REAL-TIME CONTROL OF ELECTRIC POWER SYSTEMS

REAL-TIME CONTROL OF ELECTRIC POWER SYSTEMS

PROCEEDINGS OF THE SYMPOSIUM ON REAL-TIME CONTROL OF ELECTRIC POWER SYSTEMS

BROWN, BOVERI & COMPANY LIMITED, BADEN, SWITZERLAND, 1971

Edited by

EDMUND HANDSCHIN

Brown, Boveri & Company Limited
Baden, Switzerland

ELSEVIER PUBLISHING COMPANY
AMSTERDAM – LONDON – NEW YORK
1972

ELSEVIER PUBLISHING COMPANY
335 JAN VAN GALENSTRAAT
P.O. BOX 211, AMSTERDAM, THE NETHERLANDS

AMERICAN ELSEVIER PUBLISHING COMPANY, INC.
52 VANDERBILT AVENUE
NEW YORK, NEW YORK 10017

LIBRARY OF CONGRESS CARD NUMBER: 75–190677

ISBN 0–444–41045–7

WITH 113 ILLUSTRATIONS AND 12 TABLES

PRINTED IN THE NETHERLANDS

Foreword

Shortly after the inception of the Brown Boveri Research Center in 1966, plans were developed to organize a series of Brown Boveri scientific symposia, to be held at proper intervals in Baden—on a different subject each time. The subject for a symposium should be chosen with the following requirements in mind:

—It should characterize a part of a thoroughly scientific discipline; in other words, it should describe an area of scholarly study and research.

—It should be of current interest in the sense that important results have recently been obtained and considerable research effort is under way in the world's scientific community.

—It should bear some relation to the scientific and technological activity of the Company.

These symposia will be intimately related to one of the very basic concepts which have governed the work of many modern manufacturing companies: *close coupling between science and engineering*. It is to this coupling that we owe the technical standard of our products and it is this coupling which we hope will be furthered by our symposia.

It is often said that the important technological innovations come from the basic sciences, and the *transistor* is taken as a brilliant example. Indeed, the transistor is based on quantum mechanics, a thoroughly fundamental scientific structure which in turn predicts the behaviour of electrons in a crystal lattice and explains the existence of energy bands. Without a knowledge of quantum mechanics, the invention of the transistor would have been impossible. This is a model case for the following sequence of events: first, scientific discovery; then, engineering invention. *Nuclear energy* is another case of this kind.

It is good to note that this course, although frequent, is not the only path along which technological progress takes place. In fact, the opposite sequence is quite common: first, engineering invention; then, discovery of the underlying scientific principles. The subject of the first Brown Boveri symposium which took place two years ago in 1969—fluid dynamics—is almost a model for this process. The first steam turbines were designed and successfully built by people who were only very vaguely acquainted with the scientific laws underlying their machines, and in fact some of these laws were not even known at the time. The fundamental concept of entropy was just about being introduced into the engineers' language, and much time elapsed before turbine designers were willing to adopt this concept in their daily work. The fact that these men were able to design very satisfactory turbines of 10,000 kilowatts in this manner is an impressive illustration of their deep intuitive insight. For further progress, however, science was essential. We are now considering turbines that are hundred times larger, or even more. The inventor still plays a key part in this process; but he must interact closely with the scientist, and he must frequently ask himself: How far from the optimum is my invention? Am I expecting something that is against the fundamental laws of nature? If not, what can be ultimately achieved with my idea? Without answers to these questions, technical progress would be almost impossible beyond a certain point, and this, of course, holds not only for turbines, but for all fields of technology.

As we have said, the coupling between science and engineering is essential in today's technological world. Yet the scholars and scientists on the one side, the inventors and engineers on the other are two very different breeds of people. Their mode of work is very different also. Inventors and engineers, by the very nature of their task, have to perform most of their duties behind closed doors. They are barred from excessive publicity, and, in fact, it is doubtful whether they would seek it even if they were allowed. The scientists on the other hand form one large, world-wide community with many ties and links. To keep scientific results secret would be useless. The body of scientific knowledge is a single, large pool to which many sources contribute. It is unwise for anyone to draw from this pool without making a contribution towards replenishing it. Our Company has throughout the years of its existence drawn liberally from this pool, and we may say that it has also done its share of replenishment throughout its history of 80 years. In the year in which Brown Boveri was founded (1891) one of the co-founders, Charles E. L. Brown, published a paper entitled "High voltages, their generation, transmission and distribution" in the Proceedings of the German Electrotechnical Society. In his paper he points out that voltages of 30 or 40 kV are quite feasible, as against 2–4 kV, which was generally considered to be the upper limit. The key to this big step forward is seen in the use of transformers which, although of course not new in principle, were apparently not generally known at that time; Brown found it necessary in his paper to describe in some detail what a transformer is and how it works. Among many other achievements, Charles Brown pioneered the design of transformers for higher voltages and thus opened the door to a new era in power networks. This paper had far-reaching effects and made possible transmission of power over long distances and therefore electrification on a large scale. Scientific and technical papers published by Brown's successors in Switzerland, Germany and France in the following eight decades have been numerous and significant. The Company's management is determined to continue this tradition. Currently, a new building for the Corporate Research Center is under construction. It is being erected in the pleasant surroundings of Baden, about four kilometers from the town center, and will eventually house 300 scientists and auxiliary personnel. We expect to move into the new building in 1972.

The second symposium was devoted to "Real-Time Control of Electric Power Systems." The choice of this subject was indeed not incidental. Brown Boveri has a long history in the field of control of power systems, manufacturing a range of equipment which includes: load frequency controllers; speed governors; excitation systems; protective devices; turbine supervisory systems (Turbomat); data links, including power line carrier and micro-wave; time division and frequency division multiplexed telemetry systems; computer-based load dispatching systems.

With present trends in dispatching requirements, we find that process computers are required to perform more and more tasks in real time. To enable us to design a system which can undertake these tasks we require a better understanding of the power system behaviour, and more advanced techniques are required for the solution of network problems. Future applications of on-line process computers to power systems will demand the development of algorithms which are economic both in storage requirements and computer loading. In our Group Research Center we have a scientific group studying advanced techniques in power system control and developing new methods for the solution of power system problems.

In the future, when more and more tasks must be performed by the control system in real time, great care will be required in modelling and in accepting approximations to arrive at a feasible but economic solution. It is from the efforts of the research teams, together with the cross fertilization of ideas which comes from symposia such as the present one, that viable solutions will be found.

This symposium has shown that better control is needed and suggested where advanced methods in achieving this objective can be applied. Nowadays, when interruptions in our power supply become more and more intolerable, a dispatcher must have pertinent information at his immediate disposal to enable him to take remedial action after a network incident. He must have adequate notice of impending problems. The computer with its high speed capacity for data reduction and algorithm solution is an appropriate tool to support him in his sometimes strenuous task. However, the economics of such systems must be carefully considered, though often it is difficult to justify a solution economically, as the real benefits are qualitative, not quantitative.

The symposium was attended by 111 participants from 16 countries. It was both an honour and a pleasure to welcome these scientists from so many parts of the world. Their willingness to travel to Baden and to spend two full days with us was a challenge as well as an obligation to us as organizers and we sincerely hope that the expectations which prompted them to attend were not disappointed.

We sincerely hope that the papers and discussions that were presented during the two days have inspired further investigations which will advance our knowledge in this sphere. The future trends in Real-Time Control of Power Systems have, we believe, been clearly indicated during this symposium.

To conclude, we should like to take this opportunity to express our sincere gratitude to every participant in the symposium. We hope that the time they spent with us was satisfying. Thanks are due primarily to the authors for having spared no effort in preparing the papers. The content of this volume reflects the excellent quality of their work. We thank also the participants in the discussions, both official and unofficial—they have enabled everyone to profit from the wealth of their experience. A sincere vote of thanks must be extended to Mr. H. Glavitsch, scientific secretary, Mr. E. Arn, administrative secretary, and Dr. E. Handschin, editor of the proceedings. These gentlemen, by their considerable efforts and close attention to detail, have contributed so much to the material content and the smooth running of the symposium.

Owing to our limited space and the desire to avoid an unwieldily large congress, it was not possible to extend invitations to a much larger circle; at the last moment, we had with much regret to disappoint many other people who wished to participate. We hope that the publication of the present volume may be a partial consolation for those whom we did not have the pleasure of greeting as our guests.

R. Schüpbach A. P. Speiser

Directors, Brown, Boveri & Co.

Preface

This proceeding volume contains all the material presented at the second Brown, Boveri Symposium, which was devoted to the subject 'Real-Time Control of Electric Power Systems' and held on 27 and 28 September 1971 at Baden, Switzerland. The symposium provided a forum for scientists to report on recent advances within the field of modern control as applied to electric power systems and on new applications of process computers for real-time control tasks.

Electricity supply systems have made increasing use in recent years of the computer, particularly for off-line studies concerning planning and design of power stations and systems, simulation of transient phenomena and evaluation of theoretical concepts. Today, increasing demands with regard to the security of the supply and the economics of generation and distribution of power requires new concepts and solutions. In their implementation they have to meet the stringent requirements of real-time operation. The content of this volume covers both the present state and the future tendencies of on-line control of electric power systems.

All the papers presented in the symposium are invited contributions. In choosing topics, the following three criteria have been observed:

1. The contributions must indicate clearly the applicability to power systems problems.
2. The work should have reached a certain maturity as far as theoretical solutions and practical applications are concerned.
3. The solutions proposed must be based on sound scientific grounds.

Despite the rapid development, the material presented in this volume represents a significant contribution toward the final goal of power system operation.

The twelve papers of this volume were presented in four sessions, the first of which was devoted to *analytical methods and algorithms.* The first paper, by Prof. K. J. Åström, deals with the theoretical and practical aspects of system identification. Indeed, the availability of a mathematical model representing the system sufficiently accurately is a fundamental prerequisite for any application of control theory. The complete and reliable knowledge of the power system state at any instant of time is essential for the secure and economic operation of the system. The paper by E. Handschin contains several real-time data processing algorithms based on state estimation theory. Under certain network configurations it has been observed that self-excited electromechanical oscillations arise in the transmission system. This problem has been investigated by Prof. G. Hosemann and R. Hartenstein and has resulted in a computer-based stability analysis program. The effects of speed governors and voltage regulators can be included for the analysis of a multimachine system.

The second part of the symposium contains three contributions to the topic of *optimal control.* The review of several minimization techniques given by A. Brameller and K. L. Lo is particularly oriented towards the solution of large scale systems. An important application for static optimization procedures is the economic load dispatch problem. The solution chosen at Electricité de France is described in the chapter written by J. G. Siroux, J. Pouget and A. Merlin. Optimal control theory applied to load-frequency control is discussed in the contribution by H. Glavitsch and F. D. Galiana. The comparison with conventional solutions shows the advantages which can be achieved by means of modern control theory.

Because of the rapidly-increasing dependency on the electric power supply the ample field of *systems security* was chosen as the subject of the third part of the symposium. The solution to be chosen for security monitoring and control depends on the time scale one is interested in. K. N. Stanton extends his previous work on power systems dynamic energy balance to multi-area systems. This approach is valid for a time scale of a few seconds up to minutes. If one is interested in the system behaviour within fraction of seconds, the details of the transient phenomena must be taken into consideration, as shown in the contribution by Prof. M. Mansour. This deals with the application of the second method of Liapunov and shows some possibilities for improving the stability by means of optimal control theory. At the other end of the time scale there is the steady-state security monitoring problem. On the basis of experience at the Bonneville Power Administration, W. F. Tinney and N. M. Peterson report on a fast load-flow algorithm used for a stability problem where the time scale ranges from minutes to hours.

In the final session of the symposium papers on three different subjects were presented. G. Quazza and E. Ferrari contributed an extensive study of problems concerning power station control. Their main interest centers around the effects of the primary control such as speed governors and voltage regulators on the system performance. During the last few years large scale system theory has been studied extensively both in Europe and USA. As shown by Prof. F. C. Schweppe and Prof. S. K. Mitter, a power system is a typical example which fits into the framework of this theory. For several hierarchical principles the information exchange and coordination are investigated and discussed in detail with the aim of improving the efficiency of the overall system operation. Although the importance of the computer has been emphasized in all contributions mentioned so far, the role of the human operator should not be underestimated. The last contribution by M. Syrbe and F. Bacher is therefore devoted to the cooperation between man and computer. The man–machine interface is analyzed in order to find ways to achieve the envisaged improvement of system economy and security by means of real-time process computers.

In editing this volume I have retained the literary style of the contributors. Whenever editorial alterations were required I discussed them with the authors and asked for their approval. Finally, I would like to acknowledge the excellent cooperation of all authors while editing the material for this volume.

Baden, May 1972 E. HANDSCHIN

Contents

List of Participants

T. O. Adielson
Allmänna Svenska Elektriska Aktiebolaget
S-72183 Västerås, Sweden

Dott.-Ing. F. Ariatti
ENEL DPT
Dir. Settore Interconnessione e Trasmissione
Via G.B. Martini 3, I-00100 Rome, Italy

E. Arn (Administrative Secretary)
Brown, Boveri & Co. Ltd.,
Group Research Center,
CH-5401 Baden, Switzerland

H. Asal
Elektrizitäts-Gesellschaft, Laufenburg AG,
CH-4335 Laufenburg, Switzerland

Prof. Dr. K. J. Åström (Chairman)
Lund Institute of Technology,
Division of Automatic Control,
P.O. Box 725, S-22007 Lund, Sweden

P. F. Bach Jensen
Det jysk-fynske Elsamarbejde,
ELSAM, DK-7000 Fredericia, Denmark

F. Bacher
Brown, Boveri & Co. Ltd.,
Department EDV, CH-5401 Baden,
Switzerland

Prof. Dr. J. G. Balchen
Division of Automatic Control,
Technical University of Norway,
Trondheim, Norway

R. Bertschi
Brown, Boveri & Co. Ltd.,
Department EIR-TE, CH-5401 Baden,
Switzerland

R. Bidard
Directeur Scientifique, Cie. Electro-Mécanique,
37, rue du Rocher, Paris, France

H. Bloch
Brown, Boveri & Co. Ltd.,
Department TGR, CH-5401 Baden,
Switzerland

Prof. Ir. H. B. Boerema
Technological University Delft,
Department of Electrical Engineering,
Mekelweg 4, Delft, Netherlands

J. P. Bowles
Hydro Quebec Institute of Research
P.O. Box 1000, Varennes, P.Q., Canada

Dr. A. Brameller
The University of Manchester
Institute of Science and Technology,
P.O. Box No. 88, Sackville Street, Manchester 1,
England

G. L. Brewer
Central Electricity Generating Board,
Burymead House, Portsmouth Road,
Guildford, Surrey, England

Prof. Dr. J. A. Bubenko
The Royal Institute of Technology,
Department of Electric Power Systems,
S-10044 Stockholm 70, Sweden

H. Bühler
Elektro-Watt, Ingenieurunternehmung AG,
CH-6914 Ciona di Carona, Switzerland

Miss R. M. Burke (representing the publisher)
Elsevier Publishing Company,
Amsterdam, Netherlands

J. A. Burley
Brown, Boveri & Co. Ltd.,
Department EDV, CH-5401 Baden,
Switzerland

J. Chapuy
Directeur Adjoint de la Production et du Transport, Electricité de France,
3, Rue de Messine, Paris 8, France

W. W. Clauson
International Business Machines
Dreikönigstr. 24, CH-8022 Zürich, Switzerland

Dott.-Ing. E. Clerici
Centro Elettrotecnico Sperimentale Italiano Giacinto Motta,
Via Rubattino 54, I-20134 Milano, Italy

Dott.-Ing. E. Codecá
Centro Elettrotecnico Sperimentale Italiano Giacinto Motta,
Via Rubattino 54, I-20134 Milano, Italy

Dr. V. Crastan
Suiselectra,
Malzgasse 32, CH-4000 Basel, Switzerland

G. Désilets
Hydro-Quebec,
17ie, Tour de la Bourse, Montreal, P.Q., Canada

Prof. Dr. T. S. Despotović
Elektrotehnički Institut Nikola Tesla
Bulevar Vojvode Mišića 43, Belgrade, Yugoslavia

J. L. Dineley
The University of Newcastle-upon-Tyne
Department of Electrical and Electronic Engineering,
Newcastle-upon-Tyne NE1 7RU, England

D. DoDai
Motor Columbus Ingenieurunternehmung AG,
Parkstrasse 27, CH-5400 Baden, Switzerland

Dr. L. S. Dzung
Brown, Boveri & Co. Ltd.,
Group Research Center,
CH-5401 Baden, Switzerland

Dr. C. J. Eaglen
Brown, Boveri & Co. Ltd.,
Group Research Center,
CH-5401 Baden, Switzerland

J. Ehlert Knudsen
NESA, System Planning & Research,
Strandvejen 102, DK-2900 Hellerup, Denmark

Prof. Dr. A. H. El-Abiad
Department of Electrical Engineering,
Purdue University,
Lafayette, Indiana 47907, U.S.A.

P. Feintuch
Electricité de France,
2, rue Louis Murat, Paris 8, France

Dr. E. Ferrari
ENEL, Centro Ricerca di Automatica,
18 via Beruto, I-20131 Milano, Italy

Dr. W. Frey
Brown, Boveri & Co. Ltd.,
Department CR, CH-5401 Baden, Switzerland

Dr. F. D. Galiana
Brown, Boveri & Co. Ltd.,
Group Research Center,
CH-5401 Baden, Switzerland

Prof. Dr. E. Gerecke
Freiestr. 212, CH-8032 Zürich, Switzerland

A. Germond
Chaire d'Installations Electriques,
Ecole Polytechnique Fédérale,
16, Chemin Bellerive, CH-1000 Lausanne, Switzerland

Dr. A. H. Glattfelder
Institut für Regelung und Dampfanlagen
Eidg. Technische Hochschule,
Sonneggstrasse 3, CH-8006 Zürich, Switzerland

J. Goossens
Directeur des Grands Réseaux de l'EBES,
Verenigde Energiebedrijven van het Scheldeland N.V.,
Vaartkaai 2, B-2060 Merksem, Belgium

H. Glavitsch (Scientific Secretary)
Brown, Boveri & Co. Ltd.,
Group Research Center,
CH-5401 Baden, Switzerland

N. Haase
NESA, Program Development,
Strandvejen 102, DK-2900 Hellerup, Denmark

H. Hager
Brown, Boveri & Co. Ltd., Department CT,
CH-5401 Baden, Switzerland

Dr. E. Handschin (Editor)
Brown, Boveri & Co. Ltd.,
Group Research Center,
CH-5401 Baden, Switzerland

Dr. H. Hartenstein
Institut für elektrische Energieversorgung
Technische Hochschule Darmstadt,
Schlossgraben 1, D-61 Darmstadt, Germany

U. Hartmann
Bernische Kraftwerke AG, Viktoriaplatz 2,
CH-3000 Bern, Switzerland

D. Hayeur
Hydro-Quebec, Stock Exchange Building,
4, Place Victoria, Montreal, P.Q., Canada

R. Hill
Hydro-Quebec, Stock Exchange Building,
4, Place Victoria, Montreal, P.Q., Canada

R. Hofmann
Bernische Kraftwerke AG,
Dr. Schneiderstrasse 14, CH-2560 Nidau,
Switzerland

Prof. Dr. G. Hosemann
Institut für elektrische Energieversorgung,
Technische Hochschule Darmstadt,
Schlossgraben 1, D-61 Darmstadt, Germany

Prof. Dr. M. W. Humphrey Davies
Queen Mary College, Department of Electrical
and Electronic Engineering,
Mile End Road, London E.1, England

E. Inderbitzin
Elektro-Watt Ingenieurunternehmung AG,
Selnaustrasse 16, CH-8022 Zürich, Switzerland

W. B. Jervis
Merz and McLellan, Amberley, Killingworth,
Newcastle-upon-Tyne NE12 0RS, England

T. Johansson
Statens Vattenfallsverk SKN,
S-162 87 Vällingby, Sweden

A. Jud
Nordostschweiz, Kraftwerke AG.
Parkstrasse 23, CH-5400 Baden, Switzerland

T. Kierulf
A/S Norsk Elektrisk & Brown, Boveri,
Messeveien 18, Oslo 2, Norway

Dr. W. Klapdor
Brown, Boveri & Cie AG, Department ZEK-EJ,
Postfach 351, D-68 Mannheim 1, Germany

Dr. U. G. Knight
Central Electricity Generating Board
Operations Dept.,
20, Newgate St., London E.C.1, England

Dr. H.-J. Koglin
Institut für elektrische Energieversorgung
Technische Hochschule Darmstadt,
Schlossgraben 1, D-61 Darmstadt, Germany

Dr. J. Kohlas
Brown, Boveri & Co. Ltd.
Group Research Center,
CH-5401 Baden, Switzerland

Dr. W. Latzel
Brown, Boveri & Cie AG, Dept. KJ-S,
Postfach 351, D-68 Mannheim 1, Germany

Dr. M. A. Laughton
Queen Mary College, Dept. of Electrical
Engineering, University of London,
Mile End Road, London E.1., England

M. Leibundgut
S.A. L'Energie de l'Ouest Suisse,
Place de la Gare, CH-1001 Lausanne,
Switzerland

K. Linke
Vereinigte Elektrizitätswerke Westfalen AG,
Ostwall 51, D-4600 Dortmund, Germany

Dr. K. L. Lo
The University of Manchester
Institute of Science and Technology,
Power System Laboratory,
Sackville Street, Manchester M60 1QD, England

Prof. Dr. M. Mansour
Lehrstuhl für Automatik
Eidg. Technische Hochschule,
Gloriastrasse 35, CH-8006 Zürich, Switzerland

A. Merlin
Electricité de France, Direction des Etudes et
Recherches Service Etudes de Réseaux,
17, avenue de la Libération,
F-92 Clamart, France

Prof. Dr. S. K. Mitter
Dept. of Electrical Engineering,
Massachusetts Institute of Technology,
Cambridge, Mass. 02139, U.S.A.

P. Mygind
International Business Machines,
Public Utility Industry Center,
149, Avenue Louise, B-1050 Bruxelles, Belgium

L. Nevanlinna
Imatran Voima Osakeyhtiö,
P.O. Box 10138, Helsinki 10, Finland

K. E. I. Norbäck
Sydsvenska Kraft AB, Carl Gustavsväg 1,
S-200 70 Malmö 5, Sweden

M. Nordgard
A/S Norsk Elektrisk & Brown, Boveri,
Messeveien 18, Oslo 2, Norway

Dr. H. A. Nour Eldin
Lehrstuhl für Automatik
Eidg. Technische Hochschule,
Gloriastrasse 35, CH-8006 Zürich, Switzerland

D. Oeding
Brown, Boveri & Cie AG, Dept. EB,
Postfach 351, D-68 Mannheim 1, Germany

L. M. Panis
Brown, Boveri & Co. Ltd., Department TGV,
CH-5401 Baden, Switzerland

P. Péraud
Cie Electro-Mécanique,
12, Rue Portalis, F-75 Paris (8e), France

H. Peterson
Sydsvenska Kraft AB, Driftavdelningen,
S-200 70 Malmö 5, Sweden

M. Picek
Nordostschweiz. Kraftwerke AG,
Parkstrasse 23, CH-5400 Baden, Switzerland

J. Pouget
Electricité de France,
17, Avenue du Général de Gaulle,
F-92 Clamart, France

Dr. G. Quazza (Chairman)
ENEL, Centro Ricerca di Automatica,
Via Beruto 18, I-20131 Milano, Italy

Dr. K. Reichert
Brown, Boveri & Co. Ltd., Department CT,
CH-5401 Baden, Switzerland

H. Rotter-Woletz
Direktor Abt. GSJ/M,
Brown Boveri & Cie AG,
Postfach 351, D-68 Mannheim 1, Germany

Dr. B. Sakić
Brown, Boveri & Co. Ltd., Department ED,
CH-5401 Baden, Switzerland

R. Sandberg
Brown, Boveri & Co. Ltd., Department TGA,
CH-5401 Baden, Switzerland

F. Schär
Aare-Tessin AG für Elektrizität,
Bahnhofquai 12, CH-4600 Olten, Switzerland

W. S. Seifert
Brown, Boveri & Co. Ltd., Department EDV,
CH-5401 Baden, Switzerland

Dr. F. M. Schlaepfer
IBM Research, San Jose, U.S.A.,
Monterey & Cottle Roads, San Jose,
CA 95114, U.S.A.

Dr. W. Schneider
Bayernwerk AG, Blutenburgstrasse 6,
München 2, Germany

Dr. R. Schüpbach (representing the host)
Technical Director, Department E,
Brown, Boveri & Co. Ltd.,
CH-5401 Baden, Switzerland

Dr. F. Schwab
Aare-Tessin AG für Elektrizität,
Bahnhofquai 12, CH-4600 Olten, Switzerland

Prof. Dr. F. C. Schweppe (Chairman)
Electric Power Systems, Engineering Laboratory,
Massachusetts Institute of Technology,
Cambridge, Mass. 02139, U.S.A.

J. G. Siroux
Electricité de France,
Direction des Etudes et Recherches,
17, avenue de la Libération,
F-92 Clamart, France

Dr. R. Sontheim (representing the host)
Managing Director,
Brown, Boveri & Co. Ltd.,
CH-5401 Baden, Switzerland

Prof. Dr. A. P. Speiser (representing the host)
Director of Research,
Brown, Boveri & Co. Ltd.,
CH-5401 Baden, Switzerland

Dr. K. N. Stanton (Chairman)
Systems Control, Inc., 260 Sheridan Avenue,
Palo Alto, Ca. 94306, U.S.A.

B. Streckeisen
Société Générale pour l'Industrie,
Case postale 133, CH-1000 Lausanne,
Switzerland

J. Svoen
NVE, Statskraftverkene,
Middelthunsgt. 29, Oslo 3, Norway

Dr. M. Syrbe
Institut für Informationsverarbeitung in
Technik und Biologie,
Breslauer Str. 48, D-75 Karlsruhe-Waldstadt,
Germany

H. Tillberg
Statens Vattenfallsverk,
S-162 87 Vällingby, Sweden

W. F. Tinney
United States Department of the Interior,
Bonneville Power Administration,
P.O. Box 3621, Portland, Oregon 97208, U.S.A.

Dr. I. Tiroshi
Brown, Boveri & Co. Ltd.,
Group Research Center,
CH-5401 Baden, Switzerland

W. R. Tracey
Regional Manager Control Engineering,
Hoyles Niblock Overseas Limited,
P.O. Box 11-1535, Tehran, Iran

Ir. L. Verbeek
System Planning,
N.V. Electriciteitsbedrijf Zuid-Holland,
Alexanderstraat 16, Den Haag, Netherlands

Prof. E. Voipio
Imatran Voima Osakeyhtiö,
P.O. Box 10138, Helsinki 10, Finland

Dr. P. Wach
Institut für Bau und Betrieb elektrischer Anlagen,
Technische Hochschule in Graz,
Kopernikusgasse 24, A-8010 Graz, Austria

J. P. Waha
Intercom S.A.,
Service Investissements en Grands Réseaux,
1, place du Trône, B-1000 Bruxelles, Belgium

K. Wedin
Statens Vattenfallsverk, Avd. DKT,
S-162 87 Vällingby, Sweden

Dr. J. Weiss
Brown, Boveri & Co. Ltd., Department EDV,
CH-5401 Baden, Switzerland

E. Widmer
Bernische Kraftwerke AG,
Viktoriaplatz 2, CH-3000 Bern, Switzerland

Dr. K. Zollenkopf
Hamburgische Electricitäts-Werke,
Ueberseering 12, D-2000 Hamburg 39, Germany

Prof. Dr. R. Zwicky
Lehrstuhl für Industrielle Elektronik und Messtechnik,
Eidg. Technische Hochschule,
Gloriastrasse 35, CH-8006 Zürich, Switzerland

Introduction

H. Glavitsch, Scientific Secretary

A quite remarkable tendency to employ computers can be observed today in the power industry; specifically, the process computer for control, guidance, logging and monitoring of power systems and stations. There is a number of reasons for this development.

First, before the advent of computers, the power industry already relied heavily on automatic control equipment and on the basis of this experience the utilities are willing to keep up with the development.

Second, within the last ten years space technology has had a tremendous impact on modern control theory supplying at the same time new methods for the set-up of models, on-line data processing and data clean-up, observation and estimation of system states, optimal control, structuring of hierarchical systems.

Third, off-line computation has proved very successful in power system analysis and system engineering and the computer has gained a very important position in this field.

Fourth, there is an increasing demand for better control, mainly due to security and economic requirements.

However, one is still not quite satisfied with what has been actually realized in the field. Many limitations in the modelling, in the theory, in the computation, etc. have been encountered. Since it is Brown Boveri's main business to supply power equipment and complete systems to the power industry, including control and information systems, it was thought appropriate to select the topic of 'Real-Time Control of Electric Power Systems'. It is a quite timely topic, since the stage has been reached where the computer is being accepted for certain classes of problems and further applications are to be expected.

As to the choice of papers and sessions, we tried to take into account the importance of systems for monitoring and rescheduling, as well as economic dispatch, which makes use of optimization methods. The latter enter into a number of other areas as will become apparent. Identification and estimation will serve as an introduction. Last but not least, we have included the topic of the role of power station control and the role of the human operator in overall system operation. As far as the human operator is concerned it is clear that a power system cannot be run without him. Therefore, it will be wise to employ hardware and software to make his intervention more efficient. Making available to him relevant data and giving him the possibility to interact with the system on various levels is the necessary consequence of this idea.

The support and encouragement by the management of Brown Boveri represented by Dr. R. Sontheim, Dr. A. P. Speiser and Dr. R. Schüpbach in the preparation and realization of this symposium is highly appreciated.

Modelling and Identification of Power System Components

K. J. ÅSTRÖM

Division of Automatic Control, Lund Institute of Technology, Lund, Sweden

SUMMARY

There is a continuing tendency to apply many of the powerful results of modern control theory to various industrial processes. Power systems have been indicated as one area where significant progress can be expected. Practically all results of modern control theory require that models of the processes in terms of state equations are available. The need to obtain such models has been a strong motivation for research in the area of modelling and identification. Some progress made in this area is reviewed in this paper. Modelling based on physical equations and on plant experiments is discussed and compared. Particular emphasis is given to parameter estimation techniques like the maximum likelihood method which offer a possibility of combining physical *a priori* knowledge with experimental investigations. The formulation of identification problems is discussed, including the choice of criteria and model structures.

The techniques are illustrated by applications to data obtained from measurements on various components of a power system. The examples include an electric generator, a nuclear reactor and a drum boiler, and serve to illustrate the potentials and limitations of system identification and modelling techniques when they are applied to real data.

1. INTRODUCTION

The design of a control system is frequently divided into two steps: determination of a mathematical model and design of a control strategy. In fact most of the control theory that has been developed postulates that a model of the system and its environment is available. To use much of the existing control theory it is therefore necessary to have techniques to determine suitable models for the processes to be controlled.

Before the advent of modern control theory most results were restricted to linear systems, assuming a model specified by a transfer function. The modelling then reduces to the determination of transfer functions. This is conveniently done experimentally by introducing a sinusoidal variation in the input and measuring amplitude and phase relations between the input and the output. Interesting applications of this technique to power systems were soon made. In Sweden, for example, a group at the ASEA company under the direction of Dr. A. Garde made extensive measurements on power system components for the purpose of designing control systems. A typical study was the determination of the transfer function from power input to frequency variations in the Swedish power net reported by Oja[20].

A characteristic feature of many significant results in control theory that have been developed over the past 20 years is that they require other models than transfer functions. Typically, many of the results of modern control theory assume that the system is described by time domain models like

$$\begin{aligned} \frac{d\mathbf{x}}{dt} &= \mathbf{f}(\mathbf{x}, \mathbf{u}, \mathbf{v}) \\ \mathbf{y} &= \mathbf{g}(\mathbf{x}, \mathbf{u}, \mathbf{v}) \end{aligned} \tag{1}$$

where $\mathbf{u}$ is the input, $\mathbf{y}$ the output, $\mathbf{x}$ the state variable and $\mathbf{v}$ a disturbance.

To apply the results of modern control theory to industrial processes it is therefore necessary to have techniques available to determine models like eqn. (1) for the different processes. In this paper we will outline some progress towards the solution of this problem. The results, which can be applied to many industrial processes, are illustrated by computations on data obtained from experiments on power system components. Two approaches, modelling from physical *a priori* knowledge and modelling based on measurements of inputs and outputs only, are discussed in §2. Possibilities of combining the techniques are also investigated. Parameter estimation methods and formulation of identification problems are covered in §3. The choice of model structures for linear deterministic and linear stochastic systems is reviewed in §4 and §5. Selection of criteria for identification problems is the topic of §6. In §7, §8 and §9 the techniques discussed in the previous sections are applied to the modelling of power system components. The results are based on measurements on real processes. The examples include a power generator, a nuclear reactor and a drum boiler.

2. MODELLING AND IDENTIFICATION

The problem to be considered is thus how to obtain a model like the one given by eqn. (1) for industrial processes. This problem is sometimes called the inverse problem because a solution is given and the problem is to find the equation which has the given solution. Problems of this type do of course arise in many fields: biology, medicine, economy, physics and chemistry. There are certain advantages in the modelling and identification problems originating from the field of automatic control:

There is a specific purpose in doing the modelling (design of control strategies).
It is often fairly easy to do experiments. (Control systems are designed in such a

way that control variables can be manipulated and outputs measured.)

Strictly speaking, the problem may not be so well defined. Even if the design of a control system is the final goal, it is of course often very valuable to have insight and understanding of several of the system properties that do not enter the control design directly. The possibility of making experiments may be severely limited because it may be necessary to experiment under normal operating conditions, and large changes in inputs may be prohibitive for safety and economic reasons.

Process models can be obtained from basic physical laws, from pure input–output experiments or from a combination of these. The different approaches have advantages and disadvantages which are briefly discussed below.

2.1 Modelling from physical principles

The required models can in principle be derived from basic physical laws expressing conservation of mass, momentum and energy, combined with material equations like Boyle's law or Hooke's law. The models obtained in this way have the great advantage of a wide range of validity. Usually, they also provide a good insight into the behaviour of the system. The drawbacks of modelling from physical laws are that: the required knowledge is not always available; the procedure is frequently time-consuming (consider the time required to develop Newtonian mechanics); and it is often difficult to make sensible approximations. A typical difficulty is to find good approximations of distributed parameter systems. Experience has shown that the models developed from basic physical principles tend to be very complex. Since the complexity of the model indirectly implies a complex control strategy and vice versa, then if a system can be successfully controlled by a simple strategy it can probably also be modelled satisfactorily by a simple model. In the area of power systems, dynamics of components like generators, motors, transmission lines and hydroelectric stations are well understood in the sense that models can conveniently be derived from physical principles. On the other hand, components like thermal boilers and nuclear power stations are not sufficiently well known for models suitable for control to be derived from physical principles alone. In thermal boilers we have, for example, the difficulty with the two-phase flow and the uncertainty in heat transfer coefficients. Even if the basic neutron kinetics in nuclear power stations is well known it leads to very complex models if three-dimensional effects are considered. A large portion of nuclear reactor dynamics also consists of thermal dynamics and hydromechanics.

2.2 Input–output modelling

The pure case of input–output modelling consists of the determination of a model from input–output measurements only. This is often called the black box approach. The advantage of this method is that it is usually done fairly quickly. Experience has also shown that it usually leads to fairly simple models. One serious disadvantage is that in most methods it is possible to determine linear models only. This means that the validity of the model is limited. A change in operating condi-

References pp. 23–24

tions, input signals, etc., may thus lead to a different model. Another disadvantage is that available *a priori* knowledge is not used. For example, it is almost impossible to exploit *a priori* knowledge when a transfer function is determined using frequency response methods. However, in some cases the input–output approach may be the only possibility. This may be the case in the characterization of disturbances such as load variations.

Recognizing the advantages and disadvantages of modelling from physical equations and from input–output experiments alone it seems highly desirable to try to exploit both methods in order to solve the modelling problem. In the next section we will discuss techniques that can be used to do this.

3. PARAMETER ESTIMATION

Using available *a priori* knowledge about the system to be controlled, it is frequently possible to arrive at a class of models $\{M\}$ that can represent the system. The class of models can, for example, be the class of all stable linear systems having positive impulse responses, all systems described by equations like eqn. (1), where the functions **f** and **g** depend on a parameter, etc.

The problem is to design an experiment on the real system which makes it possible to select one model in the class which is a good representation of the real system. For a control system the most natural way to select a suitable model would be to compare the performances of the control strategies designed on the models when used to control the real system. Since such a selection is very difficult to do, simpler ways are often chosen. It is common to select the models by comparing the error between the model variables and the corresponding system variables. The comparison is often based on minimization of a loss function, for example of the type

$$V(y, y_m) = \int_0^\infty e^2(t)\mathrm{d}t \tag{2}$$

where y is the system output, y_m the model output and e the error. The error can for example be defined as

$$e = y - y_m \tag{3}$$

or

$$e = \mathrm{A}y - \mathrm{A}y_m \tag{4}$$

where A is some operator. Examples of this are discussed in ref. 5.

Using this formulation, the identification problem reduces to an optimization problem. Select the model in the class $\{M\}$ such that the chosen criterion is as small as possible. It is natural to ask if there is a unique minimum, if there is a natural choice of loss functions and how the results are influenced by the choice of loss functions. Some of these problems will be covered in the following sections. In particular, it will be shown that if auxiliary assumptions are made there are in fact natural loss functions and there will be unique minima.

4. LINEAR DETERMINISTIC SYSTEMS

Consider the case when the model is given by a state equation like (1) where the functions **f** and **g** depend on a set of parameters $\alpha_1, \alpha_2, \ldots, \alpha_m$ which are considered as components of a vector $\boldsymbol{\alpha}$. Also assume that the functions **f** and **g** are linear in **x** and **u** and that disturbances **v** are neglected. We thus have the standard state space description of a linear system

$$\frac{\mathrm{d}\mathbf{x}}{\mathrm{d}t} = \tilde{\mathbf{A}}\mathbf{x} + \tilde{\mathbf{B}}\mathbf{u}$$
$$\mathbf{y} = \mathbf{C}\mathbf{x} + \mathbf{D}\mathbf{u} \tag{5}$$

where the elements of the matrices $\tilde{\mathbf{A}}$, $\tilde{\mathbf{B}}$, **C** and **D** depend on the parameter $\boldsymbol{\alpha}$. It will be shown that some of the problems raised in the previous section are non-trivial even in such simple cases. Let the experiment be arranged in such a way that the input **u** is constant over sampling intervals of constant length. The values of the state variables, the inputs and the outputs at the sampling instants are then given by

$$\mathbf{x}(t+1) = \mathbf{A}\mathbf{x}(t) + \mathbf{B}\mathbf{u}(t)$$
$$\mathbf{y}(t) = \mathbf{C}\mathbf{x}(t) + \mathbf{D}\mathbf{u}(t) \tag{6}$$

where the sampling interval is chosen as the time unit and **A**, **B**, **C** and **D** are constant matrices whose elements depend on the parameter $\boldsymbol{\alpha}$. The matrices **A** and **B** are related to $\tilde{\mathbf{A}}$ and $\tilde{\mathbf{B}}$ through well-known equations.

Let n, p and r be the dimensions of **x**, **u** and **y** respectively. The descriptions (5) and (6) of the system then contain

$$N_1 = n^2 + n(r + p) + rp \tag{7}$$

parameters. It is well known that the input–output relation for eqns. (6) can be characterized by at most

$$N_2 = n(r + p) + rp \tag{8}$$

parameters. It is thus clear that in an identification experiment where the input **u** of eqns. (5) or (6) is perturbed and the output **y** is observed it is possible to determine at most N_2 parameters. Since the models (5) and (6) have N_1 coefficients, it is clear that it is not possible to determine all coefficients of eqns. (5) or (6) from an identification experiment. We can thus say that the model is not identifiable. The difficulty can be overcome by choosing a canonical structure or by using additional information about the system.

It is well known that if the coordinates in the state space eqns. (6) are changed by the linear transformation

$$\mathbf{z}(t) = \mathbf{T}\mathbf{x}(t) \tag{9}$$

References pp. 23–24

then eqns. (6) transform to

$$\begin{aligned}\mathbf{z}(t+1) &= \mathbf{TAT}^{-1}\mathbf{z}(t) + \mathbf{TBu}(t)\\ \mathbf{y}(t) &= \mathbf{CT}^{-1}\mathbf{z}(t) + \mathbf{Du}(t)\end{aligned} \tag{10}$$

which have the same input–output relation as eqns. (6). It is now natural to ask if there are transformations **T** such that the transformed eqns. (10) are characterized by fewer coefficients than eqns. (6). Many transformations with this property are known (see, for example, refs. 1, 18 and 26). One example is given below.

4.1 Canonical structures

Assume that the system (6) is completely controllable and completely observable. Let **A** be non-singular and let **C** have rank r. Then there exists a transformation **T** such that the transformed system (10) becomes

$$\mathbf{z}(t+1) = \begin{bmatrix} a_{11} & a_{12}\ldots a_{1n} \\ a_{21} & a_{22}\ldots a_{2n} \\ \vdots & \vdots \quad\quad \vdots \\ a_{r1} & a_{r2}\ldots a_{rn} \\ & \mathbf{E} \end{bmatrix}\mathbf{z}(t) + \begin{bmatrix} b_{11} & b_{12}\ldots b_{1p} \\ b_{21} & b_{22}\ldots b_{2p} \\ \vdots & \vdots \quad\quad \vdots \\ b_{n1} & b_{n2}\ldots b_{np} \end{bmatrix}\mathbf{u}(t) \tag{11}$$

$$\mathbf{y}(t) = \begin{bmatrix} 1 & 0\ldots 0\ldots 0 \\ 0 & 1\ldots 0\ldots 0 \\ \vdots & \vdots \quad \vdots \quad \vdots \\ 0 & 0\ldots 1\ldots 0 \end{bmatrix}\mathbf{z}(t) + \mathbf{Du}(t)$$

where **E** is a matrix with one non-zero element in each row. (This element can be chosen as 1.) The proof is straightforward and omitted. There are also duel representations.

The representation (11) contains N_2 parameters and it is thus a minimum parameter representation. There are, however, two difficulties with this representation. The matrix **E** reflects the way in which the state variables are coupled to the output. In general, there might be

$$N_3 = \binom{n-1}{r-1} \tag{12}$$

different matrices **E**. Hence if there is no *a priori* knowledge about the manner in which the outputs are coupled to the states there are (N_3) different models of the type (11). In practice, this means that in order to fit a general linear model (11) to experimental data it is necessary to determine the best values of the parameters a_{ij}, b_{ij} and d_{ij} for all N_3 possible **E** matrices. If $r = 1$ then $N_3 = 1$ and there are no difficulties, since there is only one alternative. The case $r = 1$ implies that the system has only one output. Completely observable and completely controllable linear systems with one output thus have only one structural parameter—namely the order of the system. For such systems it is therefore possible to obtain unique canonical representations. To identify linear systems with one output it is sufficient to choose a minimal parameter structure and determine the parameters of models with successively increasing order. For true multivariable systems the problem is much more difficult, since for each model of order n there are N_3 models with different internal couplings. This means that if no structural information is available *a priori* it is necessary to consider N_3 cases for each order of the model. Since N_3 is a fairly large number even for moderate values of r and n, investigation of all possible internal couplings is a significant burden.

There are many canonical structures similar to eqns. (11). They will, however, all suffer from the same difficulty that unless the internal couplings are known there are many different models.

If it is attempted to bypass the difficulties by identifying a multivariable system as an interconnection of single output systems, another difficulty is encountered. Owing to uncertainties, poles of the single output systems originating from the same mode of the multivariable system are easily estimated as being different. This means that the model obtained will be of too high an order due to false modes and that the internal couplings of the multivariable system will also be represented incorrectly.

It is thus crucial to exploit *a priori* information in the identification of a multivariable system.

There is also another difficulty with the representation (11). Due to its peculiar structure it can be shown to be very sensitive to parameter variations. Examples are given in ref. 15.

4.2 Using a priori *knowledge*

Having found some difficulties associated with the choice of canonical structures, we will now consider some problems associated with the use of *a priori* physical knowledge. With some knowledge about the physics of the process it may be possible to impose some conditions on the elements of the matrices $\tilde{\mathbf{A}}$, $\tilde{\mathbf{B}}$, $\mathbf{C}$ and $\mathbf{D}$ of eqns. (5) or $\mathbf{A}$, $\mathbf{B}$, $\mathbf{C}$ and $\mathbf{D}$ of eqns. (6). An example is given below.

Example 1

Consider a system described by eqns. (5) where

$$\tilde{\mathbf{A}} = \begin{bmatrix} \alpha_1 & 0 & \alpha_6 & \alpha_{10} & \alpha_{15} \\ \alpha_2 & 0 & \alpha_7 & \alpha_{11} & \alpha_{16} \\ \alpha_3 & 0 & \alpha_8 & \alpha_{12} & \alpha_{17} \\ \alpha_4 & 0 & 0 & \alpha_{13} & 0 \\ \alpha_5 & 0 & \alpha_9 & \alpha_{14} & \alpha_{18} \end{bmatrix}$$

References pp. 23–24

$$
\tilde{\mathbf{B}} = \begin{bmatrix} 0 & \alpha_{20} & \alpha_{24} \\ 0 & \alpha_{21} & \alpha_{25} \\ 0 & \alpha_{22} & \alpha_{26} \\ \alpha_{19} & 0 & 0 \\ 0 & \alpha_{23} & \alpha_{27} \end{bmatrix}
$$

$$
\mathbf{C} = \begin{bmatrix} 1 & 0 & 0 & 0 & 0 \\ 0 & 1 & 0 & 0 & 0 \end{bmatrix}
$$

$$
\mathbf{D} = \begin{bmatrix} 0 & 0 \\ 0 & 0 \end{bmatrix}
$$

This is in fact a simplified model of a drum boiler (see ref. 9). The state variables and the inputs have the following meanings:

x_1 drum pressure
x_2 drum level
x_3 drum liquid mean temperature
x_4 riser tube mean temperature
x_5 mean value of steam-to-water ratio in drum and risers
u_1 fuel flow
u_2 feedwater flow
u_3 steam flow

The fact that certain elements of the matrices $\tilde{\mathbf{A}}$, $\tilde{\mathbf{B}}$, $\mathbf{C}$ and $\mathbf{D}$ are zero is obtained from physical considerations[9].

If it is attempted to fit a model having the structure given in Example 1 it is of course of interest to know if all the parameters $\alpha_1, \alpha_2, \ldots, \alpha_{27}$ can be determined from input–output experiments. The existence of a neat criterion to decide this is still an open problem. Algorithms which show that the parameters can be determined locally are available (see ref. 6). In this particular example we have $n = 5$, $r = 2$, $p = 3$. Also notice that the matrix $\mathbf{D}$ vanishes identically. The input–output relation can be characterized by 25 parameters. We thus find that the 27 parameters of the model cannot be determined from an input–output experiment in this example.

5. LINEAR STOCHASTIC SYSTEMS

When solving control problems, the characteristics of disturbances are frequently as important as the process dynamics. A significant contribution in modern control theory has been to model disturbances as stochastic processes and to exploit the theory of stochastic processes to obtain control strategies that take into account certain characteristics of the disturbances (see, for example, ref. 2). The following model is frequently used to represent a control system subject to random disturbances:

$$\begin{aligned} \mathbf{x}(t+1) &= \mathbf{A}\mathbf{x}(t) + \mathbf{B}\mathbf{u}(t) + \mathbf{v}(t) \\ \mathbf{y}(t) &= \mathbf{C}\mathbf{x}(t) + \mathbf{D}\mathbf{u}(t) + \mathbf{e}(t) \end{aligned} \tag{13}$$

In this model $\{\mathbf{v}(t), t = 0, \pm 1, \pm 2, \ldots\}$ and $\{\mathbf{e}(t), t = 0, \pm 1, \pm 2, \ldots\}$ are sequences of independent equally distributed random vectors with zero mean values and given covariance matrices

$$\begin{aligned} \operatorname{cov}[\mathbf{v}(t), \mathbf{v}(t)] &= \mathbf{R}_1 \\ \operatorname{cov}[\mathbf{e}(t), \mathbf{e}(t)] &= \mathbf{R}_2 \\ \operatorname{cov}[\mathbf{v}(t), \mathbf{e}(t)] &= \mathbf{R}_{12} \end{aligned} \tag{14}$$

The model (13) is fairly general. It can, for example, be used to represent finite dimensional linear systems whose disturbances are weakly stationary random processes with rational spectral densities.

The problem of determining the parameters of the model (13) will now be discussed. If all the elements of the matrices **A**, **B**, **C**, **D**, $\mathbf{R}_1$, $\mathbf{R}_{12}$ and $\mathbf{R}_2$ are considered as parameters the model contains

$$N_4 = n(1.5n + p + 2r + 0.5) + r(p + 0.5r + 0.5) \tag{15}$$

parameters. Apart from the difficulties associated with the determination of the parameters of the deterministic part of the system, *i.e.* **A**, **B**, **C** and **D**, it can be shown that all elements of the matrices $\mathbf{R}_1$, $\mathbf{R}_{12}$ and $\mathbf{R}_2$ cannot be determined from the data of an input–output experiment.

Assume, for example, that the matrix $\mathbf{R}_2$ is positive definite. Then the model given by eqns. (13) is equivalent to the model

$$\begin{aligned} \hat{\mathbf{x}}(t+1) &= \mathbf{A}\hat{\mathbf{x}}(t) + \mathbf{B}\mathbf{u}(t) + \mathbf{K}\boldsymbol{\varepsilon}(t) \\ \mathbf{y}(t) &= \mathbf{C}\hat{\mathbf{x}}(t) + \mathbf{D}\mathbf{u}(t) + \boldsymbol{\varepsilon}(t) \end{aligned} \tag{16}$$

where $\{\boldsymbol{\varepsilon}(t), t = 0, \pm 1, \pm 2, \ldots\}$ is a sequence of independent equally distributed random vectors with zero mean values and covariances **R**. The models are equivalent in the sense that the input–output relations are the same and that the stochastic properties of the outputs are also the same[2].

The proof of the statement follows from the Kalman–Bucy filtering theorem. The assumption that $\mathbf{R}_2$ is positive definite can be relaxed. The model (16) is not unique. There are in general many matrices **K** for which the stochastic properties are the same. The representation is unique, however, if it is required that the matrix $\mathbf{A} - \mathbf{KC}$ has all eigenvalues inside the unit circle.

The model (16) contains

$$N_5 = n(n + 2r + p) + r(0.5r + p + 0.5) \tag{17}$$

parameters. It has several interesting properties. The state variable $\hat{\mathbf{x}}$ of eqns. (16) can be interpreted as the best linear estimate of the state **x** of eqns. (13) based on observed outputs. The quantities $\{\boldsymbol{\varepsilon}(t)\}$ are the innovations associated with the

References pp. 23–24

stochastic process $\{\mathbf{y}(t), t = 0, \pm 1, \pm 2, \ldots\}$ of eqns. (13) or (16). The matrix $\mathbf{K}$ can be interpreted as the steady state gain of the Kalman filter associated with the model (13). The Kalman filter for estimating the state of eqns. (13) is thus given by

$$\hat{\mathbf{x}}(t+1) = \mathbf{A}\hat{\mathbf{x}}(t) + \mathbf{B}\mathbf{u}(t) + \mathbf{K}[\mathbf{y}(t) - \mathbf{C}\hat{\mathbf{x}}(t) - \mathbf{D}\mathbf{u}(t)] \tag{18}$$

This equation is trivially obtained by eliminating $\boldsymbol{\varepsilon}$ in eqns. (16).

Hence if the purpose of the identification is to design a predictor or a regulator based on linear stochastic control theory the model (16) has several advantages over the model given by eqns. (13). If the model (13) is determined it is necessary to solve a Riccati equation in order to obtain the gain of the Kalman filter. It also turns out that the algorithms for the identification problem are simpler for the model structure (16). Notice, however, that even with eqns. (16) there are the difficulties with ambiguities discussed previously for deterministic systems.

In the special case of systems with one output the redundancy in the general model can be reduced by first transforming the state variables so that $\mathbf{TAT}^{-1}$ becomes a matrix on companion form and then using the transformation given by eqns. (16). We will then obtain the following model:

$$\mathbf{z}(t+1) = \begin{bmatrix} -a_1 & 1 & 0\ldots 0 \\ -a_2 & 0 & 1\ldots 0 \\ \vdots & \vdots & \vdots \quad \vdots \\ -a_{n-1} & 0 & 0\ldots 1 \\ -a_n & 0 & 0\ldots 0 \end{bmatrix} \mathbf{z}(t) + \begin{bmatrix} b_{11} & \ldots b_{1p} \\ b_{21} & \ldots b_{2p} \\ \vdots & \vdots \\ b_{n-1,1} & \ldots b_{n-1,p} \\ b_{n1} & \ldots b_{np} \end{bmatrix} \mathbf{u}(t) + \begin{bmatrix} k_1 \\ k_2 \\ \vdots \\ k_{n-1} \\ k_n \end{bmatrix} \varepsilon(t)$$

$$y(t) = [1 \quad 0 \quad 0\ldots 0]\mathbf{z}(t) + \varepsilon(t) \tag{19}$$

By eliminating the state variable $\mathbf{z}$ the following input–output relation is obtained:

$$\begin{aligned} y(t) + a_1 y(t-1) + \cdots + a_n y(t-n) &= b_{11}u_1(t-1) + \cdots + b_{n1}u_1(t-n) + \cdots \\ &+ b_{1p}u_p(t-1) + \cdots + b_{np}u_p(t-n) + \varepsilon(t) + c_1\varepsilon(t-1) + \cdots + c_n\varepsilon(t-n) \end{aligned} \tag{20}$$

where

$$c_i = a_i + k_i \quad \text{for} \quad i = 1, 2, \ldots, n \tag{21}$$

The model (20) is thus a canonical representation of a linear stochastic system with one output and several inputs.

The model (20) can be written in a slightly more compact form if the polynomials

$$\begin{aligned} A(z) &= z^n + a_1 z^{n-1} + \cdots + a_n \\ B(z) &= [b_{11} \ldots b_{1p}]z^{n-1} + [b_{21} \ldots b_{2p}]z^{n-2} + [b_{n1} \ldots b_{np}] \\ C(z) &= z^n + c_1 z^{n-1} + \cdots + c_n \end{aligned} \tag{22}$$

and the shift operator q defined by

$$\mathrm{q}x(t) = x(t + 1) \tag{23}$$

are introduced. The model (20) then becomes

$$A(\mathrm{q})y(t) = B(\mathrm{q})\mathbf{u}(t) + C(\mathrm{q})\varepsilon(t) \tag{24}$$

6. CRITERIA

Having chosen a model whose parameters can be determined from input–output data it remains to find the parameters of the model such that the model fits the experimental data. The crucial problem is then to find suitable criteria. Having the control application in mind, it would be natural to evaluate the model on the basis of the performance of the control strategies designed from it. This is generally very difficult to do, and it is thus necessary to use other criteria. The criteria can be chosen ad hoc as, for example, in eqn. (2).

If statistical assumptions are made, the parameters can also be determined using statistical parameter estimation techniques. This leads frequently to an optimization problem with a given criterion. It is thus often possible to give statistical interpretations to many criteria.

6.1 Maximum likelihood estimates

Assuming that the process is governed by the model (16) where $\{\boldsymbol{\varepsilon}(t)\}$ is a sequence of independent gaussian random variables, the parameters can be determined using the method of maximum likelihood. It can be shown that the likelihood function for estimating the model parameters is given by

$$-\ln L = \frac{1}{2}\sum_{t=1}^{N} \boldsymbol{\varepsilon}^T(t)\mathbf{R}^{-1}\boldsymbol{\varepsilon}(t) + \frac{N}{2}\ln \det \mathbf{R} + \text{const} \tag{25}$$

where N is the record length and $\mathbf{R}$ the covariance of $\boldsymbol{\varepsilon}(t)$. The identification problem then reduces to the problem of minimizing the function (25) with respect to the unknown parameters. The matrix $\mathbf{R}$ is frequently not known. This means that it is necessary to minimize eqn. (25) with respect to $\mathbf{R}$ also. Notice that this can be done analytically. We have

$$\mathrm{Min}_{\mathbf{R}}\left[\frac{1}{2}\sum_{t=1}^{N} \boldsymbol{\varepsilon}^T(t)\mathbf{R}^{-1}\boldsymbol{\varepsilon}(t) + \frac{N}{2}\ln \det \mathbf{R}\right] = \frac{rN}{2} + \frac{N}{2}\ln \det \frac{1}{N}\sum_{t=1}^{N} \boldsymbol{\varepsilon}(t)\boldsymbol{\varepsilon}^T(t) \tag{26}$$

The minimum is assumed for

$$\mathbf{R} = \mathbf{R}_0 = \frac{1}{N}\sum_{t=1}^{N} \boldsymbol{\varepsilon}(t)\boldsymbol{\varepsilon}^T(t) \tag{27}$$

References pp. 23–24

This follows from the identity

$$\sum_{i=1}^{N} \mathbf{x}_i^T \left(\sum_{j=1}^{N} \mathbf{x}_j \mathbf{x}_j^T \right)^{-1} \mathbf{x}_i = n \tag{28}$$

where n is the dimension of the vectors $\mathbf{x}_i$.

The maximum likelihood identification thus reduces to the minimization of the loss function

$$V = \det \sum_{t=1}^{N} \boldsymbol{\varepsilon}(t)\boldsymbol{\varepsilon}^T(t) \tag{29}$$

and conversely an identification problem with the criterion (29) can be interpreted as a parameter estimation problem. An estimate of the covariance $\mathbf{R}$ is then given by eqn. (27).

In the special case of systems with one output, ε is a scalar and the loss function reduces to

$$V = \sum_{t=1}^{N} \varepsilon^2(t) \tag{30}$$

Under the restrictive assumption that the data were actually generated by a model (16) where $\{\boldsymbol{\varepsilon}(t)\}$ is a sequence of independent gaussian random variables it is possible to pose and answer several statistical problems. With minor additional assumptions it can be shown that:

(a) the estimates converge to the true parameters with probability one as the record length N increases (consistency);

(b) for large N there are no other estimation procedures that give estimates with smaller variances (asymptotic efficiency);

(c) an estimate of the covariance of the estimate $\hat{\boldsymbol{\alpha}}$ is given by

$$\operatorname{cov}[\boldsymbol{\alpha}, \boldsymbol{\alpha}] = \frac{2V}{N} V_{\alpha\alpha}^{-1} \tag{31}$$

(asymptotic normality).

The precise statements of these results are given in ref. 3 for the single output case and in refs. 7 and 27 for the multivariable case.

Many other statistical problems can also be posed. For example, the determination of the order of a model can be approached as a statistical problem. Let V_n denote the loss function obtained for a model with n parameters. The function V_n decreases with increasing n. The problem is to decide if the decrease in V is significant or not. For systems with one output the test quantity

$$F_{n_1,n_2} = \frac{V_{n_1} - V_{n_2}}{V_{n_2}} \frac{N - n_2}{n_2 - n_1} \quad \text{for} \quad n_2 > n_1 \tag{32}$$

where N is the number of observations of the output, can be shown to be asymptotically F-distributed. The quantities F_{n_1,n_2} can thus be used to test whether the loss function is reduced significantly when the number of parameters in the model is increased from n_1 to n_2.

Notice that one can test whether the residuals are gaussian and uncorrelated. It is, however, virtually impossible to assert that the data were actually generated by a model (16) with specific parameters. In practice, this is of course never true, since the model (16) is only an approximation of a complex process. Thus, the results obtained by using the statistical theories must be handled with great care when applied to real data. Notice that when the methods are tried on simulated data it is always possible to assert that the assumptions required by the statistical theory are fulfilled!

6.2 *Other interpretations*

The algorithm obtained by minimizing the criterion (29) can be given a physical interpretation, even in the case when no statistical assumptions are made. Equation (16) can be rewritten as

$$\begin{aligned} \hat{\mathbf{x}}(t+1) &= \mathbf{A}\hat{\mathbf{x}}(t) + \mathbf{B}\mathbf{u}(t) + \mathbf{K}[\mathbf{y}(t) - \mathbf{C}\hat{\mathbf{x}}(t) - \mathbf{D}\mathbf{u}(t)] \\ \boldsymbol{\varepsilon}(t) &= \mathbf{y}(t) - \mathbf{C}\hat{\mathbf{x}}(t) - \mathbf{D}\mathbf{u}(t) \end{aligned} \tag{33}$$

The quantity $\mathbf{C}\hat{\mathbf{x}}(t) + \mathbf{D}\mathbf{u}(t)$ can be interpreted as a one-step prediction of $\mathbf{y}(t)$ based on $\mathbf{y}(t-1), \mathbf{y}(t-2), \ldots$. The quantity $\boldsymbol{\varepsilon}(t)$ can thus be interpreted as the error in the prediction of the output. To minimize the loss function (29) thus means that the parameters of the model are changed in such a way that the error in predicting the output one step ahead is as small as possible according to the criterion (29). In the case of single output systems the criterion (30) is simply the mean squares prediction error.

7. POWER GENERATOR DYNAMICS

In this section, the techniques described in the previous sections will be applied to power generator modelling. The results of this section are based on ref. 17.

The analysis is based on experiments made by Dr. Stanton[24, 25] on a 50 MW turboalternator. The experiments consisted of recording the variations in the terminal voltage V, the active and reactive components of the armature current, I_r and I_q respectively, and the angular velocity ω during normal operation. The instrumentation used is described in ref. 23. This report also contains a description of the experimental procedure and the difficulties with this type of experiment. Several experiments were performed by Stanton. In this example, we will base computations on an experiment performed with the governor blocked, since we are interested in open loop dynamics. The record length was 600 sec and the sampling interval was 0.5 sec for the angular velocity and 0.125 sec for the other variables.

In this example, the dynamics relating angular velocity to electric torque $M = VI_r/\omega$ will be considered. A plot of these two variables is shown in the upper part of Fig. 1. The generator is frequently described by the simplified model

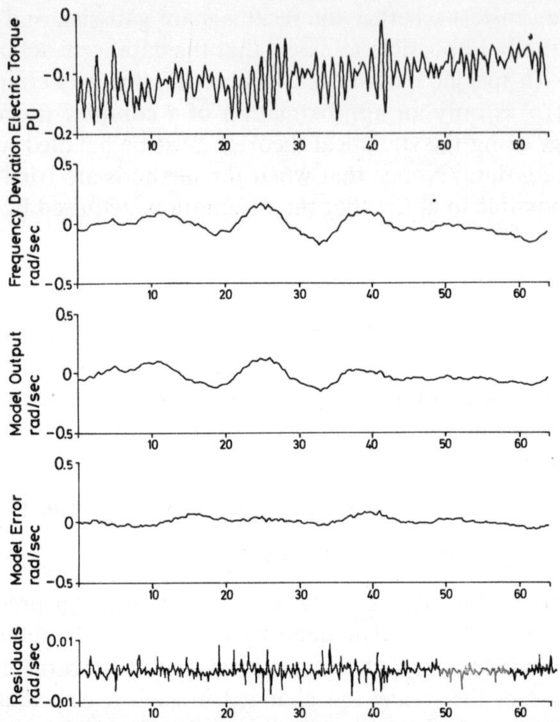

Fig. 1. The results of identification of power generator dynamics. The uppermost curve shows the measured input (electric torque $\delta M = I_r V/\omega$ in per unit and measured output (angular velocity ω in rad/sec). The model output computed from the fifth order model with coefficients given in Table 1 and disturbances neglected, the model error and the residuals $\{\varepsilon(t)\}$ are also shown in the diagram.

$$J\frac{d\omega}{dt} + D\omega = M_m - \frac{VI_r}{\omega} \tag{34}$$

where ω is the angular velocity of the rotor, J the moment of inertia of the rotor, D the damping coefficient, V the terminal voltage, I_r the active component of the armature current and M_m the mechanical torque. There are also more elaborate descriptions leading to models of higher order.

Since it is not known *a priori* that a model like eqn. (34) is compatible with the data, one could first attempt to fit general linear models with the canonical structure (20) and different orders *n*. This will give an indication of the complexity required to describe the data. The loss functions obtained from a maximum likelihood identification of models of different order are shown in Table 1. To save computer time, the computations are based on 1000 input–output pairs only. Since the sampling interval of the angular velocity is 0.5 sec and the sampling interval of the electric power is 0.125 sec, a compatible data set is created by interpolating the angular velocity. The initial state of the model (20) is also determined.

In Table 1 the test quantities (32) for testing the order of the system are also

TABLE 1

MINIMAL VALUES OF THE LOSS FUNCTION FOR MAXIMUM LIKELIHOOD IDENTIFICATION OF MODELS OF DIFFERENT ORDER RELATING ANGULAR VELOCITY ω TO ELECTRIC TORQUE $I_r V/\omega$ FOR THE ELECTRIC GENERATOR
The computations are based on 1000 data points

n	V_n	$F_{n,n+1}$
1	0.01442	126
2	0.009562	26
3	0.008560	73
4	0.006656	17
5	0.006230	2.0
6	0.006180	

TABLE 2

COEFFICIENTS OF FIFTH ORDER MODEL RELATING ANGULAR VELOCITY ω TO ELECTRIC POWER FOR ELECTRIC GENERATOR

N	1000	2400
a_1	-2.628 ± 0.031	-2.581 ± 0.024
a_2	2.290 ± 0.074	2.204 ± 0.049
a_3	-0.715 ± 0.082	-0.640 ± 0.048
a_4	0.078 ± 0.067	0.024 ± 0.044
a_5	-0.025 ± 0.026	-0.004 ± 0.019
b_1	-0.835 ± 0.092	-0.836 ± 0.061
b_2	1.120 ± 0.244	1.016 ± 0.016
b_3	-0.805 ± 0.308	-0.529 ± 0.197
b_4	0.599 ± 0.250	0.198 ± 0.162
b_5	-0.080 ± 0.100	0.150 ± 0.075
c_1	-0.899 ± 0.036	-0.854 ± 0.027
c_2	-0.005 ± 0.039	-0.011 ± 0.024
c_3	-0.013 ± 0.036	-0.027 ± 0.024
c_4	-0.624 ± 0.034	-0.617 ± 0.021
c_5	0.579 ± 0.029	0.541 ± 0.021
λ	0.0025	0.0025

evaluated. Assuming that all assumptions required for the order test are fulfilled, a straightforward application of the order test would thus indicate that a fifth order model is appropriate. The coefficients of the fifth order model are shown in Table 2. The accuracy estimates are obtained from eqn. (31). The results obtained by performing the identification on a longer data set $N = 2400$ are also shown in Table 2. A comparison of the results obtained for $N = 1000$ and $N = 2400$ shows that it is not unreasonable to assume that the system is time-invariant. A comparison of the accuracy estimates also indicates that they decrease as $1/\sqrt{N}$, as can be expected from the theory. Notice in Table 2 that there is a significant difference between the relative accuracies of the parameters b_i and the relative accuracies of a_i and c_i. The results are quite typical of those obtained in other cases when no perturbations in the input are used.

References pp. 23–24

The results of the identification are illustrated in Fig. 1, which shows the measured inputs and outputs, the model output and the difference between the measured output and the model output. The model output is computed from the model (20) with the coefficients given in the first column of Table 2 and disturbances $\{e(t)\}$ neglected. The model output thus explains how much of the actual output that can be explained from the input. Figure 1 reveals that only about half of the observed output can be related to the input. The signal-to-noise ratio is thus fairly low. The situation is quite typical for data obtained during normal operation with no extra perturbations introduced. Also notice from Table 2 that with a record of length 125 sec it is possible to get reasonable parameter estimates of the parameters a_i and c_i.

The residuals can be interpreted as the one-step prediction errors obtained from a predictor determined from the model (20). The residuals thus show how well the output can be predicted one step ahead. The standard deviation of the residuals is $\lambda = 0.0025$, which means that it is possible to predict the angular velocity one sampling interval ahead with a standard deviation of 0.0025 rad/sec using the model obtained.

If the identification procedure is to have a nice statistical interpretation the residuals should be independent stochastic variables. In Fig. 2 the covariance of the residuals is shown. This diagram indicates that the residuals at least are uncorrelated. The residuals are, however, not normally distributed, as is seen by the diagram of cumulative frequencies in Fig. 3. This means, for example, that the results of the order test can be questioned in a case like this.

The results obtained clearly indicate that the simple first order model (34) is not compatible with the data.

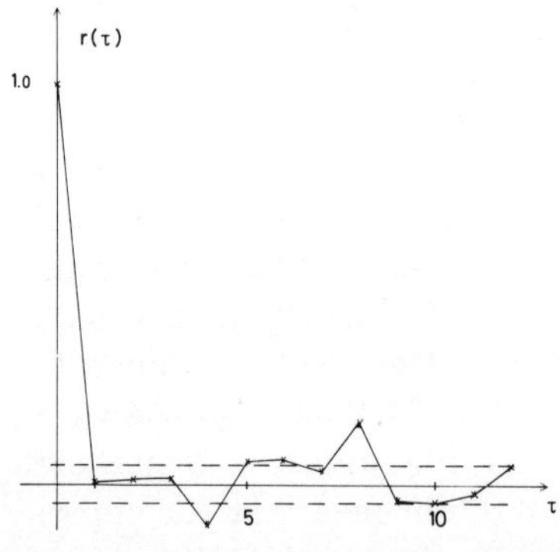

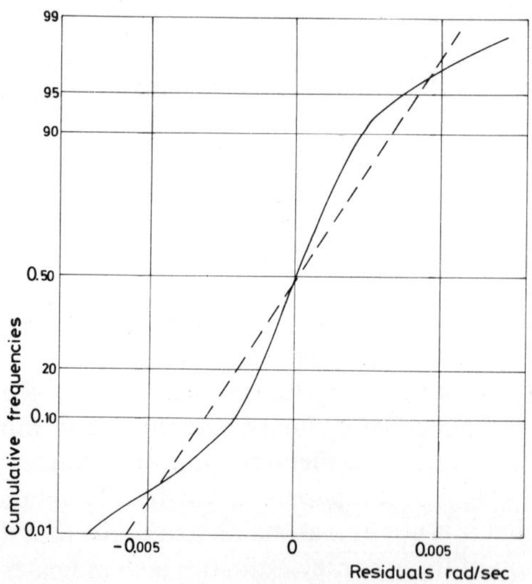

Fig. 2. Sample covariance function for the residuals of the model (20) with parameters according to the first column of Table 2.

Fig. 3. Cumulative frequencies of residuals of the model (20). The scales of the diagram are such that a normal distribution corresponds to a straight line.

8. NUCLEAR REACTOR DYNAMICS

The experiments were done by AB Atomenergi on the ÅGESTA reactor, located in a suburb of Stockholm. A schematic picture of the reactor is shown in Fig. 4. The results of this section are based on ref. 12.

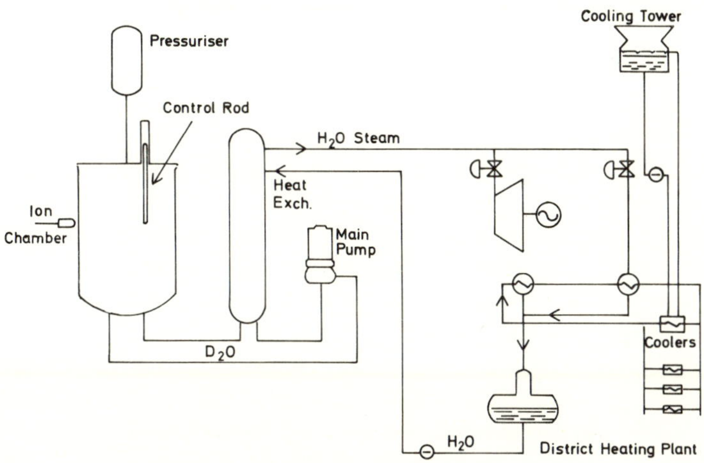

Fig. 4. Schematic diagram of the Ågesta nuclear reactor.

In the experiments the control rod position was perturbed and the nuclear power was measured. The input was chosen as a PRBS* sequence with period 127. The shortest pulse length was 20 sec and the sampling interval was 5 sec. The input-output signals from the experiment are shown in Fig. 5. As an initial attempt, a maximum likelihood identification is carried out using models (20) of different orders. The values of the loss functions obtained are shown in Table 3.

A straightforward application of the order test indicates that the model has to be of at least fourth order. At a risk level of 5% the test limit is 3, and the decrease in the loss function obtained when going from a fourth order system to a fifth order system is therefore not significant. The coefficients of the fourth order model are shown in Table 4.

The results of the identification are illustrated in Fig. 5. It is clear from this diagram that a major part of the observed output is caused by the input. The magnitude of the model error never exceeds 0.8 MW. The residuals have a standard deviation of 0.113 MW, which means that the output can be predicted 5 sec ahead with a standard deviation of 0.113 MW. The large model error and the large residual at time 20.2 min were traced to a malfunction of the control rod servo.

By comparing the results given in Table 4 with those in Table 2 we find that the relative accuracies of the parameters associated with the nuclear reactor dynamics are significantly higher than those associated with the dynamics of the electric

* PRBS = pseudo random binary sequence.

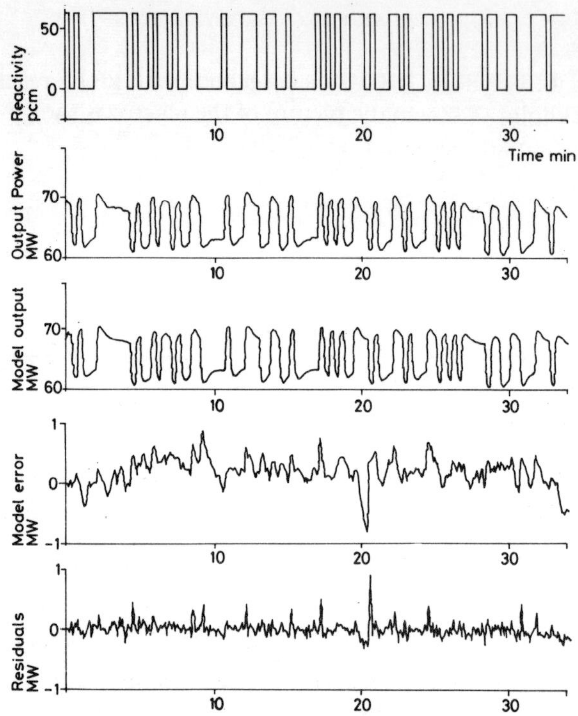

Fig. 5. The identification of nuclear reactor dynamics. The figure shows the measured input, reactivity in pcm (per cent mille $= 10^{-5}$) (control rod position), the measured output in MW (nuclear power), the model output, the model error and the residuals.

TABLE 3

MINIMAL VALUES OF THE LOSS FUNCTION FOR MAXIMUM LIKELIHOOD IDENTIFICATION OF MODELS OF DIFFERENT ORDER RELATING NUCLEAR POWER TO CONTROL ROD POSITION

The computations are based on 1125 input–output pairs

n	V_n	$F_{n,n+1}$
1	203.344	2502
2	26.378	218
3	16.642	61.2
4	14.284	1.2
5	14.238	0.2
6	14.229	

TABLE 4

PARAMETERS OF A FOURTH ORDER MODEL RELATING NUCLEAR POWER TO CONTROL ROD POSITION

The estimates are based on 1125 input–output pairs with a sampling interval of 5 sec

a_1	-2.442 ± 0.023
a_2	2.096 ± 0.047
a_3	-0.679 ± 0.029
a_4	0.032 ± 0.004
b_1	0.1013 ± 0.0002
b_2	-0.2377 ± 0.0024
b_3	0.1884 ± 0.0045
b_4	-0.0517 ± 0.0023
c_1	-1.479 ± 0.038
c_2	0.502 ± 0.061
c_3	0.122 ± 0.055
c_4	-0.072 ± 0.030
λ	0.113

generator. Compare in particular the coefficients b_i. This is due to the fact that perturbation signals were used in the reactor experiment, while no perturbations were introduced in the generator experiment. Similar observations have been made in many other situations.

9. DRUM BOILER DYNAMICS

In this section, applications to drum boiler modelling are discussed. The results are based on refs. 10 and 4. The experiments were performed on the P16–G16 power plant of the Öresundsverket of Sydkraft AB in Malmö, Sweden. The unit consists of a Steinmüller drum boiler and a Stal–Laval turbine. It has a power of 160 MW. A schematic picture of the boiler is shown in Fig. 6.

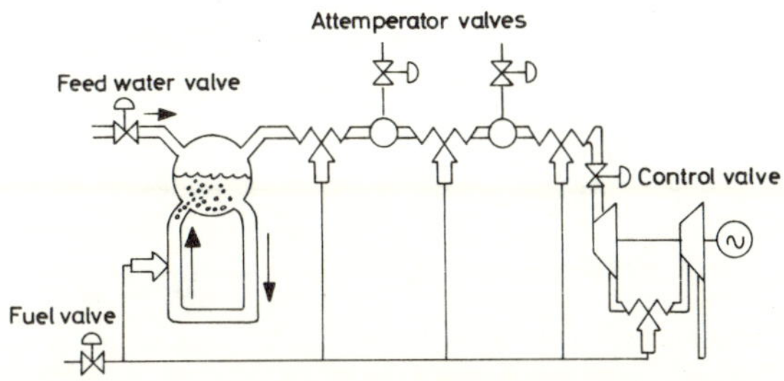

Fig. 6. Schematic diagram of the boiler–turbine unit.

In the experiment, the input variables fuel flow, feedwater flow and control valve position were perturbed and the relevant process variables were recorded. Since the open loop dynamics were of major interest all major regulators except the fuel–air regulator were removed. For safety reasons only one process variable at a time was changed. The determination of linear multivariable models was based on superposition of the results of several experiments.

Many experiments were performed, both for the purpose of determining linear steady state models and for the purpose of determining non-linear models. A detailed discussion of the measurements including identification and modelling is given by Eklund[10].

In this example, the modelling of the drum boiler only will be discussed. The inputs are taken as fuel flow, feedwater flow and steam flow. The output variables are chosen as drum level and drum pressure. A set of experimental data is shown in Fig. 7. The sampling interval was chosen as 10 sec.

9.1 Linear models

Maximum likelihood identification of single output models for drum pressure and drum levels leads to models of orders 3 and 2. It is thus not unreasonable that a model like the one given in Example 1, §4, is compatible with the data. The model given in Example 1 was obtained from physical considerations. Since it contains

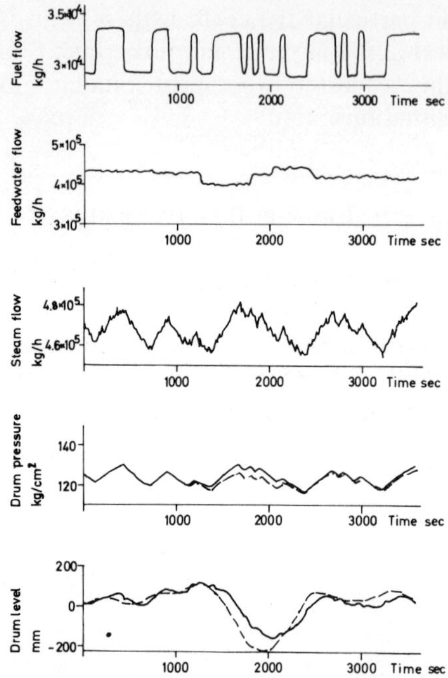

Fig. 7. Results of linear modelling of a drum boiler. The full lines are the measured data and the broken lines are the model outputs.

27 parameters, while at most 25 parameters can be determined from input–output data, some parameters must be fixed. There are disturbances in the data which are not white measurement noise. It is then necessary to choose the model structure (16) which allows for a more general noise model. Thus the 10 elements of the **K**-matrix must also be included as parameters. Since a model (16) was determined from physical considerations initial estimates of all parameters except those of the **K**-matrix were available. Several attempts were made to identify different parameter sets. A typical result obtained is listed in Table 5. Notice the significant reduction in the value of the loss function. The model obtained with the parameters of Table 5 is compared with the measurements in Fig. 7.

The estimated covariance matrix is

$$\hat{\mathbf{R}} = \begin{bmatrix} 0.0240 & 0.016 \\ 0.016 & 12.4 \end{bmatrix}$$

This means that the drum pressure can be predicted 10 sec ahead with a standard deviation of 0.16 bar and the drum level with a standard deviation of 3.5 mm.

9.2 *Non-linear models*

For power system analysis it is highly desirable to have fairly simple models available which give the gross behaviour of the different system components.

TABLE 5

INITIAL GUESSES AND PARAMETER ESTIMATES OBTAINED WHEN FITTING A MODEL WITH THE STRUCTURE OF EXAMPLE 1 TO DRUM BOILER MEASUREMENTS

	Initial	*Identified*
V	0.2567×10^{17}	0.3850×10^{5}
α_1	-0.02835	-0.02052
α_3	0.01382	0.00896
α_5	-0.1341×10^{-3}	-0.0738×10^{-3}
α_{10}	0.0406	0.1678
α_{13}	-0.0454	-0.3295
α_{16}	-0.2266	-0.3622
α_{19}	0.1162×10^{-4}	0.1446×10^{-4}
α_{24}	-0.01575	-0.00496
k_{11}	0	1.271
k_{12}	0	-0.1041×10^{-2}
k_{21}	0	0.01027
k_{22}	0	0.1279×10^{-2}
k_{31}	0	-0.1214
k_{32}	0	-0.06576
k_{41}	0	0.5432
k_{42}	0	0.01398
k_{51}	0	-0.7702×10^{-3}
k_{52}	0	0.3585×10^{-4}

From studies of linear models like the one just discussed a great deal of insight into the behaviour of boiler–turbine units was obtained. It was found that the gross behaviour of a boiler–turbine unit could be described by the equations

$$\begin{aligned} \frac{dp}{dt} &= \alpha[-f(p, u_2) + g(u_1, u_3)] \\ P &= f(p, u_2) \end{aligned} \tag{35}$$

where p is the drum pressure, P the output power, u_1 the fuel flow, u_2 the control valve setting and u_3 the feedwater flow. A model of the structure (35) can also be derived from an energy balance if several (crude) approximations are done. Such a derivation will also give the structure of the functions f and g. Details of this are given in ref. 4. It is shown that a possible choice is

$$f(p, u_2) = \alpha_4[u_2 p^{5/8} - \alpha_5] \tag{36}$$

$$\alpha g(u_1, u_3) = \alpha_2 u_1 - \alpha_3 u_3 \tag{37}$$

$$\alpha = \alpha_1/\alpha_4 \tag{38}$$

The boiler–turbine unit can thus be represented by a non-linear model with five parameters $\alpha_1, \alpha_2, \ldots, \alpha_5$. Determining these parameters from several experiments covering a wide operating range the following values are obtained for the unit P16–G16.

$\alpha_1 = 0.00305$
$\alpha_2 = 0.02$
$\alpha_3 = 4.4 \times 10^{-2}$
$\alpha_4 = 11.45$
$\alpha_5 = 8.2$

The performance of the model at two different operating conditions is illustrated in Figs. 8 and 9. Notice in particular that the model agrees well with experiments over a wide operating range.

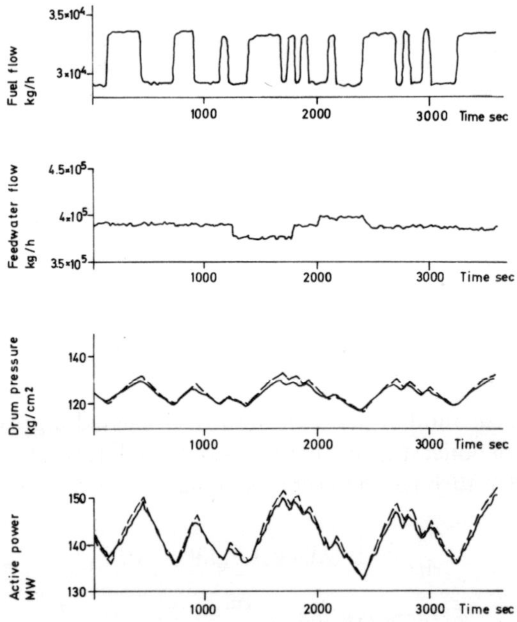

Fig. 8. Comparison of measured boiler–turbine data (full lines) and the responses of the non-linear model (35) (broken lines).

ACKNOWLEDGEMENTS

It is my pleasure to acknowledge my gratitude to several persons and institutions. My research in the field of system identification has been partially supported by the Swedish Board for Technical Development (contract 71-50/U33). Data from the power generator measurement were kindly provided by Dr. K. N. Stanton. The results of experiments on the nuclear reactor were provided by AB Atomenergi, Stockholm, Sweden. The experiments on the thermal power plant were done by Dr. K. Eklund as part of his thesis in collaboration with Sydkraft AB, Malmö, Sweden. The computations reported have been performed by my collaborators at the Lund University, Dr. K. Eklund and Mr. S. Lindahl. Special thanks are due to I. Gustavsson with whom I have enjoyed extensive discussions on applications of system identification over many years.

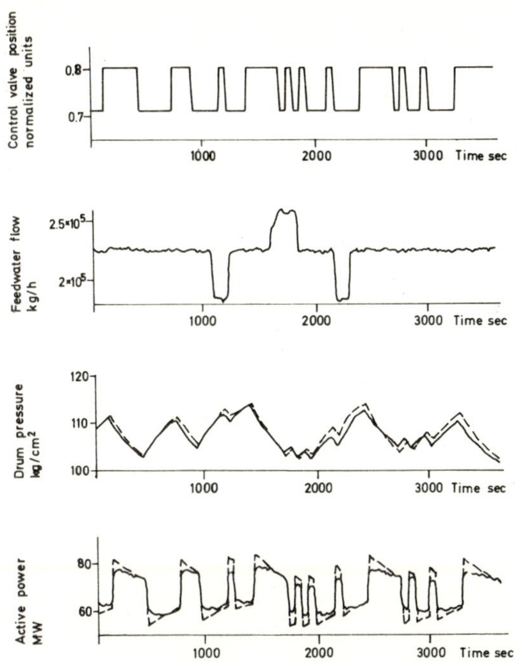

Fig. 9. Comparison of measured boiler–turbine data (full lines) and the responses of the non-linear model (35) (broken lines).

NOTES AND REFERENCES

System modelling and identification is discussed in many recent articles. Many references are given in the recent survey articles[5,19] and in the book *System Identification*[22]. There have also been two IFAC symposia in Prague in 1967 and 1970 entirely devoted to this field. The use of maximum likelihood techniques is discussed in refs. 3, 7, 8, 11, 13, 14, 16 and 27. Canonical structures for linear multivariable systems are treated in refs. 1, 18 and 26. A more detailed discussion of the nuclear reactor identification is given in ref. 13. The same techniques have also been applied to the modelling of the Halden reactor in Norway (see ref. 21). The section on boiler modelling is based on refs. 4 and 10 which include detailed analyses of many measurements.

REFERENCES

1 I. ACKERMANN, Die minimale Ein-Ausgangs-Beschreibung von Mehrgrössen-Systemen und ihre Bestimmung aus Ein-Ausgangs-Messungen, *Regelungstechnik und Prozessdatenverarbeitung*, **5** (1971) 203–206.

2 K. J. ÅSTRÖM, *Introduction to Stochastic Control Theory*, Academic Press, New York, 1970.

3 K. J. ÅSTRÖM AND T. BOHLIN, Numerical identification of linear dynamic systems from normal operating records, paper at the IFAC Symposium on Theory of Self-Adaptive Systems, Teddington, England, in P. H. HAMMOND (Ed.), *Theory of Self-Adaptive Control Systems*, Plenum Press, New York, 1966.

4 K. J. ÅSTRÖM AND K. EKLUND, A simplified nonlinear model of a drum-boiler turbine unit, *Report 7104*, Division of Automatic Control, Lund Institute of Technology, April 1971, to appear in *Intl. J. Control.*

5 K. J. ÅSTRÖM AND P. EYKHOFF, System identification—a survey, *Automatica*, **7** (1971) 123–162.

6 R. BELLMAN AND K. J. ÅSTRÖM, On structural identifiability, *Mathematical Biosciences*, **7** (1970) 329–339.

7 P. E. CAINES, The parameter estimation of state variable models of multivariable linear systems, *Control Systems Centre Report No. 146*, The University of Manchester Institute of Science and Technology, April 1971.

8 J. EATON, Identification for control purposes, *IEEE Winter Meeting, New York, 1967.*

9 K. EKLUND, Linear mathematical models of the drum–downcomer–riser loop of a drumboiler, *Report 6809*, Division of Automatic Control, Lund Institute of Technology, November 1968.

10 K. EKLUND, Linear drum boiler–turbine models, *Report 7117*, Division of Automatic Control, Lund Institute of Technology, November 1971.

11 C. A. GALTIERI, The problem of regulation in the computer control of industrial processes, Part I: Theory, *IBM Res. Rept RJ382*, San Jose, California, 1966.

12 I. GUSTAVSSON, Maximum likelihood identification of dynamics of the Ågesta reactor and comparison with results of spectral analysis, *Report 6903*, Division of Automatic Control, Lund Institute of Technology, February 1969.

13 I. GUSTAVSSON, Parametric identification of multiple input single output dynamic systems, *Report 6907*, Division of Automatic Control, Lund Institute of Technology, July 1969.

14 I. GUSTAVSSON, Comparison of different methods for identification of linear models for industrial processes, *2nd IFAC Symposium on Identification and Process Parameter Estimation, Prague, 1970*, Paper 11.4.

15 P. HAGANDER, C. KÄLLSTRÖM AND K. J. ÅSTRÖM, Real time computing: Implementing linear filters, *Report 7106B*, Division of Automatic Control, Lund Institute of Technology, May 1971.

16 R. L. KASHYAP, Maximum likelihood identification of stochastic linear systems, *IEEE Trans. Autom. Control*, **AC-15** (1970) 25.

17 S. LINDAHL, Identification of power generator dynamics from normal operating data, Term paper in course on System Identification Techniques, Division of Automatic Control, Lund Institute of Technology, 1971.

18 D. G. LUENBERGER, Canonical forms for linear multivariable systems, *IEEE Trans.*, **AC-12** (1967) 290–293.

19 R. E. NIEMAN, D. G. FISHER AND D. E. SEBORG, A review of process identification and parameter estimation techniques, *Intl. J. Control*, **13** (1971) 209.

20 V. OJA, Frequency-response method applied to the study of turbine regulation in the Swedish power system, in R. OLDENBURGER (Ed.), *Frequency Response*, Macmillan, New York, 1955.

21 G. OLSSON, Maximum likelihood identification of the Halden boiling water reactor, unpublished.

22 A. P. SAGE AND J. L. MELSA, *System Identification*, Academic Press, New York, 1971.

23 K. N. STANTON, An incremental wattmeter for use in interconnected power systems, *Trans. Instn Engrs. Aust.*, May 1964, p. 20.

24 K. N. STANTON, Estimation of turboalternator transfer functions using normal operating data, *Proc. IEE*, **122** (1965) 1713–1720.

25 K. N. STANTON, Use of normal operating data from a power system to estimate turboalternator transfer functions, Report, Purdue University, West Lafayette, Indiana, 1969.

26 W. G. TUEL, JR., Canonical forms for linear systems—I, *IBM Res. Rept RJ375*, San Jose, California, 1966.

27 K. T. WOO, Maximum likelihood identification of noisy systems, *2nd IFAC Symposium on Identification and Process Parameter Estimation, Prague, 1970*, Paper 3.1.

Discussion

J. P. WAHA (*INTERCOM, S.A., Brussels, Belgium*)

In non-linear identification it is felt that quasi-linearization is useful. Work done in

Belgium on a thermal model of an 80 MVA 150 kV/70 kV transformer has shown its possibilities. The theoretical model used was of the first order, with non-linear forcing functions of the form U^m. Using the m exponent from constructive data, errors in temperature of more than 10°C were obtained. Identifying by quasi-linearization and using the new coefficients, including a new m exponent, led to an identification error of less than 2°C. The measurement length was approximately one week, the sampling period was of the order of minutes, and the dominant time constant of the transformer was about 4 hours.

What method was used to identify the non-linear model of the drum boiler?

K. J. ÅSTRÖM
The particular values of the coefficients were obtained by curve fitting to several different experiments. The full details are given in ref. 4 of my paper.

L. S. DZUNG (*Brown, Boveri & Co. Ltd., Baden, Switzerland*)
The difference between the measured behaviour and that predicted by the theoretical model may be due to insufficiency of the model, or to errors of measurement or of the predictor. Professor Åström tries to minimize the deviation by successively increasing the number of free parameters of his model. This is certainly in order if the aim is to simulate the system behaviour as accurately as possible. Now, in the case of a small number of model parameters, it is possible to ascribe physical meaning to each parameter, which may then be modified to obtain the desired system behaviour. If the number of parameters is increased, it is not always possible to interpret the proper physical meaning of each parameter. In fact, some of these parameters may have no physical significance. A model with too many parameters may therefore be less useful for the designer of the system, although it may be helpful to the designer of control apparatus. I think there should be some criterion for the optimal number of parameters for this aspect also.

K. J. ÅSTRÖM
The loss function will always decrease with an increasing number of parameters. The decrease in the loss function, however, may not be significant. As discussed in §6 of my paper, I have found it useful to approach the problem of order as a statistical hypothesis test using the test quantity (32). I do not claim that this gives *the optimal* number of parameters. It is, however, one way to get an indication of the order. The possibility of giving a physical interpretation of the model parameters depends on the model structure.

C. J. EAGLEN (*Brown, Boveri & Co. Ltd., Baden, Switzerland*)
How does the sampling interval influence parameter estimation for control and simulation?

K. J. ÅSTRÖM
The selection of the sampling interval is very important. If the sampling interval is h, it is virtually impossible to estimate parameters of modes with time constants less

than, say, $0.3h$ with a reasonable accuracy. Hence, the selection of sampling interval puts an upper limit to the frequency range over which the model is valid. This problem is further discussed in the following references.

K. J. ÅSTRÖM, On the choice of sampling rates in parametric identification of time series, *Information Sciences,* **1** (1969) 273–278.

I. GUSTAVSSON, Choice of sampling interval for parametric identification, *Report 7103,* Division of Automatic Control, Lund Inst. of Technology, April 1971.

H. GLAVITSCH (*Brown, Boveri & Co. Ltd., Baden, Switzerland*)
Referring to the first practical case studied, I would like to ask if the power generator supplied an isolated load or if it was connected to an infinite bus.

K. J. ÅSTRÖM
The measurements were done by Dr. Stanton. Since he is in this room I would kindly ask him to answer this question.

K. N. STANTON (*Systems Control, Inc., Palo Alto, California, U.S.A.*)
The power generator data were collected from a 50 MW machine operating normally and interconnected with the rest of the power system through high voltage cables, but the coupling and the data are typical of those experienced in power systems.

H. GLAVITSCH
From Dr. Stanton's answer I gather that the generator was connected to the rest of the system. Hence, the model in this case has to be at least of second order because of the interaction with other generators.

K. J. ÅSTRÖM
The order test does in fact indicate that a model of fifth order would be consistent with the experimental data.

W. B. JERVIS (*Merz and McLellan, Newcastle upon Tyne, England*)
I should like to confine my remarks to the random sampling measurements which the author referred to in the nuclear boiler studies. I note that the author employed a large signal level into the reactor controls which gave an electrical power excursion of approximately 10 MW. This value seems to be an extremely high error level for injection and the effect of non-linearities could lead to serious error if linearized theory was assumed for the model. This influence could be shown by injecting a small level perturbation, *e.g.* 1 MW or smaller, and comparing the results obtained.

What precautions would the author take to ensure that site measurements and model results give comparable results?

Regarding the length of time employed for random sampling, I noticed that the author gave an illustration for a 30 min period. My personal experience with sinusoidal sampling has shown that significant error can be introduced if the length of time employed is sufficient to allow load conditions on the machine and network to change between start and finish of the test. It would seem that short-time sampling should be the aim, provided that accuracy is not lost by losing the effects of the largest time lags. I would like to mention that practical models, particularly for boilers, have an economic incentive to a user to shorten the commissioning time at site by predicting the correct setting for the automatic control settings.

K. J. Åström

The experiments on the nuclear reactor were performed by AB Atomenergi. According to them (see ref. below) the ÅGESTA reactor can be modelled by linear models for the power excursions used. Our models also agree well with those obtained by other techniques and smaller excursions. I do not know enough about nuclear reactors to judge whether the ÅGESTA reactor is non-typical in this respect. I appreciate your remark concerning the use of models to shorten commissioning.

P. Å. Bliselius, H. Vollmer and F. Åkerhjelm, Experimental and theoretical dynamic study of the Ågesta Nuclear Power Station, *Report AE-376*, AB Atomenergi, Stockholm, 1969.

A. H. Glattfelder (*Eidg. Technische Hochschule, Zürich, Switzerland*)

I would like to ask a few questions on the last part of your paper—trying to link modelling and identification (input–output only methods).

Professor Profos has shown that certain characteristic coefficients of boiler dynamics can easily be calculated from geometrical and physical data. What are the main connections between these coefficients and the matrix elements given in your paper? How well do identified and precalculated values agree?

In the normal operation of a power plant, roughly stationary conditions exist for only about 30–60 min. How large will confidence intervals for the matrix elements be after this time? Do you feel that this would be a sufficient basis on which to construct an optimal control scheme giving better results than conventional linear control?

K. J. Åström

The model like the one in Example 1, §4.2, and the non-linear boiler model in §9 are in fact based on physical considerations following work by Professor Profos and other related results. Thus the parameters can also be determined from construction data. A detailed comparison of identified parameter values with construction data is given in Dr. Eklund's thesis (ref. 10 of my paper). If some crucial parameters were drifting so much that the performance of the control scheme deteriorated, I would propose that you tracked the parameters in real time and incorporated these estimates in your control strategy.

G. Quazza (*ENEL, Milan, Italy*)

One difficulty with statistical identification of not entirely stationary systems, as our experience in trying to determine the transfer function $\Delta f/\Delta P$ of an electric power system has clearly shown, is associated with the conflicting requirements of stationariness and low frequency accuracy in the choice of the recording sample length. Reasonably good confidence limits in the transfer function parameters affecting low frequency response could be obtained only after several hours of recording samples—in the case of the electric power system, when spontaneous random disturbances are used as the stimulating input—while power system structure changes are likely to occur and become relevant in 2 or 3 hours, with consequent modifications in the transfer function itself. On the other hand, the typical spectrum of spontaneous random load changes is approximately Poissonian, with decreasing magnitudes for increasing frequencies. Hence, the "stimulation" of high frequencies is scarce, with consequently a poor approximation of the identifiable transfer function in that range.

For these reasons, statistical identification of electrical power systems by spontaneous signals usually yields less accurate results than deterministic identification. However, it has the advantage of not necessarily requiring isolation of the network from interconnected neighbours.

Could the author comment upon such problems and on the comparison of statistical and deterministic identification?

K. J. Åström

The first statistical identification techniques relied on the spontaneous random disturbances as stimulating inputs. This is not, however, a prerequisite for using statistical identification techniques. One great advantage of the methods I have discussed today is that they make few assumptions on the input. In fact I strongly recommend that external perturbations are used. In the examples presented in my paper only the power generator experiment was done using natural perturbations. In all the other cases external perturbations were used. A comparison of the relative accuracies of the parameters of the power generator with, for example, those of the nuclear reactor clearly shows the importance of using external perturbations.

Real-time Data Processing Using State Estimation in Electric Power Systems

E. HANDSCHIN

Brown, Boveri Group Research Center, Baden, Switzerland

SUMMARY

The measurement and telemetering equipment in an electric power system is always subject to random and systematic instrument and phenomenon errors. The crude information about the system's behaviour is unsatisfactory for assessing its state and performance. The estimation problem consists in determining an approximation to the time history of the system's response variables from the erroneous measurements. Its solution implies the development of algorithms for processing the measurements in real time.

Estimation theory has been used extensively for navigation of aircraft and spacecraft, as well as for post-experimental data analysis. Only recently, it has been shown that estimation theory is well suited for on-line data processing in electric power systems. In this application there are two specific features. Firstly, the optimality of the estimator is primarily related to the reliability of the estimated quantities. Secondly, the estimation algorithm must be simple enough to be implemented on a small computer and to work in real time.

Several estimation methods are summarized for both the static and dynamic cases. While the static estimator deals only with a snapshot of the system state, the dynamic estimator yields the state as a function of time. In both cases, the estimation problem is inherently a non-linear one whose solution requires suitable approximation methods.

The applicability of state estimation techniques depends very much on the preprocessing of the crude measurements. Bad data and anomalies in the system have to be detected automatically in order to improve the reliability of the estimates. Simulation studies are indispensable for the development of estimation algorithms, yet only the test with real data finally proves the validity of the solution.

The use of state estimation in electric power systems may proceed in three stages. Firstly, the estimator is used for building up a reliable data base giving information about the system structure and state. Suitable display techniques combined with estimation algorithms effectively facilitate the human operator's task of supervising a complex, large system. In a further stage, the estimator may be used for security assessment and contingency evaluation. Finally, in the third stage, the combination of estimation and control theory may contribute towards improving load-frequency control, economic load dispatch, on-line stability analysis and power station control. In this way state estimation becomes an integral part of automatic power system control.

1. INTRODUCTION

The influence of modern control theory on power systems operation has grown rapidly during the last few years. Two reasons for this development are as follows.

(1) The expansion of today's power systems requires more and more sophisticated monitoring and control schemes, in order to help in maintaining the system in a secure and efficient mode of operation by facilitating the operator's difficult tasks.

(2) Large computer installations allow the implementation of advanced control schemes in real time. Due to programming facilities, a wide range of different algorithms can be used on the same computer in a time-sharing mode.

The availability of reliable system data is a necessary step in the satisfaction of these aims. The complete knowledge of the past and present state of the system is an indispensable prerequisite for the human operator and the computer, to guarantee reliable and economic generation and distribution of electric power. Due to the inherent uncertainty and crude measurements, system data may not be reliable enough for immediate use. Data processing using a digital computer is therefore required, in order to present to the operator only that information which is relevant for the problem he has to solve at each instant of time. It is the object of this paper to discuss the problems of data processing in power systems.

1.1 The framework of modern control theory

In this paper the problem of data processing is put into the framework of modern control theory. This theory is characterized by two main features:

(1) the state variable approach, where the minimal number of variables is selected in order to describe the system completely; and

(2) the mathematical modelling of internal and external system uncertainties.

Figure 1 serves as an illustration of the problems solved by modern control theory. The physical system is modelled together with its environment. The system state variables $\mathbf{x}$ are affected by random disturbances $\mathbf{w}$. In general, the state variables $\mathbf{x}$ are not all directly accessible. The measurement system only yields a selection of the quantities required to describe the system state. Moreover, the measurements $\mathbf{z}$ are contaminated by observation errors $\mathbf{v}$. Thus, the output quantities $\mathbf{z}$ may be unsatisfactory for assessing the system's performance. As an illustration, reference is made to the phase angles of the complex bus voltages in a power system. No satisfactory method exists to measure these quantities directly.

From the control engineer's point of view there are two basic problems to be solved.

(1) Estimation theory provides the background for determining an estimate $\hat{\mathbf{x}}$ of the system state $\mathbf{x}$.

(2) Control theory deals with the problem of computing the control variables $\mathbf{u}$ so that the system behaves in a prescribed manner.

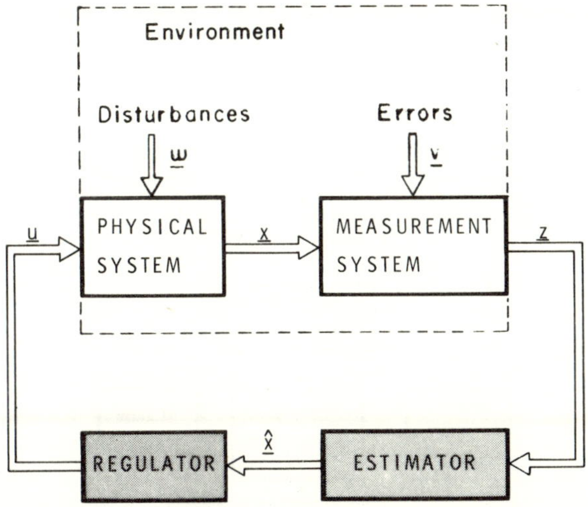

Fig. 1. General problem of estimation and control: $\mathbf{x}$ = true system state vector, $\mathbf{z}$ = measurement vector, $\hat{\mathbf{x}}$ = estimated state vector and $\mathbf{u}$ = control vector.

1.2 Estimation

According to Fig. 1, the estimator yields an approximate value $\hat{\mathbf{x}}$ of the system state $\mathbf{x}$ using noisy observations $\mathbf{z}$ as input variables. The following definition[1] may help to clarify the concept of the estimator.

Definition: A static state estimator is a data processing algorithm for use on a digital computer to transform meter readings and other information up to the present time into an estimate of the system's state or static state vector at the present time.

Estimation theory[2] has been used intensively for navigation of aircraft and spacecraft, as well as for post-experimental data analysis. Only recently, it has been found that estimation theory is well suited for on-line data processing in electric power systems[3–12]. As a specific feature of this application, it is worth mentioning that the main object is not to obtain a very accurate estimate of the state. It is more important to obtain reliable estimates even for much disturbed measurement data. Moreover, the estimation algorithms must be simple enough to be used on a process computer with limited storage capacity and computing speed.

Of course, load flow calculations[13] are very often performed for obtaining the system state. The main difference between load flow calculations and state estimation methods is shown in Fig. 2. Load flow calculations rely on the assumption

References p. 56

that the input data (*i.e.* the n load injections) are without any observation error. The number of given quantities equals the number of unknown bus voltages. Estimation theory, on the other hand, takes the observation errors into account. In order to obtain an improved solution, the input data should contain redundancy, *i.e.* the number of input data m should be larger than the number of unknown variables n. The output $\hat{\mathbf{x}}$ of the estimator gives complete information about the state of the system.

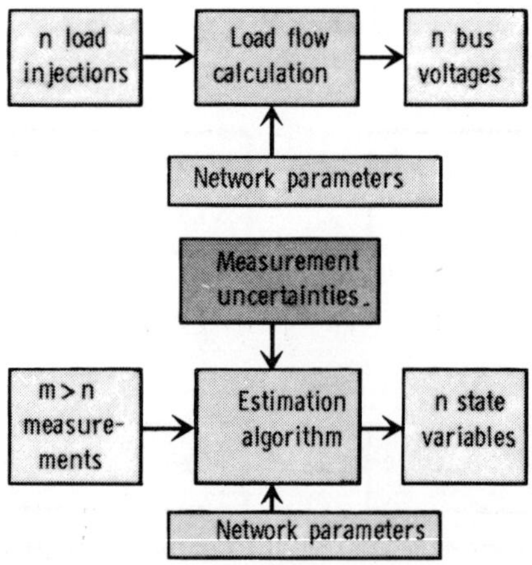

Fig. 2. Comparison between the load flow calculation and the state estimation algorithm.

1.3 Optimal regulator

The object of the optimal regulator is the computation of a control signal $\mathbf{u}$ such that some estimate $\hat{\mathbf{x}}$ of the system state $\mathbf{x}$ satisfies the optimality condition of a prespecified mathematical criterion. This aspect will not be discussed in this paper. As shown in Fig. 1, the estimator provides the input data for the regulator. Some of the control functions in a power system are solved in a closed-loop manner. It is, however, very important to study the role of the human operator in the over-all control system as he always forms an integral part in any power system control scheme[14].

1.4 Outline of the paper

The mathematical description of an electric power system is given in §2 in terms of state variables. The measurement uncertainties are discussed and their mathematical model formulated.

In §3 the estimation procedure based on the least-squares method is described. A concise review of different estimation algorithms is contained in §3.2. It is shown in §3.3 that the tracking estimator deals with the time behaviour of slowly varying quantities. A dynamic estimator, taking the time argument as independent variable into account, is contained in §3.4.

Off-line studies using several state estimation algorithms are described in §4. Two important off-line applications are discussed:

(1) detection of bad data, and

(2) meter placement.

Although off-line simulations give valuable insight into the estimation techniques, the final evaluation is only possible with on-line studies (see §5). In the first stage of on-line data processing and display, the two main aspects are the preprocessing of the crude measurements and the determination of the sampling rate. In the next stage (§5.2), the estimator is used for security monitoring, a task for which a reliable data base is essential. Finally, optimal control problems such as load-frequency control might be based on the estimator (see §5.3).

Several conclusions related to the present and future use of state estimation algorithms are given in §6.

2. MATHEMATICAL DESCRIPTION OF ELECTRIC POWER SYSTEMS

The state vector **x** of a system contains the minimal amount of information about the past history of the system which is required to predict the future behaviour. Referring to electric power systems, the following questions arise.

(1) What are the components of the state vector **x**? (§2.1.)

(2) How is the state vector **x** related to the measurements **z**? (§2.2.)

(3) What are the sources for the measurement uncertainties **v**? (§2.3.)

2.1 The state variables x

In agreement with the above definition of the state vector **x**, the state of the power system is defined as the vector containing the complex voltages of all N network buses. It has been found advantageous to work with the real, e_i, and imaginary, f_i, parts of the complex voltage V_i in bus i:

$$V_i = e_i + \mathrm{j}f_i \qquad i = 1, 2, \ldots, N \tag{1}$$

The polar coordinates

$$V_i = U_i \angle \theta_i \tag{2}$$

where U_i is the voltage magnitude and θ_i the phase angle, are sometimes used too.

For physical considerations, the phase angle at one bus (the swing bus) has to be set equal to zero. Thus, in a network with N buses, the dimension n of the state vector **x** is given by

$$\dim[\mathbf{x}] = 2N - 1 = n \tag{3}$$

Denoting the swing bus as bus number one, the components of **x** are defined as

$$\begin{aligned} \mathbf{x}^T &\triangleq [e_1, e_2, \ldots, e_N, f_2, \ldots, f_N] \\ &\triangleq [x_1, x_2, \ldots, x_n] \end{aligned} \tag{4}$$

References p. 56

A similar definition of $\mathbf{x}$ is obtained when use is made of eqn. (2) instead of eqn. (1):

$$\mathbf{x}^T \triangleq [U_1, U_2, \ldots, U_N, \theta_2, \ldots, \theta_N] \tag{5}$$

In the following, only definition (4) will be used.

Knowing $\mathbf{x}$ and the network parameters, it is possible to evaluate all other quantities in the power system, such as currents, line flows, etc.

2.2 *The measurements* z

Referring to Fig. 1, the state vector $\mathbf{x}$ is not directly accessible. It has to be estimated from the m-dimensional observation vector $\mathbf{z}$. In Table 1 five basic relations between the complex bus voltages $\mathbf{x}$ and the measurements $\mathbf{z}$ are summarized. In general, they are non-linear functions denoted by

$$\mathbf{z} = \mathbf{h}(\mathbf{x}) + \mathbf{v} \tag{6}$$

TABLE 1

FIVE BASIC MEASUREMENT SCHEMES FOR A SYSTEM WITH N BUSES AND M LINES

Case	*Components of* z	h(x)	*dim*[z]
1	Real P_i and reactive Q_i injections at all nodes (except P_1 at swing bus): $\mathbf{z} = [\ldots, P_i, Q_i, \ldots]$	Equations (7) and (8)	$2N - 1$
2	Same as case 1 plus voltage magnitudes U_i at all nodes: $\mathbf{z} = [\ldots, P_i, Q_i, \ldots, P_N, Q_N, U_1 \ldots U_N]$	Equations (7), (8) and (11)	$3N - 1$
3	Active and reactive line flows P_{ik} and Q_{ik} at both ends of each line: $\mathbf{z} = [\ldots, P_{ik}, Q_{ik}, \ldots]$	Equations (9) and (10)	$4M$
4	Same as case 3 plus voltage magnitude at all nodes: $\mathbf{z} = [\ldots, P_{ik}, Q_{ik}, \ldots, U_1, \ldots, U_N]$	Equations (9), (10) and (11)	$4M + N$
5	Full measurement system $\mathbf{z} = [\ldots P_i, Q_i, \ldots, P_{ik}, Q_{ik}, \ldots, U_i, \ldots, \theta_i, \ldots]$	Equations (7), (8), (9), (10), (11) and (12)	$4N - 1 + 4M$

where $\mathbf{h}(\mathbf{x})$ is an m-dimensional vector function and $\mathbf{v}$ is the m-dimensional observation error (see §2.3). While case 1 corresponds to an ordinary load flow computation with no redundancy, the full metering system of case 5 yields the maximal realizable redundancy. From an economic point of view the investments required for case 5 may not be realistic. In a real power system the measuring system will be a combination of the cases shown in Table 1.

The following non-linear equations define the components of **z** in Table 1.

Real injection P_i into bus i

$$P_i = e_i \sum_{k=1}^{N} (e_k G_{ik} - f_k B_{ik}) + f_i \sum_{k=1}^{N} (f_k G_{ik} + e_k B_{ik}) \qquad (7)^*$$

Reactive injection Q_i into bus i

$$Q_i = f_i \sum_{k=1}^{N} (e_k G_{ik} - f_k B_{ik}) - e_i \sum_{k=1}^{N} (f_k G_{ik} + e_k B_{ik}) \qquad (8)^*$$

Real load flow P_{ik} from bus i to k

$$P_{ik} = -[e_i(e_i - e_k) + f_i(f_i - f_k)]G_{ik} + [e_i(f_i - f_k) - f_i(e_i - e_k)]B_{ik} \qquad (9)$$

Reactive load flow Q_{ik} from bus i to k

$$Q_{ik} = [e_i(e_i - e_k) + f_i(f_i - f_k)]B_{ik} + [e_i(f_i - f_k) - f_i(e_i - e_k)]G_{ik} - (e_i^2 + f_i^2)Y_{ik}/2 \qquad (10)$$

Voltage magnitude U_i at bus i

$$U_i = (e_i^2 + f_i^2)^{1/2} \qquad (11)$$

Voltage angle θ_i at bus i

$$\theta_i = \arctan \frac{f_i}{e_i} \qquad (12)$$

As shown by eqns. (7)–(12) the components $h_i(\mathbf{x})$ of $\mathbf{h}(\mathbf{x})$ are analytic functions in **x**. The network parameters G_{ik}, B_{ik} and Y_{ik} are assumed to be known from the line model using a π-section network element (see Fig. 3).

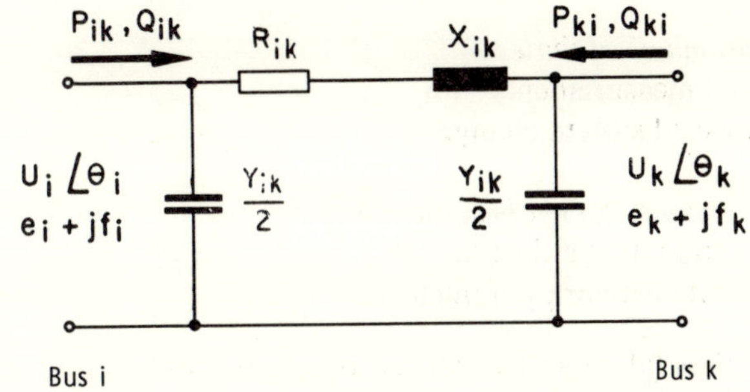

Fig. 3. Line model using a π-section network element.

* In eqns. (7), (8) and (13) the summation over *k* implies only those buses which are connected to bus *i* through a line.

References p. 56

Case 1: $k = i$

$$G_{ii} = \sum_{k=1}^{N} \frac{R_{ik}}{R_{ik}^2 + X_{ik}^2} \qquad B_{ii} = \sum_{k=1}^{N} \frac{-X_{ik}}{R_{ik}^2 + X_{ik}^2} + \frac{Y_{ik}}{2} \tag{13}^*$$

Case 2: $i \neq k$

$$G_{ik} = -\frac{R_{ik}}{R_{ik}^2 + X_{ik}^2} \qquad B_{ik} = \frac{X_{ik}}{R_{ik}^2 + X_{ik}^2} \tag{14}$$

A final remark concerns the (m, n)-dimensional Jacobian matrix $\mathbf{H}$ of $\mathbf{h}(\mathbf{x})$. It plays a central role in §3. Its elements are defined as in eqn. (15):

$$\mathbf{H} \triangleq \begin{bmatrix} \frac{\partial h_1(\mathbf{x})}{\partial x_1} & \cdots & \frac{\partial h_1(\mathbf{x})}{\partial x_n} \\ \vdots & \ddots & \vdots \\ \frac{\partial h_m(\mathbf{x})}{\partial x_1} & \cdots & \frac{\partial h_m(\mathbf{x})}{\partial x_n} \end{bmatrix} \tag{15}$$

2.3 *The measurement uncertainties*

The measurement and telemetering equipment in an electric power system is always subject to random and systematic instrument and phenomenon errors. They are modelled by the random vector $\mathbf{v}$ appearing in eqn. (6) and may be classified into three different categories.

(a) *Instrumentation errors*
Meter and communication errors
Incomplete instrumentation
Bad data points

(b) *Operation uncertainties*
Delayed measurements
Unexpected system changes

(c) *Incompleteness of mathematical model*
Errors in mathematical model
Inaccurate network parameters

The errors of class (a) could be reduced by improved metering equipment at the expense of higher installation costs. The operating uncertainties correspond to reality and are not avoidable. Although the mathematical model could be improved,

* see footnote p. 35.

the resulting loss in computing speed and the increased storage requirement limit the feasibility of such efforts.

The instrumentation errors can be modelled[8, 11] by the following random variable v_i:

$$v_i = (\alpha z_i^{\max} + \beta z_i)v_i \qquad i = 1, 2, \ldots, m \tag{16}$$

where α = percentage error of full scale value $z_i^{\max}$, *e.g.* $\alpha = 0.01$
β = percentage error of actual value z_i; β is a function of z_i of the order of 0.01
v_i = uniformly distributed random number in the interval $[-1, +1]$
For estimation purposes it is more advantageous to model all system uncertainties by a gaussian random vector. The corresponding m-dimensional density is

$$p(\mathbf{v}) = \frac{1}{[(2\pi)^m \det \mathbf{R}]^{1/2}} \exp(-\tfrac{1}{2}\mathbf{v}^T\mathbf{R}^{-1}\mathbf{v}) \tag{17}$$

where $\mathbf{R}$ is the (m, m)-dimensional covariance matrix of $\mathbf{v}$.

3. LEAST-SQUARES ESTIMATION

3.1 *Mathematical background*

In view of the system uncertainties discussed in §2.3, the stochastic approach of estimation theory provides a useful tool for determining the system state. This section deals with the data processing algorithms based on least-squares estimation theory[2].

As shown in §2.2 the relationship between the state vector $\mathbf{x}$ and the measurement vector $\mathbf{z}$ is non-linear:

$$\mathbf{z} = \mathbf{h}(\mathbf{x}) + \mathbf{v} \tag{18}$$

The optimal estimate $\hat{\mathbf{x}}$ is given by that value of $\mathbf{x}$ for which the scalar sum of the weighted squares

$$J = [\mathbf{z} - \mathbf{h}(\mathbf{x})]^T\mathbf{R}^{-1}[\mathbf{z} - \mathbf{h}(\mathbf{x})] \tag{19}$$

takes its minimum. The weighting matrix $\mathbf{R}^{-1}$ is the inverse covariance matrix of the observation noise $\mathbf{v}$ introduced in eqn. (17).

The problem to be considered is to find the minimum of the function $J(\mathbf{x})$ of n variables $\mathbf{x}$. The following quantities derived from $J(\mathbf{x})$ are needed in the various solutions of the estimation problem. The gradient of $J(\mathbf{x})$, denoted by $\mathbf{g}(\mathbf{x})$, is readily obtained from eqns. (19) and (18):

$$\mathbf{g}(\mathbf{x}) = \frac{\partial J}{\partial \mathbf{x}} = -2\mathbf{H}^T\mathbf{R}^{-1}[\mathbf{z} - \mathbf{h}(\mathbf{x})] \tag{20}$$

where $\mathbf{H}$ is the Jacobian matrix of eqn. (15). The Hessian matrix of $J(\mathbf{x})$ is denoted by $\mathbf{G}(\mathbf{x})$, *i.e.*

References p. 56

$$\mathbf{G}(\mathbf{x}) = \left[\frac{\partial^2 J}{\partial x_i \partial x_j}\right] \qquad i,j = 1,2,\ldots,n \tag{21}$$

The method of solving this problem is to compute a root of the equation $\mathbf{g}(\mathbf{x}) = \mathbf{0}$ by some iterative procedure such that

$$\mathbf{x}^{(\nu+1)} = \mathbf{x}^{(\nu)} - \mathbf{A}^{(\nu)}\mathbf{g}(\mathbf{x}^{(\nu)}) \qquad \nu = 0,1,2,\ldots \tag{22}$$

and

$$\lim_{\nu\to\infty} \mathbf{g}(\mathbf{x}^{(\nu)}) = \mathbf{0} \tag{23}$$

If the algorithm (22) converges and if $\mathbf{A}^{(\nu)}$ is non-singular for every ν it is clear that the limit value of $\mathbf{x}^{(\nu)}$, $\nu \to \infty$, satisfies eqn. (23).

The problem to be solved next is the determination of suitable matrices $\mathbf{A}^{(\nu)}$. Clearly, if $\mathbf{A}^{(\nu)}$ equals a multiple of the unit matrix, the algorithm (22) is the steepest descent method. If $\mathbf{A}^{(\nu)}$ equals the inverse of the Hessian matrix $\mathbf{G}(\mathbf{x}^{(\nu)})$ then algorithm (22) is known as Newton's method[15].

An indirect method[2] of deriving a matrix $\mathbf{A}^{(\nu)}$ is based on the Taylor series expansion of $\mathbf{h}(\mathbf{x})$ around a nominal point $\mathbf{x}^0$:

$$\mathbf{h}(\mathbf{x}) = \mathbf{h}(\mathbf{x}^0) + \mathbf{H}(\mathbf{x}^0)(\mathbf{x} - \mathbf{x}^0) + \text{higher order terms} \tag{24}$$

The original criterion (19) is now replaced by the following linearized risk function:

$$J_l = [\mathbf{z} - \mathbf{h}(\mathbf{x}^0) - \mathbf{H}(\mathbf{x}^0)(\mathbf{x} - \mathbf{x}^0)]^T\mathbf{R}^{-1}[\mathbf{z} - \mathbf{h}(\mathbf{x}^0) - \mathbf{H}(\mathbf{x}^0)(\mathbf{x} - \mathbf{x}^0)] \tag{25}$$

The linear least-squares estimate $\Delta\mathbf{x} = \mathbf{x} - \mathbf{x}^0$ satisfies the optimality condition

$$\frac{\partial J_l}{\partial \mathbf{x}} = -2\mathbf{H}^T\mathbf{R}^{-1}[\mathbf{z} - \mathbf{h}(\mathbf{x}^0) - \mathbf{H}(\mathbf{x}^0)(\mathbf{x} - \mathbf{x}^0)] = \mathbf{0} \tag{26}$$

Thus the linear estimate $\Delta\hat{\mathbf{x}}$ of $\Delta\mathbf{x}$ becomes

$$\Delta\hat{\mathbf{x}} = [\mathbf{H}^T\mathbf{R}^{-1}\mathbf{H}]^{-1}\mathbf{H}^T\mathbf{R}^{-1}[\mathbf{z} - \mathbf{h}(\mathbf{x}^0)] \tag{27}$$

It can be verified that eqn. (27) yields an absolute minimum of J_l in eqn. (25). On the other hand, $\mathbf{x}^0 + \Delta\hat{\mathbf{x}}$ does not yield an absolute minimum of the original risk function (19) unless $\mathbf{h}(\mathbf{x}) = \mathbf{H}\mathbf{x}$.

Equation (27) leads to the following iterative procedure*:

$$\hat{\mathbf{x}}^{(\nu+1)} = \hat{\mathbf{x}}^{(\nu)} + [\mathbf{H}^T\mathbf{R}^{-1}\mathbf{H}]^{-1}\mathbf{H}^T\mathbf{R}^{-1}[\mathbf{z} - \mathbf{h}(\mathbf{x}^{(\nu)})] \tag{28}$$

i.e. $\mathbf{A}^{(\nu)} = -[\mathbf{H}^T\mathbf{R}^{-1}\mathbf{H}]^{-1}$ in eqn. (22).

The covariance matrix of the estimation error $\mathbf{x} - \hat{\mathbf{x}}$ may be approximated by the matrix $(\mathbf{H}^T\mathbf{R}^{-1}\mathbf{H})^{-1}$. The initial starting point $\mathbf{x}^0$ either is the solution of a previous load flow computation or is chosen as a flat voltage profile:

$$V_i = e_i + jf_i \tag{29}$$

* It must be kept in mind that $\mathbf{H}$ is a function of $\mathbf{x}^{(\nu)}$.

with $e_i = 1$ and $f_i = 0$ for $i = 1, 2, \ldots, N$

The direct implementation of eqn. (28) is not suitable for on-line data processing*. The Jacobian matrix $\mathbf{H}$ has to be computed for each $\hat{\mathbf{x}}^{(\nu)}$. Each iteration, therefore, requires a matrix inversion. Algorithms which are designed for real-time state estimation will be discussed in the following section.

3.2 *Estimation algorithms for static state estimation*

3.2.1 *Recursive measurement processing*[6, 7]

Instead of processing the entire observation vector $\mathbf{z}$, the following two-step procedure converts the matrix inversion into divisions by scalar numbers.

Step 1: The m-dimensional observation vector $\mathbf{z}$ is partitioned into an n-dimensional vector $\mathbf{z}_0$ and an $(m - n)$-dimensional vector $\boldsymbol{\zeta}$. The vector $\mathbf{z}_0$ contains the n real and reactive load injections. Similarly, the observation equations $\mathbf{h}(\mathbf{x})$ and the Jacobian matrix $\mathbf{H}$ are partitioned such that the linearized observation eqn. (24) takes the following form:

$$\mathbf{h}(\mathbf{x}) = \begin{bmatrix} \mathbf{h}_0(\mathbf{x}^0) \\ h_1(\mathbf{x}^0) \\ \vdots \\ h_{m-n}(\mathbf{x}^0) \end{bmatrix} + \begin{bmatrix} \mathbf{H}_0(\mathbf{x}^0) \\ \mathbf{H}_1(\mathbf{x}^0) \\ \vdots \\ \mathbf{H}_{m-n}(\mathbf{x}^0) \end{bmatrix} \Delta\mathbf{x} \tag{30}$$

The load flow problem $\mathbf{z}_0 = \mathbf{h}_0(\hat{\mathbf{x}}^0)$ yields the starting value for step 2.

Step 2: The remaining $(m - n)$ observations of vector $\boldsymbol{\zeta}$ are sequentially used to update the initial estimate $\hat{\mathbf{x}}^0$:

$$\hat{\mathbf{x}}^{(\nu+1)} = \hat{\mathbf{x}}^{(\nu)} + \mathbf{P}^{(\nu)}\mathbf{H}_{\nu+1}^T[\mathbf{H}_{\nu+1}\mathbf{P}^{(\nu)}\mathbf{H}_{\nu+1}^T + R_{\nu\nu}]^{-1}[\zeta_{\nu+1} - \mathbf{H}_{\nu+1}\hat{\mathbf{x}}^{(\nu)}] \tag{31}$$

$$\text{for } \nu = 0, 1, 2, \ldots, m - n$$

where $\mathbf{H}_{\nu+1}$ is the $(n + \nu)$th row vector of the full Jacobian matrix $\mathbf{H}$, and $R_{\nu\nu}$ is the $(n + \nu)$th diagonal element of the covariance matrix $\mathbf{R}$.

The matrix $\mathbf{P}^{(\nu)}$ is updated according to

$$\mathbf{P}^{(\nu+1)} = \mathbf{P}^{(\nu)} - \mathbf{P}^{(\nu)}\mathbf{H}_{\nu+1}^T[\mathbf{H}_{\nu+1}\mathbf{P}^{(\nu)}\mathbf{H}_{\nu+1}^T + R_{\nu\nu}]^{-1}\mathbf{H}_{\nu+1}\mathbf{P}^{(\nu)} \tag{32}$$

This step only requires scalar divisions to update $\hat{\mathbf{x}}^{(\nu)}$ and $\mathbf{P}^{(\nu)}$. The repeated use of the algorithms (31) and (32) yields the general recursive estimator of eqn. (28).

The initial value of $\mathbf{P}^{(\nu)}$ is given by

$$\mathbf{P}^0 = [\mathbf{H}_0^T\mathbf{R}_0^{-1}\mathbf{H}_0]^{-1} \tag{33}$$

where $\mathbf{R}_0$ is that part of the covariance matrix $\mathbf{R}$ which contains the noise vector $\mathbf{v}_0$

* It has been pointed out by Schweppe[5, 30] that eqn. (28) may be rewritten as

$$(\mathbf{H}^T\mathbf{R}^{-1}\mathbf{H})[\mathbf{x}^{(\nu+1)} - \mathbf{x}^{(\nu)}] = \mathbf{H}^T\mathbf{R}^{-1}[\mathbf{z} - \mathbf{h}(\mathbf{x}^{(\nu)})]$$

The estimation problem is now a problem of solving a system of linear equations. Exploiting the sparsity of $\mathbf{H}^T\mathbf{R}^{-1}\mathbf{H}$ this approach may be used for on-line data processing.

References p. 56

of the measurements $\mathbf{z}_0$. Similarly, the matrix $\mathbf{H}_0$ is that part of the Jacobian $\mathbf{H}$ which is obtained by linearizing the observation equations of $\mathbf{z}_0$. Additional computing time can be saved by working only with the diagonal elements of the matrix $\mathbf{P}^{(\nu)}$.

This algorithm is particularly suitable if the measurements $\mathbf{z}$, all taken at the same time, are transmitted to the control centre in a sequential manner. The recursive algorithm (30) is based on the assumption that the components of the noise vector $\mathbf{v}$ are not mutually correlated.

3.2.2 *Constant gain algorithms*[1, 5]

In the recursive measurement processing, the row vectors $\mathbf{H}_\nu$ have to be computed after each iteration for the updated estimated value $\hat{\mathbf{x}}^{(\nu)}$. Another simplification is obtained when the matrix inversion $[\mathbf{H}^T\mathbf{R}^{-1}\mathbf{H}]^{-1}$ of eqn. (28) is only performed for the initial value $\hat{\mathbf{x}}^0$. The resulting matrix

$$\mathbf{P}_0 = [\mathbf{H}_0^T\mathbf{R}_0^{-1}\mathbf{H}_0]^{-1} \tag{34}$$

is then held constant for all further iterations of the algorithm

$$\hat{\mathbf{x}}^{(\nu+1)} = \hat{\mathbf{x}}^{(\nu)} + \mathbf{P}_0\mathbf{H}^T(\mathbf{x}^{(\nu)})\mathbf{R}^{-1}[\mathbf{z} - \mathbf{h}(\mathbf{x}^{(\nu)})] \tag{35}$$

A further reduction is obtained when the Jacobian matrix $\mathbf{H}$ is computed only for $\hat{\mathbf{x}}^0$ and not for every $\hat{\mathbf{x}}^{(\nu)}$. Thus, eqn. (35) takes the form

$$\hat{\mathbf{x}}^{(\nu+1)} = \hat{\mathbf{x}}^{(\nu)} + \mathbf{P}_0\mathbf{H}_0\mathbf{R}^{-1}[\mathbf{z} - \mathbf{h}(\mathbf{x}^{(\nu)})] \tag{36}$$

In order to obtain a satisfactory convergence behaviour it is recommended that the result of a load flow calculation, as in §3.2.1, is used as a starting value $\mathbf{x}^0$.

The reduced convergence speed of eqns. (35) and (36) compared with eqn. (28) has to be seen in view of the greatly reduced computing time.

3.2.3 *Line flow measurements*[8–10]

An interesting solution of the non-linear estimation problem can be derived if the observation vector $\mathbf{z}$ contains only the line flow measurements at both ends of each line. This corresponds to case 3 in Table 1.

Firstly, the non-linear problem is transformed into a linear problem by introducing the branch voltages $\mathbf{y}$ across each line. The vector $\mathbf{y}$ is related to the state vector $\mathbf{x}$ through a linear transformation

$$\mathbf{y} = \mathbf{B}\mathbf{x} + \mathbf{v}' \tag{37}$$

where $\mathbf{B}$ is the element–node incidence matrix[13]. The noise vector $\mathbf{v}'$ represents the system uncertainties. The risk function $J(\mathbf{x})$ pertinent to eqn. (37) is quadratic, assuming $\mathbf{y}$ is observed:

$$J(\mathbf{x}) = (\mathbf{y} - \mathbf{B}\mathbf{x})^T(\mathbf{y} - \mathbf{B}\mathbf{x}) \tag{38}$$

The optimality condition

$$\frac{\partial J(\mathbf{x})}{\partial \mathbf{x}} = -2\mathbf{B}^T(\mathbf{y} - \mathbf{B}\mathbf{x}) = \mathbf{0} \tag{39}$$

yields as an estimate $\hat{\mathbf{x}}$ of $\mathbf{x}$

$$\hat{\mathbf{x}} = (\mathbf{B}^T\mathbf{B})^{-1}\mathbf{B}^T\mathbf{y} \tag{40}$$

The second step involves the determination of the non-linear relation between the branch voltages $\mathbf{y}$ and the line flows $\mathbf{z}$

$$\mathbf{y} = \mathbf{f}(\mathbf{x}, \mathbf{z}) \tag{41}$$

The pth component of the vector valued function $\mathbf{f}(\mathbf{x}, \mathbf{z})$ takes the form*

$$y_p = (R_{ik} + \mathrm{j}X_{ik})\left[\frac{P_{ik} - \mathrm{j}Q_{ik}}{e_i - \mathrm{j}f_i} - (e_i + \mathrm{j}f_i)Y_{ik}\right] \tag{42}$$

As $\mathbf{y}$ is not directly measured and depends on $\mathbf{x}$, the estimator (40) has to be iterated starting with a value $\hat{\mathbf{x}}^0$ as

$$\hat{\mathbf{x}}^{(\nu+1)} = (\mathbf{B}^T\mathbf{B})^{-1}\mathbf{B}^T\mathbf{f}(\hat{\mathbf{x}}^{(\nu)}, \mathbf{z}) \qquad \nu = 0, 1, 2, \ldots \tag{43}$$

until the convergence criterion, eqn. (49) of §3.5, is satisfied.

If convergence to some limit value $\hat{\mathbf{x}}$ occurs, it follows from eqn. (43) that $\hat{\mathbf{x}}$ satisfies the equation $\hat{\mathbf{x}} = (\mathbf{B}^T\mathbf{B})^{-1}\mathbf{B}^T\mathbf{f}(\hat{\mathbf{x}}, \mathbf{z})$.

If the non-linear transformation (41) is used in the risk function (38), the optimality condition of $J(\mathbf{x})$ with respect to $\mathbf{x}$ is

$$\frac{\partial J}{\partial \mathbf{x}} = 2\left[\frac{\partial \mathbf{y}}{\partial \mathbf{x}} - \mathbf{B}\right]^T(\mathbf{y} - \mathbf{B}\mathbf{x}) = \mathbf{0} \tag{44}$$

It is obvious that $\hat{\mathbf{x}}$ does not satisfy eqn. (44). Therefore $\hat{\mathbf{x}}$ is not the optimal value of the risk function (38) if $\mathbf{y}$ is considered to be a function of $\mathbf{x}$ and $\mathbf{z}$, as one might have suspected originally. Nevertheless, practical experience has shown that algorithm (43) yields satisfactory results.

The algorithm (43) is well-suited to on-line implementation because the matrix inversion need not be performed as long as the structure of the network remains unchanged.

3.2.4 Estimation using Fletcher–Powell algorithm

The iteration scheme (22) is known as Newton's method if $\mathbf{A}$ equals the inverse of the Hessian matrix $\mathbf{G}$. The main feature of this method is the fact that the curvature of $J(\mathbf{x})$ is taken into account. The curvature can be obtained through the calculation of $\mathbf{G}$. Sometimes it is preferable to avoid the difficulties of working out and programming explicit expressions for $\mathbf{G}$. Instead, methods have been devised that use only $J(\mathbf{x})$ and its first derivatives $\mathbf{g}(\mathbf{x})$. The Davidon's method[16], as described and extended by Fletcher and Powell[17], is a procedure in which the matrix $\mathbf{G}^{-1}$ is not evaluated directly; instead, a matrix $\mathbf{A}^{(\nu)}$ is used which is initially chosen as unit matrix and subsequently modified by the algorithm shown in Table 2. If convergence occurs, $\mathbf{A}^{(\nu)}$ approaches $\mathbf{G}^{-1}$ for $\nu \to \infty$.

When the iteration counter ν exceeds the integer n_1 the estimation procedure is stopped. As the function $J(\mathbf{x})$ has n variables the limit n_1 of step (5) must be

* The notation of eqn. (42) is introduced and explained in §2.

References p. 56

somewhat bigger than n. Indeed, a fairly good approximation of $\mathbf{G}^{-1}$ has been found experimentally after $n_1 = n + 3$ iterations. The main feature of the algorithm in Table 2 is the fact that it is directly based on the original risk function (19).

The Fletcher–Powell algorithm still yields a full $\mathbf{A}$ matrix. As an alternative approach a conjugent gradient technique[15] could be used instead[30].

TABLE 2

STATIC STATE ESTIMATION USING THE FLETCHER–POWELL ALGORITHM

(1) Given $\mathbf{x}^0$, $\mathbf{A}^0 = \mathbf{I}$, $v = 0$ and $n_1 = n + 3$

(2) Evaluate $J(\mathbf{x}^{(v)})$ and $\mathbf{g}(\mathbf{x}^{(v)}) = -2\mathbf{H}^T\mathbf{R}^{-1}[\mathbf{z} - \mathbf{h}(\mathbf{x}^{(v)})] = \mathbf{g}^{(v)}$

(3) Find $\mathbf{x}^{(v+1)}: J(\mathbf{x}^{(v+1)}) = \min\limits_{\alpha} J(\mathbf{x}^{(v)} - \alpha\mathbf{A}^{(v)}\mathbf{g}^{(v)})$

(4) Evaluate $\Delta\mathbf{x}^{(v)} = \mathbf{x}^{(v+1)} - \mathbf{x}^{(v)}$

$$\Delta\mathbf{g}^{(v)} = \mathbf{g}^{(v+1)} - \mathbf{g}^{(v)}$$

(5) If v is less than n_1 go to (6); otherwise *stop*

(6) Evaluate $\mathbf{M}_1^{(v)} = \dfrac{\Delta\mathbf{x}^{(v)}\Delta\mathbf{x}^{T(v)}}{\Delta\mathbf{x}^{T(v)}\Delta\mathbf{g}^{(v)}}$

$$\mathbf{M}_2^{(v)} = \frac{-\mathbf{A}^{(v)}\Delta\mathbf{g}^{(v)}\Delta\mathbf{g}^{T(v)}\mathbf{A}^{(v)}}{\Delta\mathbf{g}^{T(v)}\mathbf{A}^{(v)}\Delta\mathbf{g}^{(v)}}$$

$$\mathbf{A}^{(v+1)} = \mathbf{A}^{(v)} + \mathbf{M}_1^{(v)} + \mathbf{M}_2^{(v)}$$

(7) Set v to $v + 1$; return to step (2)

3.3 Tracking state estimation

So far only static state estimation has been considered. By introducing the time k as independent variable, tracking and dynamic estimators are obtained. Dynamic state estimation requires good representative models for the time behaviour of the state $\mathbf{x}$ and observation error $\mathbf{v}$. The time dependency is denoted by $\mathbf{x}(k)$ and $\mathbf{v}(k)$ where k is the discrete time argument. This approach will be further discussed in §3.4.

The tracking state estimator is based on the assumption that the state vector $\mathbf{x}$ does not change or changes only very little between two consecutive instants of time, k and $k + 1$ respectively. The initial value $\mathbf{x}^0$ at time $k + 1$ is replaced by the estimate $\hat{\mathbf{x}}(k)$ of the previous time instant k based on $\mathbf{z}(k)$. Static and tracking state estimation are compared in Fig. 4.

If the observation error $\mathbf{v}$ is not too large *one* iteration is sufficient to compute $\hat{\mathbf{x}}(k + 1)$ from $\mathbf{z}(k + 1)$ and $\hat{\mathbf{x}}(k)$. Then any static estimation algorithm of §3.2 is readily altered into a tracking algorithm by changing the iteration counter v into the time argument k. The general algorithm (28) becomes

$$\hat{\mathbf{x}}(k + 1) = \hat{\mathbf{x}}(k) + [\mathbf{H}^T(k)\mathbf{R}^{-1}(k)\mathbf{H}(k)]^{-1}\mathbf{H}^T(k)\mathbf{R}^{-1}(k)[\mathbf{z}(k + 1) - \mathbf{h}(\hat{\mathbf{x}}(k))] \quad (45)$$

The simplifications discussed in §3.2 have been used[1,18,28,29] to reduce the computation effort caused by direct implementation of the algorithm in Table 2. As they affect the convergence speed it is sometimes necessary to use more than one iteration per time step; *i.e.* after each increment of the time argument k several iterations ν, as discussed in §3.2, must be performed keeping the observation vector $\mathbf{z}(k)$ constant.

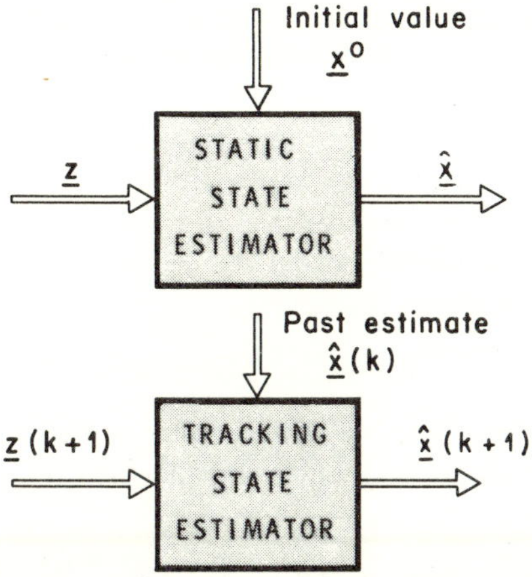

Fig. 4. Comparison between static and tracking state estimation.

The direct algorithm in Table 2 is well suited for tracking estimation, too. Provided $\mathbf{x}(k)$ does not change too much with the time k, the matrix $\mathbf{A}^{(\nu)}$ can be stored from one time instant k to the next. This reduces the computation time very much because the time-consuming task of computing $\mathbf{A}^{(\nu)}$ need only be done infrequently. A flow diagram of the tracking estimator based on the algorithm of Table 2 is shown in Table 3.

TABLE 3

TRACKING STATE ESTIMATION USING THE FLETCHER–POWELL ALGORITHM

(I) Set time argument $k = 1$ and read observation vector $\mathbf{z}(k)$

(II) Execute one complete Fletcher–Powell computation as shown in Table 2

(III) Store $\hat{\mathbf{x}}^{(\nu)}(k)$ and $\mathbf{A}^{(\nu)}(k)$ for $\nu = n_1$

(IV) Change time k to $k + 1$ and read new observation vector $\mathbf{z}(k)$

(V) Process new observation $\mathbf{z}(k)$ using algorithm in Table 2 with its steps (1) and (5) replaced by:

(1) $\mathbf{x}^0 = \hat{\mathbf{x}}^{(n_1)}(k-1)$; $\mathbf{A}^0 = A^{(n_1)}(k-1)$; set $n_1 = 3$

(5) If ν is less than n, go to (6) of algorithm in Table 2; otherwise continue with step (VI) of this algorithm

(VI) Compute the squared residual $r(k)$:

$$r(k) = [\mathbf{z}(k) - \mathbf{h}(\hat{\mathbf{x}}(k))]^T[\mathbf{z}(k) - \mathbf{h}(\hat{\mathbf{x}}(k))]$$

(VII) If $r(k) < \varepsilon$, go to step (IV); otherwise go to step (II)

The test value ε in step (VII) for restarting the Fletcher–Powell algorithm has to be determined experimentally. The iteration limit $n_1 = 3$ in step (V) is an empirically chosen value and is based on the quadratic convergence behaviour which guarantees fast convergence as soon as $\mathbf{A}^{(\nu)}$ is a good approximation of the inverse Hessian matrix.

3.4 Dynamic state estimation

Dynamic state estimation is based on a mathematical model for the time behaviour of the state vector $\mathbf{x}(k)$. From the operation point of view, it is well known that the bus voltages do not change much over a short period of time. As the sampling time between k and $k + 1$ is of the order of 5–60 sec the above assumption can be used to postulate the following model[18]:

$$\mathbf{x}(k+1) = \mathbf{x}(k) + \mathbf{w}(k) \tag{46}$$

In eqn. (46) any changes of the bus voltages are modelled by the n-dimensional white gaussian random vector sequence $\mathbf{w}(k)$ specified by

$$p[\mathbf{w}(k)] = \frac{1}{[(2\pi)^n \det \mathbf{Q}]^{1/2}} \exp -[\tfrac{1}{2}\mathbf{w}^T(k)\mathbf{Q}^{-1}(k)\mathbf{w}(k)] \tag{47}$$

The choice of the covariance matrix $\mathbf{Q}(k)$ may be based on past data analysis. It is reasonable to keep $\mathbf{Q}(k)$ constant over a long period of time. As an alternative approach an adaptive estimation scheme[19] could be used for the computation of $\mathbf{Q}(k)$, keeping in mind that the computational load of such a scheme is quite severe.

Due to the non-linear observation eqn. (18), the extended Kalman filter[20] has to be used. It is based on the Taylor series expansion of eqn. (18). The particular form of recursive measurement processing (§3.2.1) can readily be combined with the general dynamic solution[20]. The resulting algorithm is shown in Table 4. Due to the recursive processing of the observation vector $\mathbf{z}$ no matrix inversion is required. Further simplifications are achieved when only the diagonal elements of the matrix $\mathbf{P}(k)$ are considered.

Comparing the tracking with the dynamic estimator in Table 4 the following points are worth mentioning.

(a) All estimation algorithms in this paper are based on algebraic, static models of power systems. The repeated use of static estimators in the tracking mode is well suited for on-line data processing in electric power systems.

(b) The dynamic estimator in Table 4, based on the unstable model (46), can be termed a tracking estimator with a memory. It cannot estimate the transient behaviour of the system state $\mathbf{x}(k)$ because the model (46) is not sufficiently accurate under rapidly changing conditions. The choice of the covariance matrix $\mathbf{Q}(k)$ in eqn. (47) may be difficult and may require complicated additional computations.

(c) Dynamic state estimation in power systems must be based on dynamic models. To this end it has been proposed to use the magnetic flux linkages in all the synchronous generators in the network as state vector[21].

TABLE 4

DYNAMIC STATE ESTIMATION USING THE EXTENDED KALMAN FILTER

(1) Set the time argument $k = 1$

(2) Store the filter state $\hat{\mathbf{x}}^0(k) = \hat{\mathbf{x}}^{(m)}(k-1)$ and set the approximate covariance matrix $\mathbf{P}^0(k) = \mathbf{P}^{(m)}(k-1) + \mathbf{Q}(k)$; set the iteration counter $v = 1$

(3) Compute the new matrix $\mathbf{P}^{(v)}(k)$:

$$\mathbf{P}^{(v)}(k) = [\mathbf{I} - \mathbf{K}^{(v)}\mathbf{H}_v(k)]\mathbf{P}^{(v-1)}(k)$$

(4) Compute the filter gain vector $\mathbf{K}^{(v)}(k)$:

$$\mathbf{K}^{(v)}(k) = \mathbf{P}^{(v-1)}(k)\mathbf{H}_v^T(k)[R_{vv}(k) + \mathbf{H}_v(k)\mathbf{P}^{(v-1)}(k)\mathbf{H}_v^T(k)]^{-1}$$

(5) Process the vth component $z_v(k)$ of the observation vector $\mathbf{z}(k)$:

$$\hat{\mathbf{x}}^{(v)}(k) = \hat{\mathbf{x}}^{(v-1)}(k) + \mathbf{K}^{(v)}(k)[z_v(k) - \mathbf{H}_v(k)\hat{\mathbf{x}}^{(v-1)}(k)]$$

(6) If v is less than m, set v to $v + 1$ and return to step (3); otherwise set $k = k + 1$ and return to step (2)

3.5 Convergence problems

The performance of an estimator is usually measured by means of the covariance matrix of the estimation error

$$\mathbf{e}_x = \mathbf{x} - \hat{\mathbf{x}} \tag{48}$$

Due to the non-linearity of the observation eqn. (18), the error covariance matrix cannot be computed analytically. A simple method to detect convergence consists of comparing estimates of two consecutive iterations:

$$\mathbf{\Delta} = \hat{\mathbf{x}}^{(v+1)} - \hat{\mathbf{x}}^{(v)} \tag{49}$$

This scheme suffers from the drawback that slow convergency may lead to a premature iteration stop. In order to overcome this difficulty, the convergence test can be based on the size of the risk function (19), provided a direct method is used. A third possibility, which requires all load flow and load injection measurements, is the evaluation of the power balance in each node. For on-line applications this convergence test is in general not possible for lack of additional data.

4. OFF-LINE SIMULATIONS

4.1 The simulation procedure

For a preliminary evaluation, off-line simulation studies yield valuable insight into the feasibility of state estimation. The block diagram of the simulation procedure is shown in Fig. 5. The voltage profile $\mathbf{x}$ may either be assumed arbitrarily or be based on a previous load flow calculation. The observation vector $\mathbf{z}$ is obtained as

References p. 56

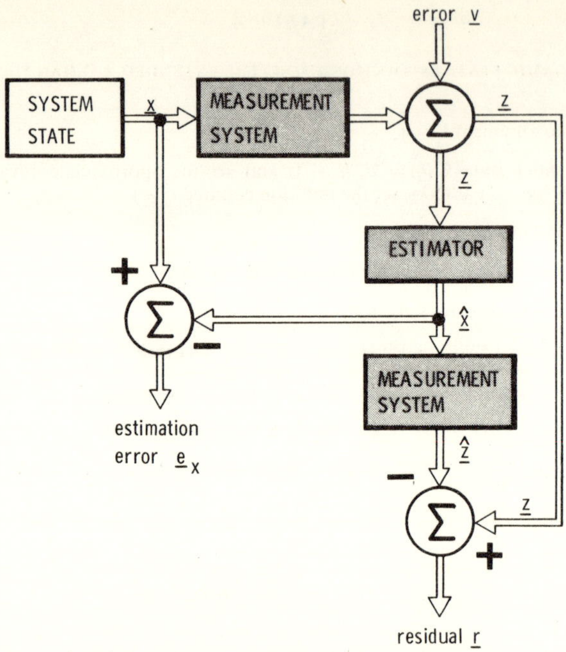

Fig. 5. Block diagram of simulation procedure.

the sum of the non-linear transformation $\mathbf{h}(\mathbf{x})$ and the observation error $\mathbf{v}$. Several types of observation equations $\mathbf{h}(\mathbf{x})$ are given in Table 1. For the simulation studies reported in this section, the line flow measurements (case 3) are used. The observation error $\mathbf{v}$ is obtained from a gaussian random number generator.

In simulation studies the estimated state vector $\hat{\mathbf{x}}$ can be compared with the true state vector $\mathbf{x}$. Thus, the estimation error $\mathbf{e}_x$, defined as

$$\mathbf{e}_x \triangleq \mathbf{x} - \hat{\mathbf{x}} \tag{50}$$

may be computed*. Repeating the simulation p times with the same $\mathbf{x}$ value but different observation errors $\mathbf{v}$ a random sample $\{\mathbf{e}_x\}_p$ of size p is obtained. This sample may be used for estimating the average scalar sum of the squared estimation error:

$$s_x = \frac{1}{p} \sum_{i=1}^{p} (\mathbf{e}_x)_i^T (\mathbf{e}_x)_i \tag{51}$$

The residual vector $\mathbf{r}$, defined as

$$\mathbf{r} \triangleq \mathbf{z} - \hat{\mathbf{z}} = \mathbf{z} - \mathbf{h}(\hat{\mathbf{x}}) \tag{52}$$

may also be used for evaluating the performance of the estimator. Repeating the

* In the on-line application the estimation error $\mathbf{e}_x$ is unknown because $\mathbf{x}$ is not accessible. An approximate covariance matrix of the estimation error $\mathbf{e}_x$ is given by $[\mathbf{H}^T\mathbf{R}^{-1}\mathbf{H}]^{-1}$.

simulation p times allows the estimation of the scalar sum of the squared residuals:

$$s_z = \frac{1}{p} \sum_{i=1}^{p} (\mathbf{r})_i^T (\mathbf{r})_i \tag{53}$$

The squared estimation error (51) or the squared residuals (53) may be computed for each iteration, using the notation $s_x^{(v)}$ and $s_z^{(v)}$, respectively.

The performance of the estimator is much affected by the redundancy η, defined as

$$\eta = \frac{\dim[\mathbf{z}]}{\dim[\mathbf{x}]} = \frac{m}{n} \tag{54}$$

The five-bus sample network[13] shown in Fig. 6 is used for the results shown in Figs. 7–12. The redundancy $\eta = 3.11$. All results are given in per unit on a 100 000 kVA base.

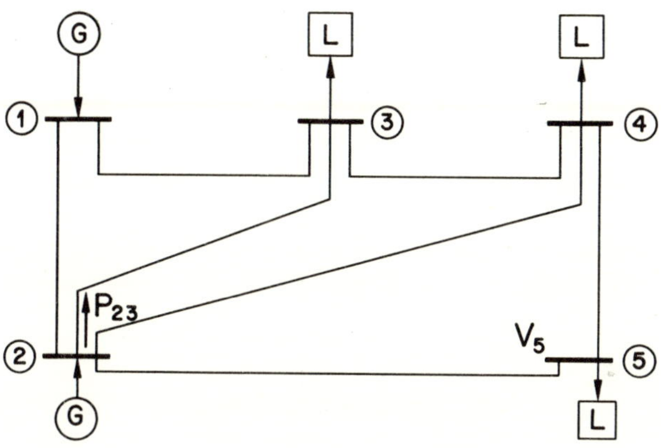

Fig. 6. Five-bus sample network.

4.2 *Static state estimation*

Using the line flow measurement algorithm of §3.2.3 the normalized estimation error $s_x^{(v)}/s_x^0$ is shown in Fig. 7 for several standard deviations $\sigma_v = \sqrt{R_{ii}}$ of the observation error $\mathbf{v}$.

In the deterministic case ($\sigma_v = 0$) the convergence criterion (49) is satisfied for $\varepsilon = 10^{-4}$ after three iterations. The final error is due to numerical inaccuracies of the input data. For the stochastic cases with $\sigma_v \neq 0$ the convergence criterion (49) is satisfied after five iterations. The resulting curves are based on $p = 20$ simulations.

Figure 8 shows the standard deviation $\sigma_{\hat{z}} \triangleq (\text{var}[\hat{z}_i])^{1/2}$ of one component $\hat{z}_i$ of $\hat{\mathbf{z}}$ as a function of σ_z. If the redundancy $\eta = 1$, no accuracy improvement is obtained, *i.e.* $\sigma_{\hat{z}} = \sigma_z$. For $\eta > 1$ the accuracy improvement of the output data $\hat{\mathbf{z}}$ compared with the input data $\mathbf{z}$ depends on η.

References p. 56

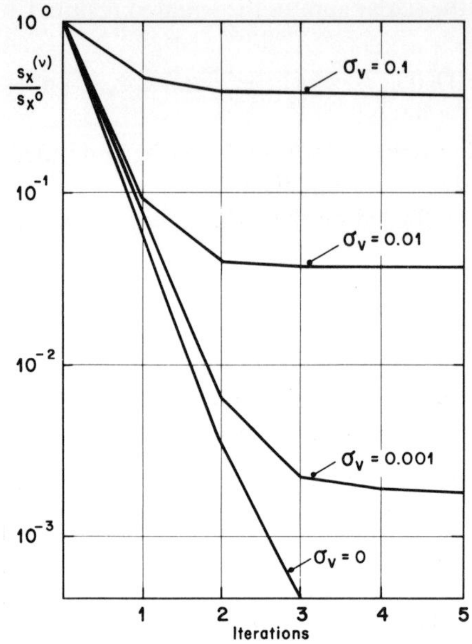

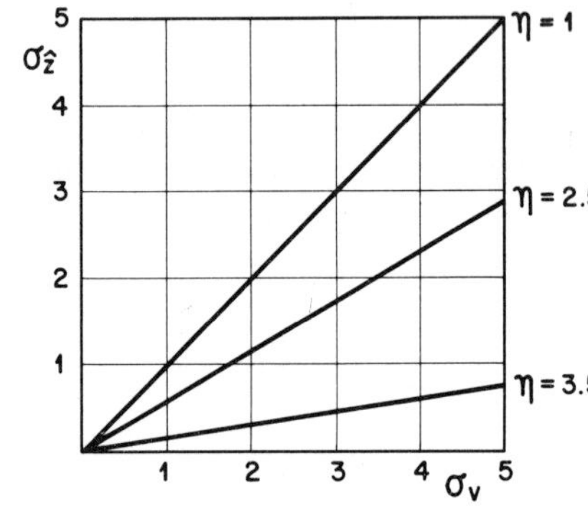

Fig. 7. Convergence behaviour of the static estimator. The normalized estimation error is defined as $s_x^{(v)}/s_x^0$.

Fig. 8. Standard deviation $\sigma_{\hat{z}}$ of output data $\hat{\mathbf{z}}$ *versus* the standard deviation σ_z of observation $\mathbf{z}$ with η as parameter.

4.3 Tracking state estimation

For the on-line use of state estimation algorithms the tracking mode is very important (see §3.3). For simulation purposes the state vector $\mathbf{x}$ is changed from one time instant, k, to the next, $k + 1$, by a fixed constant $\boldsymbol{\varepsilon}(k)$:

$$\mathbf{x}(k+1) = \mathbf{x}(k) + \boldsymbol{\varepsilon}(k) \tag{55}$$

The elements of $\boldsymbol{\varepsilon}(k)$ are such that the voltage changes by $\pm 1\%$ to $\pm 2\%$ in ten time samples k. In Figs. 9 and 10 the sign of $\boldsymbol{\varepsilon}(k)$ is changed after $k = 20$. The total simulation lasts until $k = 40$. The diagonal elements of the noise covariance matrix $\mathbf{R}$ are all equal to 10^{-3}.

If the state vector $\mathbf{x}(k)$ in the network of Fig. 6 is changed according to eqn. (55), the tracking algorithm in Table 3 yields the results shown in Figs. 9 and 10.

4.4 Dynamic state estimation

In order to illustrate the difference between the tracking and dynamic estimator the estimation error s_x as defined by eqn. (51) is shown in Fig. 11 for the five-bus network. The diagonal elements of the covariance matrix $\mathbf{R}(k)$ of the observation

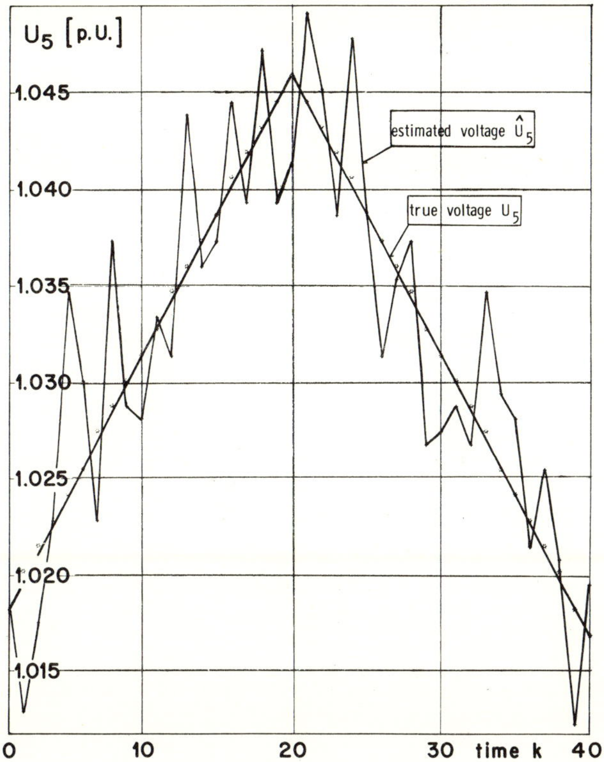

Fig. 9. True and estimated voltage magnitude U_5 and $\hat{U}_5$ respectively of bus 5 as a function of time k.

noise $\mathbf{v}$ are equal to 10^{-3}. The mean values of the observations $\mathbf{z}(k)$ are held constant.

The size of the estimation error s_x of the static estimator used in the tracking mode is given by the convergence criterion (49). It remains constant for all time arguments k. The estimation error s_x of the dynamic estimator is time dependent. Starting from an initial condition the error $s_x(k)$ decreases with a certain time constant. The dynamic estimator has a memory.

An important consequence of this behaviour applies when processing disturbed data. Figure 11 indicates that the dynamic estimator is less sensitive to individual bad data then the tracking estimator. On the other hand, the reaction of the dynamic estimator to a true step change is delayed by the time constant shown in Fig. 11.

The results discussed so far are all obtained with the small network of Fig. 6. In the engineering computing centre of Brown Boveri, Baden, the estimation programs have been expanded for real case studies with networks up to 70 nodes and 100 lines. These studies have shown that no major difference exists between the results from the sample network and the real systems, provided the algorithms are judiciously modified to deal with large networks.

References p. 56

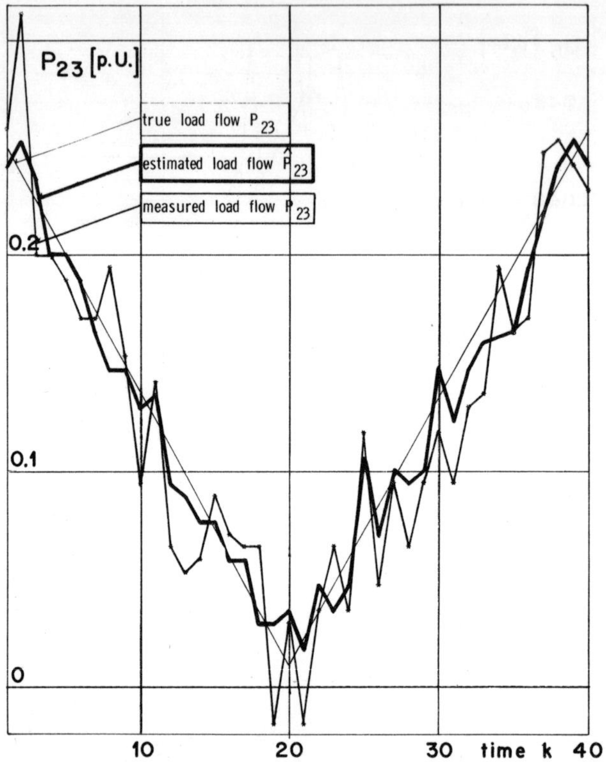

Fig. 10. True, measured and estimated real power flow from bus 2 to bus 3 as a function of time *k*.

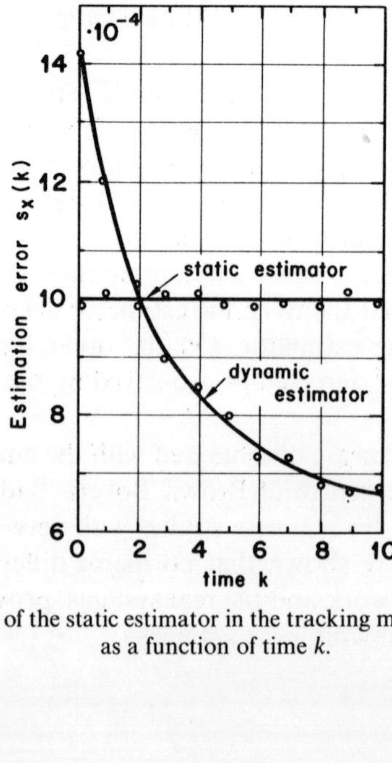

Fig. 11. Estimation error $s_x(k)$ of the static estimator in the tracking mode and of the dynamic estimator as a function of time *k*.

4.5 *Detection of wrong information*

The application of static state estimation to actual measurements **z** may give erroneous results because some components of **z** may lie outside the limits given by probability theory. Such information has to be eliminated or corrected before the estimation algorithm can give plausible results. To this end the residual vector **r** of eqn. (52) may be used. The element z_i corresponding to the largest residual component r_i is replaced by a corrected value or, if no additional information is available, z_i is eliminated. Further state estimations are performed until all large residuals are corrected, one in each run.

A second off-line application of static state estimation is the supervision of the switch position information, *i.e.* of the system structure. As shown in Table 5 the switch position information and the real load flow measurements need not be consistent. In the ambiguous cases either of the two sets of information may be wrong. The question of which information is wrong may be solved by two estimation runs. Firstly, it is hypothesized that the measurement of the real load flow is wrong and the switch position information correct. The second time the contrary is assumed. The solution which yields the smaller value for the risk function $J(\mathbf{x})$ is accepted as being correct.

TABLE 5

SUPERVISION OF NETWORK STRUCTURE

		Switch	
Line flow P_{ik}		open	closed
	$= 0$	consistent	ambiguous
	$\neq 0$	ambiguous	consistent

4.6 *Meter placement*

The problem of optimal meter placement can successfully be tackled by static state estimation. Given a certain network configuration the question arises: What is the most economical meter placement giving reliable system information? Using engineering experience the location of most instruments is straightforward. The decision about where additional instruments should be placed and what improvement they yield can be assisted by using the state estimation algorithm. Comparing the minimum value of the risk function $J(\mathbf{x})$ of eqn. (19) before and after additional instruments are assumed gives an accurate measure of the realizable accuracy gain. This trial and error procedure has given satisfactory results in a real case study.

Instead of comparing the values of $J(\mathbf{x})$ for different meter arrangements the instrument placement problem may also be based on the estimation error covariance matrix or some function thereof[30].

References p. 56

5. ON-LINE APPLICATION OF STATE ESTIMATION

The complete implementation of on-line estimation algorithms in the control centre of a power company is quite demanding. Before proceeding to more advanced applications it is suggested that some operating experience should be gained in simpler tasks, followed by the implementation of more complicated solutions. It is suggested, therefore, to proceed via the following three stages.

Stage 1: Data processing and display
- Bad data detection
- Sampling rate

Stage 2: Security monitoring
- Overload limits
- Re-scheduling
- Switching
- Load shedding

Stage 3: Optimal control
- Load-frequency control
- Economic load dispatching

5.1 Data processing and display

In the first stage of developing an on-line state estimator two main problems to be solved are:

(1) bad data detection, and
(2) sampling rate.

The state estimator which is implemented in an area or regional control centre receives the crude measurements through telemetering equipment. Practical experience shows that some of the incoming data may contain unusually large errors. As the estimator is sensitive to bad data it is important to use a scheme which detects bad data. A bad data suppression may be incorporated into the estimator itself[22]. As an alternative possibility the incoming data may be preprocessed before being used in the estimator. In the preprocessor the crude measurements $\mathbf{z}(k+1)$ at time $k+1$ are compared with the past estimate $\hat{\mathbf{z}}(k)$. Unless a switching operation takes place the two values $\hat{\mathbf{z}}(k)$ and $\mathbf{z}(k+1)$ should not differ by more than 3% up to a maximum of 10%. Of course these values depend on the sampling interval between two measurements. The object of the preprocessor is to eliminate those components of $\mathbf{z}(k+1)$ which are identified as bad data. Such an identification is not ambiguous because an unreported line switching is readily detected in the comparison between $\mathbf{z}(k+1)$ and $\hat{\mathbf{z}}(k)$. Any change in the structure of the system affects all components of $\mathbf{z}(k+1)$ adjacent to the fault location. On the other hand, individual large differences between several components of $\mathbf{z}(k+1)$ originating from different parts of the network are readily interpreted as bad data.

The important aspect of sampling rate must also be studied carefully. For data processing a sampling rate for the measurements of 5–60 sec seems appropriate.

As indicated in Fig. 12 it is not necessary to process new data as long as the difference between the last estimate $\hat{\mathbf{z}}(k-1)$ and the current measurement $\mathbf{z}(k)$ is less than a preset value $\boldsymbol{\varepsilon}$. As long as this difference is small enough the time counter k (in Fig. 12 time is denoted by t and $t + \Delta t$) is advanced without further computation. In other words, it is possible that the on-line estimation algorithm may take 10–60 sec for one complete estimation run. A schematic solution for the problems encountered in stage 1 is shown in Fig. 12. Appropriate display methods are required for transmitting the entire data base to the operator. Metering, logging and event recording may readily be based on the state estimator output.

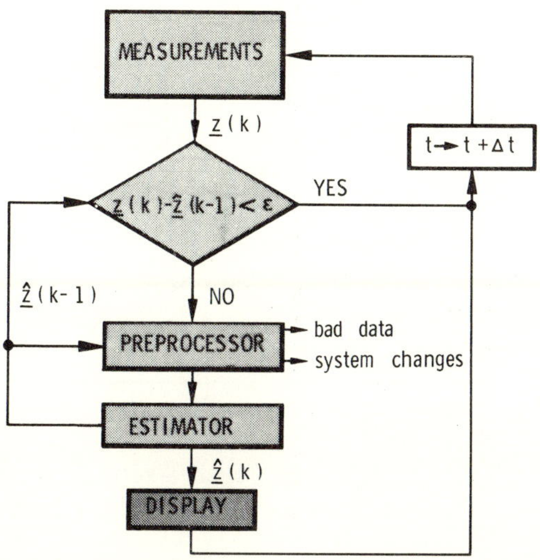

Fig. 12. Data processing and display using state estimation techniques.

5.2 Security monitoring

The second stage of on-line state estimation is concerned with the important problem of security monitoring.

(1) The estimated quantities must be compared against overload limits and the results indicated by suitable display methods.

(2) If limits are approached the operator must be able to choose between re-scheduling, switching operations and load shedding.

(3) Contingency evaluations are based on the data basis of the estimator.

Figure 13 shows the role of the estimator for security monitoring. The inputs to the estimator include the network topology, the system parameters and the various limits on power generation, transmission and system voltages, and scheduled line switching operations.

At each instant of time the measurements $\mathbf{z}(k)$ are processed and displayed as described in §5.1. The estimated values are then checked against overload limits. If no limit is violated, the output of the estimator may be used for contingency

References p. 56

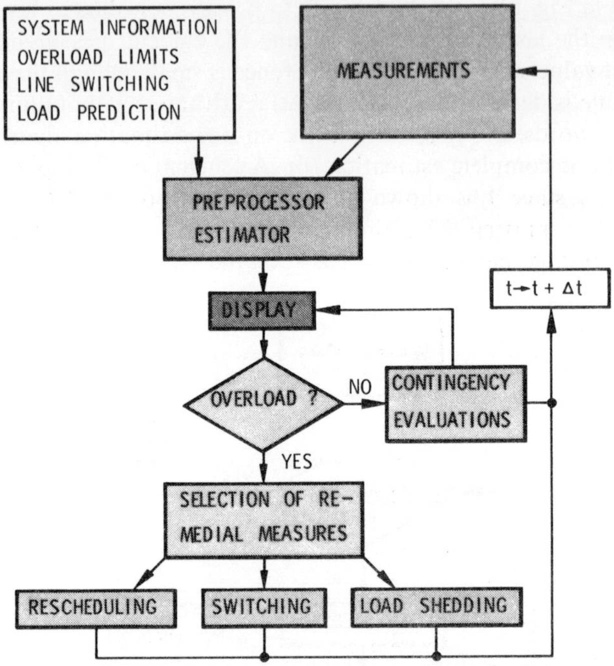

Fig. 13. Security monitoring using state estimation techniques.

evaluations[23]. When overload limits are reached suitable remedial measures help the operator to choose between re-scheduling, line switching and load shedding.

5.3 Optimal control

Optimal control theory is studied at present in connection with the following problems:

(1) Emergency dispatching
(2) Short-term forecasting
(3) Load-frequency control
(4) Economic dispatching
(5) Unit commitment

It is believed that state estimation techniques may successfully be used for problems (2), (3) and (4), while the problems (1) and (5) are only indirectly affected by state estimation.

Load forecasting plays an important part in the reliable and economic operation of a power system. State estimation techniques have been used for developing forecasting models[24–26, 31].

The load-frequency control (LFC) operates on a time scale comparable with state estimation. The main task of LFC is to make sure that load disturbances do not cause an imbalance of power and therefore a frequency deviation. Estimates for tie-line flows and power generations can be obtained from the state estimator[27].

6. CONCLUSIONS

A state estimator is a data processing algorithm to transform meter readings and other sets of information into an estimate of the system state. The estimator based on the least-squares approach minimizes a weighted sum of the errors between measured and calculated system variables. Conceptually, state estimation algorithms are closely related to load flow computations. The main difference is that state estimation takes the system uncertainties into consideration by using redundant measurements.

Due to the non-linear nature of the problem one has to resort to approximation techniques. In view of on-line applications several suboptimal solutions are discussed. Starting from a flat voltage profile the solution is obtained after three to five iterations.

The repeated use of static estimators in the tracking mode is well suited to on-line data processing in electric power systems. Assuming that the observation error $\mathbf{v}$ is not too large any static estimation algorithm is readily altered into a tracking algorithm by changing the iteration counter ν into the discrete time argument k.

The concept of state estimation techniques has been evaluated with off-line simulation studies. It has been found that the increase of redundancy effectively reduces the system uncertainties.

The input data of the estimator are the network parameters, the specifications of the measurement errors $\mathbf{v}$, the system configuration and the measured quantities $\mathbf{z}$.

Before the measurements $\mathbf{z}$ are processed in the estimator they are passed through a preprocessor in order to detect bad data and unreported system changes.

The output data of the estimator are the complex voltages of all buses, the real and reactive power flows in all lines, and possibly the real and reactive power injections at all nodes.

An interesting application of static state estimation is the important problem of meter placement. Using a trial and error procedure a satisfactory solution has been obtained in a real case study.

The complete implementation of on-line estimation algorithms in the control centre of a power company is quite demanding. It is suggested, therefore, to proceed via the following three stages: data processing and display, security monitoring, and optimal control. The first stage of state estimation will be completed in the next few years. At present, work is in progress to use state estimation techniques for the second and third stages.

ACKNOWLEDGEMENTS

The author wishes to express his gratitude to his colleagues Dr. K. Reichert and Mr. J. Caneel, both from the engineering computing centre of Brown Boveri, Baden, for many interesting discussions on state estimation. The case studies mentioned in this paper have been carried out in the computing centre in Baden.

References p. 56

REFERENCES

1 R. D. Masiello and F. C. Schweppe, A tracking static state estimator, *IEEE Trans.*, **PAS-90** (1971) 1025–1033.
2 R. Deutsch, *Estimation Theory*, Prentice-Hall, Englewood Cliffs, N.J., 1965.
3 F. C. Schweppe and J. Wildes, Power system static-state estimation—Part I: Exact model, *IEEE Trans.*, **PAS-89** (1970) 120–125.
4 F. C. Schweppe and D. B. Rom, Power system static-state estimation—Part II: Approximate model, *IEEE Trans.*, **PAS-89** (1970) 125–130.
5 F. C. Schweppe, Power system static-state estimation—Part III: Implementation, *IEEE Trans.*, **PAS-89** (1970) 130–135.
6 R. E. Larson, W. F. Tinney and J. Peschon, State estimation in power systems—Part I: Theory and feasibility, *IEEE Trans.*, **PAS-89** (1970) 345–352.
7 R. E. Larson, W. F. Tinney, L. P. Hajdu and D. S. Piercy, State estimation in power systems—Part II: Implementation and applications, *IEEE Trans.*, **PAS-89** (1970) 353–363.
8 G. W. Stagg, J. F. Dopazo, O. A. Klitin and L. S. VanSlyck, Techniques for the real-time monitoring of power system operations, *IEEE Trans.*, **PAS-89** (1970) 545–555.
9 J. F. Dopazo, O. A. Klitin, G. W. Stagg and L. S. VanSlyck, State calculation of power systems from line flow measurements I, *IEEE Trans.*, **PAS-89** (1970) 1698–1708.
10 J. F. Dopazo, O. A. Klitin and L. S. VanSlyck, State calculation of power systems from line flow measurements II, *Proc. 7th PICA Conference, Boston, Mass., May, 1971.*
11 F. Ariatti, Selection of telemetered quantities for on-line supervision and control of an electric power system with the aid of a real-time computer, *Symposium on the application of operations research methods in Varna, Bulgaria, 25–27 May, 1970.*
12 O. J. M. Smith, Power system state estimation, *IEEE Trans.*, **PAS-89** (1970) 363–375.
13 G. W. Stagg and A. H. El-Abiad, *Computer Methods in Power System Analysis*, McGraw-Hill, New York, 1968.
14 M. Syrbe and F. Bacher, The role of the human operator in system control, This Symposium, p. 279.
15 D. A. Pierre, *Optimization Theory with Applications*, Wiley, New York, 1969.
16 W. C. Davidon, Variable metric method for minimization, *AEC Research and Development Report ANL-5990*, 1959.
17 R. Fletcher and M. J. D. Powell, A rapidly convergent descent method for minimization, *Computer J.*, **6** (1963) 163–168.
18 A. S. Debs and R. E. Larson, A dynamic estimator for tracking the state of a power system, *IEEE Trans.*, **PAS-89** (1970) 1670–1677.
19 L. J. Levy, Adaptive estimation algorithms, *Technical Report ERI 93400*, Engineering Research Institute, Iowa State University, Ames, Iowa, 1970.
20 H. W. Sorenson, Kalman filtering techniques, in C. T. Leondes (Ed.), *Advances in Control Systems*, Vol. 3, Academic Press, New York, 1966, Ch. 5.
21 W. L. Miller and J. B. Lewis, Dynamic state estimator in power systems, *IEEE Trans.*, **AC-16** (1971) 841–846.
22 H. M. Merrill and F. C. Schweppe, Bad data suppression in power system static state estimation, *IEEE Trans.*, **PAS-90** (1971) 2718–2725.
23 W. F. Tinney and N. M. Peterson, Steady-state security monitoring, This Symposium, p. 191.
24 J. Toyoda, M. S. Chen and Y. Inoue, An application of state estimation to short term load forecasting—Part I: Forecasting modeling, *IEEE Trans.*, **PAS-89** (1970) 1678–1682.
25 J. Toyoda, M. S. Chen and Y. Inoue, An application of state estimation to short term load forecasting—Part II: Implementation, *IEEE Trans.*, **PAS-89** (1970) 1683–1688.
26 J. Toyoda and M. Chen, A state estimation type approach to practical load forecasting, *Proc. 7th PICA Conference, Boston, Mass., May, 1971.*
27 H. Glavitsch and F. D. Galiana, Load-frequency control with particular emphasis on thermal power stations, This Symposium, p. 115.
28 A. S. Debs and L. P. Hajdu, A tracking state-estimator algorithm and security-related display concepts for the BPA system control center, *Final report to the BPA, Contract 14-03-79968 (Exhibit E)* by SCI, Palo Alto, Calif. 1970.
29 A. S. Debs and W. H. Litzenberger, Experimental evaluation of tracking state estimation, *Proc. 7th PICA Conference, Boston, Mass., May, 1971.*
30 F. C. Schweppe, personal communication, December 1971.

31 F. D. GALIANA, An application of system identification and state prediction to electric load modelling and forecasting, *Report 28,* Electric Power Systems Engineering Laboratory, MIT, March 1971.

Discussion

G. HOSEMANN (*Technische Hochschule Darmstadt, Germany*)
Referring to "state estimation", I would like to ask our British colleagues about their plans to use such algorithms in their new Central Load Dispatch Centre in London. Why are you using it, or not using it? What has been your experience?

U. G. KNIGHT (*CEGB, London, England*)
Professor Hosemann has asked about our approach to state estimation in the National Control Centre of the CEGB. Data are displayed and logged as received over the telemetry, with, however, an indication if the format is incorrect. On the other hand, the security assessment programs are going to use information which has been checked for consistency—for instance, between circuit state and flow, between busbar selector isolators, etc. The computed flow will be compared circuit by circuit with the telemetered information and any significant differences displayed to the control engineer, enabling him to judge the viability of subsequent outage calculations.

As a general comment, I suggest we should consider the use to which the state estimation may be put when considering need and method. An experienced man is probably the best over-all state estimator we have. Is it right to process raw data to the extent of removing bad and missing data points, with consequent corrections to the remainder, when display to the man and not computation is concerned? May we not run the risk of "smearing" good data and of possibly misleading him, particularly if information on system structure is concerned?

E. HANDSCHIN
The object of state estimation is to provide the operator with a reliable data base. Bad data contained in the measurement vector $\mathbf{z}$ may give very misleading results. The quality of the estimation depends on the ability to detect and eliminate bad data. The operator must know whether any information is removed, and which particular information is removed. In this way the state estimation algorithm helps the control engineer to locate faulty meters, transmission systems or any other cause of large, unmodelled disturbances in a systematic manner.

J. SVOEN (*NVE-Statskraftverkene, Oslo, Norway*)
The reduction of the amount of metering points and transmitted data is important for economical and technical reasons. Did you perform any general study in order to optimize the needed redundancy, keeping in mind a "significant" accuracy and reliability of the estimated data?

E. HANDSCHIN
For building up a reliable data base the problem of meter placement is very important. No solution has been found so far to solve this problem in an algorithmic way. However, state estimation is a powerful tool for the quantitative assessment

of several meter arrangements. The line flow method of §3.2.3 has been used to find the minimum redundancy η in one particular network. With a redundancy $\eta \geq 2.0$ and a judicious meter location satisfactory results are obtained.

W. R. TRACEY (*Hoyles Niblock Overseas Ltd., Tehran, Iran*)
Have you made any analysis of the sensitivity of your state estimator to the selection of measurements for raw data, *i.e.* measurement of real and reactive line flow *versus* injections and loads, or bus voltages and angles?

E. HANDSCHIN
The numerical results reported in this paper are all based on the line flow method (§3.2.3) in which there is no choice with respect to the measurement variables $\mathbf{z}$. Other algorithms also mentioned in this paper can make use of any of the basic measurement schemes of Table 1. In ref. 28 there is a comparison between several measurement systems. Little difference exists between line flow measurements at both ends of each line and the measurements of injections and voltages at generator nodes plus line flows at one end of each line.

B. SAKIĆ (*Brown, Boveri & Co. Ltd., Baden, Switzerland*)
With respect to the required computing time, I would like to ask you whether state estimation methods could be used for protection purposes. In this case the execution time should be of the order of 10–20 msec. Does this require different algorithms or a large computer?

E. HANDSCHIN
Adopting a hierarchical structure for power system control, I think that all the protection devices operate on a low level because they must react very quickly. State estimation belongs to a higher level and we do not intend to use it for protection purposes on the low level. There is no direct interference of state estimation in protection devices.

F. ARIATTI (*ENEL, Rome, Italy*)
I would like to ask the author whether he made any comparisons between the least-squares state estimation method and load flow techniques which use load injections and load flow measurements.

E. HANDSCHIN
So far we have not made such a comparison, the main reason being that no programs of comparable efficiency are available for the same network. I hope that the questionnaire on power system state estimation proposed by Dr. Ariatti to the Working Group A1(C.32) of CIGRE may help to assess the quality of the various estimation methods and algorithms.

K. J. ÅSTRÖM (*Lund Institute of Technology, Lund, Sweden*)
(1) What are the characteristics of the measurement errors? Are they calibration errors, bias or random noise?
(2) What are typical values of redundancy used in power systems?
(3) In your diagrams you get decreases of accuracy which are much more than you get in simple situations with independent measurements, where the accuracy

decreases as the inverse of the square root of redundancy. Does this depend on the structure of the system?

E. HANDSCHIN
The measurement error $\mathbf{v}$ has many different causes. The characteristics you mention belong to the instrument errors. If one is only interested in modelling such errors I think that the error model of eqn. (16) is very useful. Although bias is not included in eqn. (16) it is not a difficult problem to estimate it by comparing the estimated quantities $\hat{\mathbf{z}}$ with the raw measurement vector $\mathbf{z}$.

A typical value of the efficiency η in a network consisting of 30 buses and 37 lines is given by $\eta = 148/59 = 2.5$ if the line flow method of §3.2.3 is used.

The accuracy gain as a function of redundancy depends on the structure of the network. Figure 8 can only serve as a general illustration of what happens when the number of meters is increased, *i.e.* when η is increased. It must be emphasized that it is not sufficient to increase the number of meters; they have to be placed in a judicious manner. The accuracy gain of Fig. 8 has nothing to do with the time independence of the measurements.

F. C. SCHWEPPE (*Massachusetts Institute of Technology, Cambridge, Mass., U.S.A.*)
The main point I would like to make is that the reason why redundancy is required is not accuracy but reliability. The state estimator should make sure that

(1) the processed information does not contain any bad data,
(2) there are no modelling errors and
(3) the structure of the network is correct.

It is important to have reliable data about what the system is doing and this is the main reason for redundancy and the use of weighted least-squares.

E. HANDSCHIN
An important application of estimation theory is concerned with navigation of aircraft and spacecraft where the quality of solution is directly related to the accuracy. This is not true for the power system application. Because there are bad data in the measurement vector $\mathbf{z}$ and because the transmission system may partially fail, the main object of processing the raw data $\mathbf{z}$ by means of state estimation techniques is to obtain a reliable data base $\hat{\mathbf{z}}$ with the properties mentioned by Professor Schweppe.

G. QUAZZA (*ENEL, Milan, Italy*)
Although reliability, more than accuracy, should be the main concern of state estimation, I would still like to pose a question regarding filtering of telemetered signals. Did the author consider the problem of rejecting the low frequency noise of measured data due to electromechanical oscillations in the system? Is the choice of sampling rates affected by such a filtering requirement, leading perhaps to different sampling periods according to the location of the measured variable in the system? Does a rational decision here imply knowledge of the system dynamics?

E. HANDSCHIN
All our experience with the tracking estimator is based on simulated data. Therefore the problem of electromechanical oscillations contained in real data has not

yet been investigated. The choice of the sampling rate is dictated by the computing time required to perform one or several iteration steps at each discrete time instant. A better knowledge of the system dynamics will help the choice of the sampling interval.

J. G. BALCHEN (*Technical University of Norway, Trondheim, Norway*)
Professor Åström discussed estimation of *parameters* in *dynamic* models of power systems. Dr. Handschin was concerned only with *states* of *static* models. Since parameters may be regarded as slowly varying states in an enlarged space and a static model is a special case of a dynamic model, it is apparent that both papers become special cases of a more general case. Would you care to make a comment on how the application of estimators would place bounds on tolerable complexities?

If you have no (or incomplete) direct measurements of the structure of your system (*e.g.* the state of circuit breakers), parameters describing this structure will have to be included in your state vector. Do you have a comment on this?

How does the fundamental requirement of observability in all estimation methods relate to your term "redundancy"?

E. HANDSCHIN
Static state estimation used in a tracking mode is well suited to generating a reliable data base and updating it every 10–30 sec. The identification procedure yields dynamic models of the energy production systems. Dynamic state estimation used for the generation systems may then be combined with the tracking estimator of the network based on algebraic equations. Such a hierarchical structure of the estimation problem may keep the over-all estimation problem in a manageable form.

Uncertainties with respect to the structure of the system are discussed in connection with Table 5. They do not have to be included in the state vector if several hypotheses of structures are tested.

In the case of the line flow method, the concept of observability requires that each line must be measured at least on one side.

J. P. WAHA (*INTERCOM, S.A., Brussels, Belgium*)
With respect to Professor Balchen's question about observability I would like to make a short comment. In the case of state estimation of a linear network (*i.e.* d.c. load flow approach), observability implies that measurements are made at least on a tree of the network graph. In the non-linear a.c. approach it is felt, but not necessarily proved, that the same is true for active and reactive measurements. An interesting case is the one in which some active and reactive injections are known, and redundancy is given by line ampère measurements. In this case it can be shown that observability is true.

For a 70 node problem an algorithm execution time of 3 sec was given by the author. What kind of computer and what core requirements were needed for this result?

E. Handschin

The power system considered has 70 nodes and 97 lines. The complete processing of one measurement vector **z** using the line flow method of §3.2.3 requires 7 sec and 50 kbytes on the IBM 370/155. In the tracking mode the matrix inversion need not be done at each instant of time. The data processing without matrix inversion takes about 3 sec.

The estimation algorithm is at present implemented on an IBM 1800 for a network consisting of 30 nodes and 37 lines. In this application only 8 k words are available for the estimation algorithm.

Self-excited Electromechanical Oscillations in Power Systems and Means for Control

G. HOSEMANN AND R. HARTENSTEIN

Institut für elektrische Energieversorgung, Technische Hochschule, Darmstadt, Germany

SUMMARY

This paper describes a new method for the investigation of electromechanical oscillations in multi-machine power systems. In addition to the complete two-axis equations of rotating machines the characteristics of speed governors and voltage regulators are considered together with the properties of the interconnecting network.

Stability criteria obtained from analysing the geometrical locus plot of a complex transfer function with the angular frequency λ as independent variable are not suitable for multi-machine systems if the entire range $0 \leq \lambda \leq \infty$ is to be investigated. As an example reference is made to the generalized Cremer–Leonhard criterion. The Hurwitz criterion is also not practicable due to the very high degree of the characteristic equation. Step-by-step methods are unsatisfactory in view of the extremely large number of steps necessary.

The authors propose to investigate the stability in the case of electromechanical oscillations by the ratio $\bar{m}_v/\bar{\delta}_v$ where $\bar{m}_v$ is the shaft torque oscillation and $\bar{\delta}_v$ the rotor angle oscillation of the vth machine. The ratio $\bar{m}_v/\bar{\delta}_v$ corresponds to a generalized "complex synchronizing coefficient" including terminal voltage oscillations in magnitude and phase as well as control characteristics and machine kinetics. It is shown that for the present stability analysis the range of interest for λ/ω_N can be limited to $0.01 \leq \lambda/\omega_N \leq 0.10$. If the locus of the complex ratio $\bar{m}/\bar{\delta}$ of a single synchronous machine intersects the positive (negative) ordinate the system is electromechanically stable (unstable), as shown in §2. This statement can be extended to multi-machine systems by means of the matrix $\bar{\mathbf{D}}$ consisting of all coupling coefficients. Only the main diagonal elements of $\bar{\mathbf{D}}$ are of interest (§3 and §4).

Several examples are calculated and discussed:

(1) A generator connected to an infinite bus by a real resistance R_v (§2); the stabilization makes use of an ideal speed governor (§5);

(2) The ratios $\bar{m}_v/\bar{\delta}_v$ are calculated for a ten-machine system for $v = 3, 7$ and 10 (§4);

(3) Forced non-harmonic oscillations caused by a synchrocyclotron in a ten-machine system (§7).

1. INTRODUCTION

Extensive investigations have been carried out on self-excited oscillations of a single generator or motor connected to an infinite bus via a feeder line (see refs. 1–3). These oscillations may occur for high real resistances of the supply line or for series capacitors. Lalander[4] and Ölwegård[5] have discussed self-excited slow oscillations observed for certain network configurations in the Scandinavian interconnected power system. In order to avoid such oscillations high voltage d.c. links have been suggested (see refs. 6 and 7).

Analytic methods for the investigation of these unstable oscillations in networks with a large number of synchronous and asynchronous machines are new. Frey[8] and El-Serafi[2] have determined the stability of a single synchronous machine in the system using the Cremer–Leonhard criterion* based on locus geometry which can be derived from the Cauchy theorem. The complete knowledge of the entire frequency response of the system is an important prerequisite for the locus geometry. In order to obtain the frequency response the system must be excited by an oscillation whose angular frequency λ ranges from zero to infinity. For these excitations the network transmissions cannot any longer be considered as frequency independent. Hartenstein[9, 10] has successfully calculated forced oscillations in multi-machine power systems even for non-sinusoidal responses, such as are sometimes caused by piston drives. He has also considered the frequency dependence of transmissions. In ref. 11 it is shown that multi-machine systems can effectively be treated with two-axis components.

Based on refs. 9–11 a method is developed to analyse self-excited electromechanical oscillations in multi-machine systems. Furthermore, possibilities are investigated of eliminating these oscillations by the proper layout of appropriate control equipment.

To this end, the frequency response of the complex transfer matrix $\bar{\mathbf{D}}(\mathrm{j}\lambda)$ of all n interconnected rotating machines is computed for the frequency range $0.01 \leq \lambda/\omega_N \leq 0.10$ where $\omega_N = 2\pi \times 50 = 314\ \mathrm{sec}^{-1}$ in Europe and $\omega_N = 2\pi \times 60 = 377\ \mathrm{sec}^{-1}$ in the U.S.A. The complex transfer matrix $\bar{\mathbf{D}}(\mathrm{j}\lambda)$ is defined as

$$\bar{\boldsymbol{\delta}} = \bar{\mathbf{D}}(\mathrm{j}\lambda)\bar{\mathbf{m}} \tag{1}$$

where $\bar{\boldsymbol{\delta}}$ = n-dimensional complex vector of all rotor angle oscillations
$\bar{\mathbf{m}}$ = n-dimensional complex vector of all shaft torque oscillations
$\bar{\mathbf{D}}(\mathrm{j}\lambda)$ = complex transfer matrix including the properties of turbines, generators, the control equipment and the network.

As shown in Fig. 1 the power supply to the system is only possible via the n shafts. The ratio $\bar{\delta}_\nu/\bar{m}_\nu$ or $\bar{m}_\nu/\bar{\delta}_\nu (\nu = 1, 2, \ldots, n)$ serves as a measure of the power supply to the network through the νth turbine-generator set. For the stability analysis it is therefore sufficient to consider only the elements on the main diagonal of $\bar{\mathbf{D}}(\mathrm{j}\lambda)$. For ease of graphical representation only the complex ratio $\bar{m}_\nu/\bar{\delta}_\nu$ is used in the following. This ratio will be referred to as the inverse interference function.

* Modified form of the Nyquist criterion.

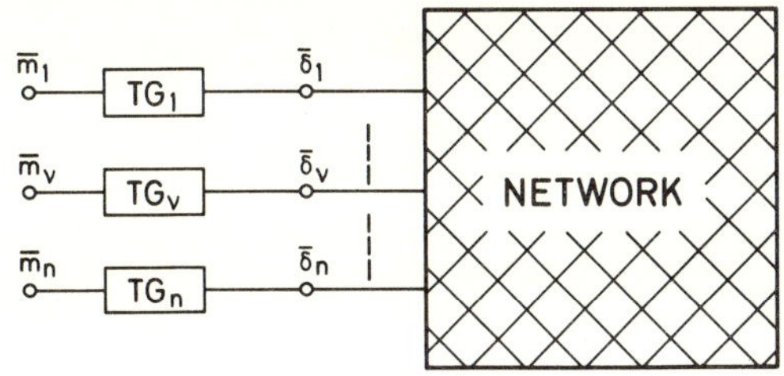

Fig. 1. Multi-machine power system oscillations $\bar{m}_\nu$ and $\bar{\delta}_\nu$ ($\nu = 1, 2, \ldots, n$) of n turbine–generator sets (TG).

2. LOCUS OF THE INVERSE INTERFERENCE FUNCTION OF ONE SYNCHRONOUS MACHINE

The torque M acting on the machine shaft is defined as

$$M = M_o + \mathrm{Re}[\bar{m}\exp(\mathrm{j}\lambda t)] \tag{2}$$

where M_o is the torque on the machine shaft under normal operating conditions. For physical reasons the use of the torque M as the input variable is very suitable for stability investigations, since when the oscillations grow the additional power must be supplied via the shaft.

In order to obtain a precise and unambiguous stability criterion the rotor angle Δ is chosen as the output variable. It is defined as

$$\Delta = \Delta_o + \mathrm{Re}[\bar{\delta}\exp(\mathrm{j}\lambda t)] \tag{3}$$

where Δ_o is the rotor angle under normal operating conditions. The magnitude of the ratio $\bar{m}/\bar{\delta}$ increases almost quadratically for $\lambda/2\pi$ greater than the electromechanical resonance frequency. This frequency is of the order of 1–6 c/s. Because self-excited electric oscillations are of higher frequencies they do not influence the stability analysis presented in this paper. As they are substantially decoupled there are simpler methods of calculation (see ref. 12) and different countermeasures apply. The transfer function $\bar{m}/\bar{\delta}$ corresponds to a generalized "complex synchronizing coefficient" which includes the terminal voltage oscillations in magnitude and phase as well as control characteristics and machine kinetics.

The rotor angle Δ of the machine is given by eqn. (4):

$$J\frac{\mathrm{d}^2\Delta}{\mathrm{d}t^2} + M_E = M_A + M_R \tag{4}$$

where J is the inertia of the machine, Δ the rotor angle, M_E the electric torque, M_A the driving torque and M_R the additional driving torque controlled by the governor.

For small alterations of angle δ and torque m a much simplified equivalent circuit diagram is shown in Fig. 2(a). It can be obtained with the following analogy based on an electrical equivalent circuit diagram with G, L and C elements. The analogy

References p. 78

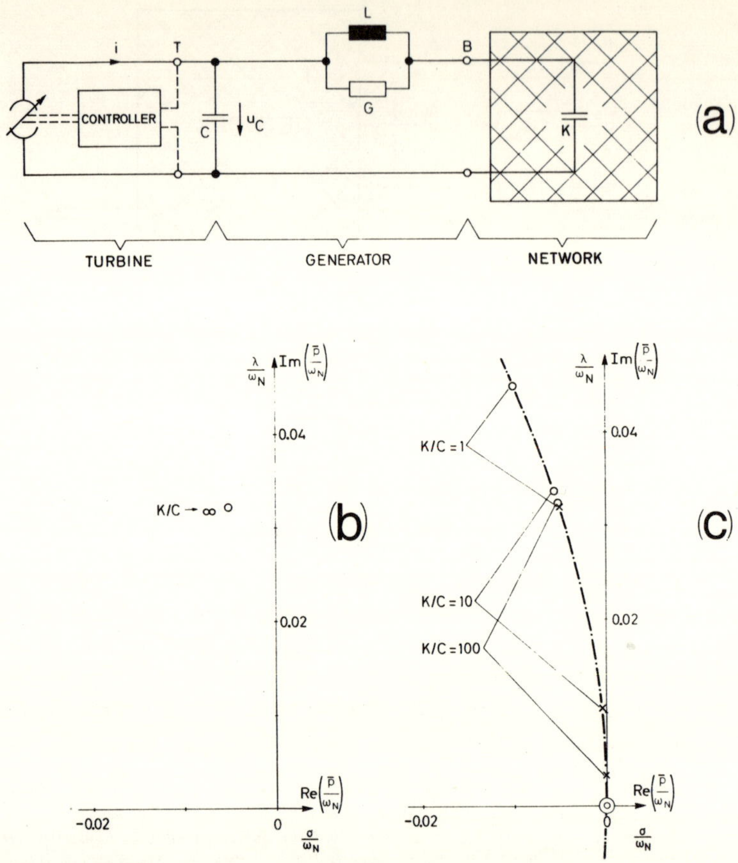

Fig. 2. (a) Equivalent circuit diagram of electromechanical system.
(b) Zeros of inverse interference function $\bar{p}\bar{y}_T \Leftrightarrow \bar{m}/\bar{\delta}$ with $K/C \to \infty$.
(c) Poles and zeros of inverse interference functions with $K/C = 1$, 10 and 100.

is chosen such that the network structure remains unchanged (isomorphic structure). In other words, parallel lines lead to parallel inductance L and serial lines lead to serial inductance L. The second characteristic of the analogy is the power invariance. The definition of the analogous capacity C results from the acceleration equation:

$$J\frac{\mathrm{d}\dot{\delta}}{\mathrm{d}t} = m \Leftrightarrow C\frac{\mathrm{d}u}{\mathrm{d}t} = i \tag{5}$$

where $\Leftrightarrow$ expresses the equivalence between the electromechanical system and the electric analogy diagram.

The definition of the inductance L results from the simplified transmission line and generator equation illustrated in Fig. 3.

The transmitted power P is given by

$$P = \frac{EU}{X} \sin \Delta \tag{6}$$

where in this analogy Δ is the angular difference between the complex input and output voltages, $\overline{E}$ and $\overline{U}$ respectively. Linearizing eqn. (6) around the operating point Δ_o yields:

$$\mathrm{d}P = \left.\frac{\partial P}{\partial \Delta}\right|_{\Delta_o} \mathrm{d}\Delta = \frac{EU}{X} \cos \Delta_o \, \mathrm{d}\Delta \tag{7a}$$

$$\mathrm{d}P = \omega_N \, \mathrm{d}M \tag{7b}$$

Putting $\mathrm{d}M \Leftrightarrow m$ and $\mathrm{d}\Delta \Leftrightarrow \delta$ and solving for m yields with eqn. (5) the following analogy relationship:

$$m = \frac{EU}{\omega_N X} \cos \Delta_o \delta \Leftrightarrow i = \frac{1}{L} \int u \, \mathrm{d}t \tag{8}$$

The analogy for small oscillations is summarized in Table 1.

The input admittance $\bar{y}_{\mathrm{T}}$ in node T of the diagram shown in Fig. 2(a) is given by

$$\bar{y}_{\mathrm{T}} = \bar{i}/\bar{u}_C \tag{9}$$

In order to use the equivalents from Table 1, eqn. (9) has to be multiplied by the Laplace operator $\bar{p} = \mathrm{j}\lambda$. Then the following equivalence relation holds:

$$\bar{p}\frac{\bar{i}}{\bar{u}_C} \Leftrightarrow \frac{\bar{m}}{\bar{\delta}} \tag{10}$$

On the border of stability there exist undamped rotor angle oscillations $\bar{\delta}$ even without torque oscillations of the turbine; *i.e.* $\bar{m}/\bar{\delta} = 0$.

According to Table 1 the dominating equivalents for the frequency of electromechanical oscillations are primarily the capacitance C and the inductance L. They correspond to the rotor inertia and the synchronizing coefficient of the machine and line. Voltage regulation and reactive power balance have only a small and indirect effect via the inductance L. Figures 2(b) and 2(c) show the poles, denoted by $\times$, and the zeros, denoted by 0, of the function $\bar{p}\bar{y}_T \Leftrightarrow \bar{m}/\bar{\delta}$ for the usual characteristics of a large turbogenerator. In Fig. 2(b) the generator is connected to an infinite bus, *i.e.* $K/C \to \infty$. In Fig. 2(c) several cases are studied for the

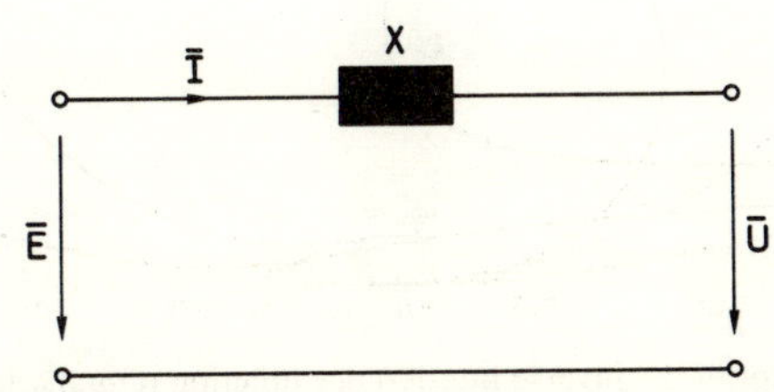

Fig. 3. Simplified transmission and generator system.

References p. 78

TABLE 1

ANALOGIES BETWEEN A POWER SYSTEM INTERFERED BY SMALL ELECTROMECHANICAL OSCILLATIONS AND AN ISOMORPHIC POWER-INVARIANT ELECTRIC CIRCUIT

Electromechanical system		*Electric circuit anology*	
Torque	m	Current	i
Angular velocity of rotor oscillations	$\dot{\delta}$	Capacitor voltages	u_C, u_K
Rotor inertia	J	Capacity	C
Equivalent network inertia	—	Capacity	K
Synchronizing torque of generator or line	$EU \cos \Delta_0/(\omega_N X)$	Inverse inductance	$1/L$
Damping torque acting on rotor	—	Conductance	G

network inertia: $K/C = 1$, 10 and 100. If the complex variable $\bar{p} = j\lambda$ of the function $\bar{p}\bar{y}_T$ is varied from $\lambda = 0$ to $\lambda \to \infty$ the locus shown in Fig. 4 is obtained. For the case when $K/C \to \infty$ the locus passes counterclockwise through two quadrants for stable operating conditions. The locus starts on the real axis at $1/L$ for $\lambda = 0$. In Fig. 2(c) an additional double zero in the origin and a pair of poles in the left half plane appear for $K/C < \infty$. At $K/C = 100$, however, they lie so closely together that their effects are nearly cancelled. This is again evident from Fig. 4 where all locus curves practically coincide in the area of technical interest, *viz.* $K/C \geq 10$.

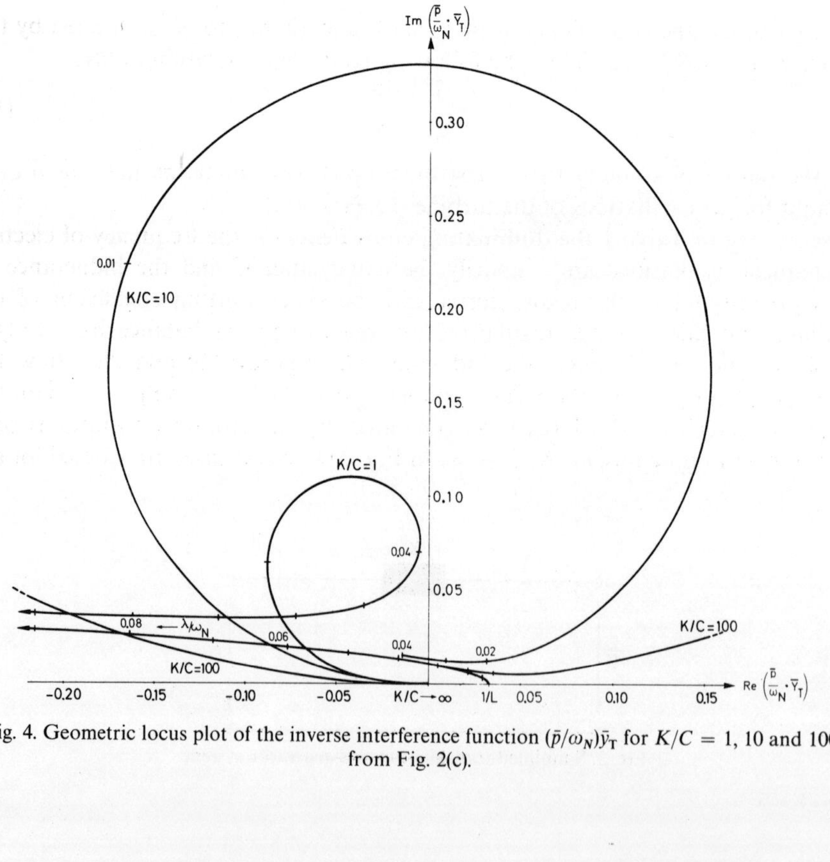

Fig. 4. Geometric locus plot of the inverse interference function $(\bar{p}/\omega_N)\bar{y}_T$ for $K/C = 1$, 10 and 100, from Fig. 2(c).

It is therefore justified to neglect interference frequencies $\lambda/\omega_N < 0.01$ for the stability analysis of electromechanical oscillations. Such extremely slow oscillations only affect the speed of synchronous clocks connected to the network and have no influence on the network stability (see ref. 10, p. 72). Hence the drift instability of the total interconnected system is neglected in the vicinity of zero frequency.

In the absence of damping, $G = 0$, the locus curves shown in Fig. 4 would coincide with the real axis for all K/C. Therefore undamped oscillations must be expected to occur at the point of intersection with the imaginary axis. For a constant driving torque $\bar{m} = 0$, the condition $\bar{m}/\bar{\delta} = 0$ to be met at the origin is satisfied for each arbitrary $\bar{\delta} \neq 0$.

The effect of large interfering angular frequencies $\bar{p} = \mathrm{j}\lambda$ on the function $\bar{p}\bar{y}_\mathrm{T} \Leftrightarrow \bar{m}/\bar{\delta}$ can also be studied with Fig. 4. In this case the admittance $\bar{y}_\mathrm{T}$ at node T is approximately equal to $\bar{p}C$. According to the analogy in Table 1 the function $\bar{p}^2C$ has to be studied. Therefore all locus curves of Fig. 4 approach the negative abscissa. Since the electromechanical resonance frequency of synchronous machines is approximately 5 c/s or less, only interfering angular frequencies $\lambda/\omega_N \leq 0.10$ are of interest.

The selection of $\bar{m}/\bar{\delta}$ as the inverse interference function yields the following criterion based on locus geometry for the operation of *one* single synchronous machine without electrical self-excitation in the network:

> Electromechanical stability exists when the locus curve $\bar{m}/\bar{\delta}$ intersects the positive ordinate axis for the range $0.01 \leq \lambda/\omega_N \leq 0.10$.

The correctness of this statement can be confirmed by extending the frequency response to the range $-\infty \leq \lambda \leq +\infty$ and complementing it at infinity into a closed curve with subsequent application of the Cauchy theorem (see refs. 2 and 8). It is a great advantage of this method that these additional portions of the curve are not required if the investigation of electromechanical oscillations is based on the use of the function $\bar{m}/\bar{\delta}$.

A relevant example is given in Fig. 5 where the connection of a synchronous machine to a grid via a large real resistance R_v is analysed. At $R_v \doteqdot 0.32\ Z_N$*, the curve $\bar{m}/\bar{\delta}$ runs through the origin so that $\bar{m} = 0$ is satisfied for any arbitrary $\bar{\delta}$. In this case undamped oscillations of the rotor angle develop. Their frequency is approximately 0.04×50 c/s $= 2$ c/s. A similar example is given in ref. 2, Fig. 10. Stabilization by means of a speed governor will be examined in §5.

3. LOCUS OF THE INVERSE INTERFERENCE FUNCTION OF *n* SYNCHRONOUS MACHINES

The results obtained so far are now generalized to systems where n generators are connected to a network (see Fig. 1). The power required for producing undamped

---

* $Z_N = U_N^2/S_N$ is the rated impedance of the generator.

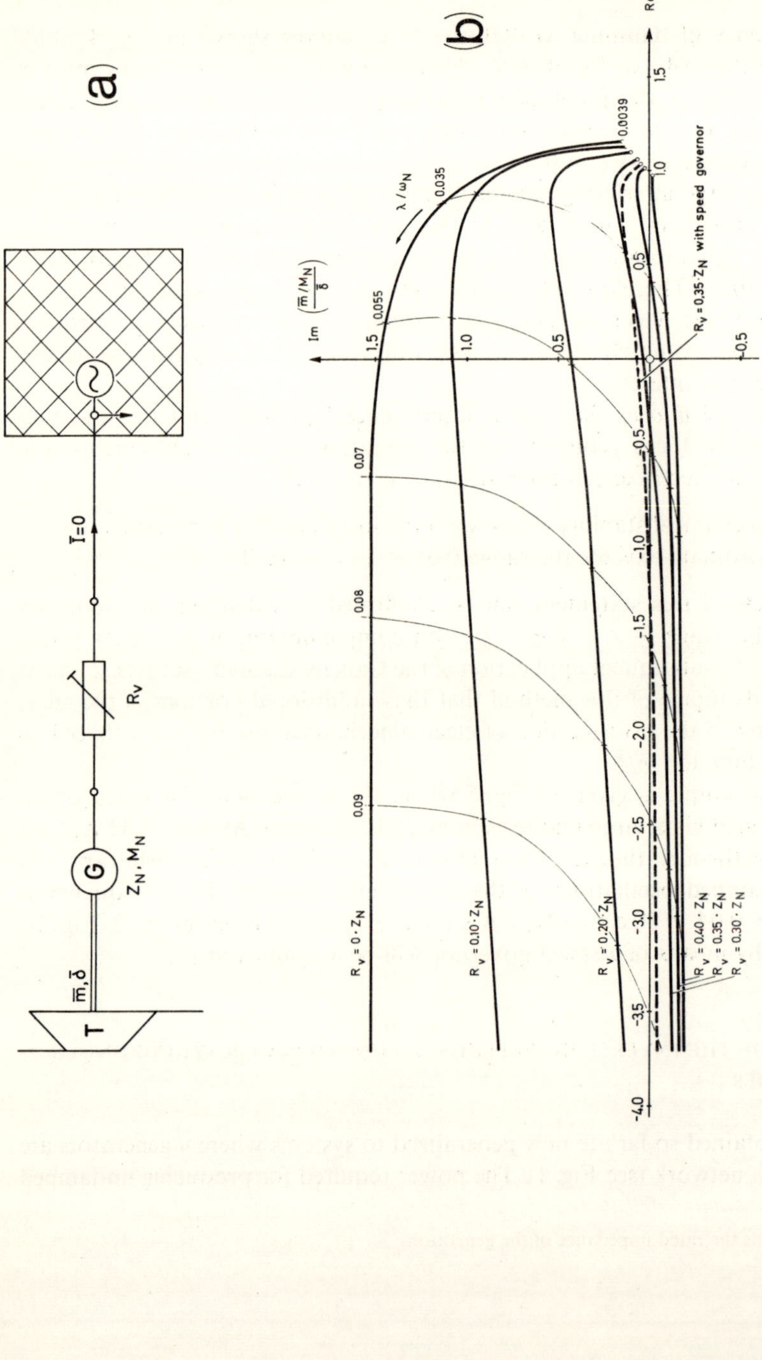

Fig. 5. (a) Circuit diagram of a generator connected to a grid by the real resistance R_v.
(b) Inverse interference function $(\bar{m}/M_N)/\delta$ for several values of R_v. The broken curve corresponds to $R_v = 0.35\, Z_N$ with a speed governor (see §5).

or exciting electromechanical oscillations must be supplied to the network via the shafts of one or several generators.

If electrical self-excitation is excluded the following statement applies:

> Electromechanical stability exists when the locus curves $\bar{m}_\nu/\bar{\delta}_\nu$ ($\nu = 1, 2, \ldots, n$) of *all* generators connected to a network intersect the positive ordinate for $0.01 \leq \lambda/\omega_N \leq 0.10$.

It is therefore sufficient to investigate the diagonal elements $\bar{D}_{\nu\nu}$ of the complex transfer matrix $\bar{\mathbf{D}}(j\lambda)$ of eqn. (1). The matrix $\bar{\mathbf{D}}$ is the matrix of the coupling coefficients.

4. MATRIX $\bar{\mathbf{D}}$ OF COUPLING COEFFICIENTS

At first only *one* generator is considered. The results are then generalized to the multi-machine case. The matrix $\bar{\mathbf{D}}$ may be calculated in the same way as shown for the investigation of forced oscillations in networks (see ref. 10). Therefore only the principle is indicated here.

The external variables of the generator are the shaft torque oscillation $\bar{m}$ and the terminal voltage oscillations $\bar{u}_d$ and $\bar{u}_q$ using the two-axis components (see ref. 11). Instead of working with $\bar{u}_d$ and $\bar{u}_q$ it may sometimes be advantageous to represent the complex terminal voltage oscillations in polar coordinates with magnitude oscillations $\bar{u}$ and phase oscillations $\bar{\gamma}$. The internal variables of the generator are the rotor angle oscillations $\bar{\delta}$ and the terminal current oscillations $\bar{i}_d$ and $\bar{i}_q$ using the two-axis components. When considering the voltage regulator, real and imaginary current oscillations $\bar{i}_r$ and $\bar{i}_i$ can be used instead. From the linearized two-axis equations (Park's equations) it is possible to formulate a matrix $\bar{\mathbf{A}}$ or a matrix $\bar{\mathbf{F}}$ for the generator. The control devices and inert masses may be included in $\bar{\mathbf{A}}$ and $\bar{\mathbf{F}}$. The complex elements of $\bar{\mathbf{A}}$ and $\bar{\mathbf{F}}$ are dependent on the interfering angular frequency λ. In rectangular coordinates the relations between the internal and external variables are given by eqn. (11):

$$\begin{bmatrix} \bar{i}_r \\ \bar{i}_i \\ \bar{\delta} \end{bmatrix} = \begin{bmatrix} \bar{A}_{11} & \bar{A}_{12} & \bar{A}_{13} \\ \bar{A}_{21} & \bar{A}_{22} & \bar{A}_{23} \\ \bar{A}_{31} & \bar{A}_{32} & \bar{A}_{33} \end{bmatrix} \begin{bmatrix} \bar{u} \\ \bar{\gamma} \\ \bar{m} \end{bmatrix} \tag{11}$$

Using the two-axis components the matrix $\bar{\mathbf{A}}$ in eqn. (11) is replaced by $\bar{\mathbf{F}}$:

$$\begin{bmatrix} \bar{i}_d \\ \bar{i}_q \\ \bar{\delta} \end{bmatrix} = \begin{bmatrix} \bar{F}_{11} & \bar{F}_{12} & \bar{F}_{13} \\ \bar{F}_{21} & \bar{F}_{22} & \bar{F}_{23} \\ \bar{F}_{31} & \bar{F}_{32} & \bar{F}_{33} \end{bmatrix} \begin{bmatrix} \bar{u}_d \\ \bar{u}_q \\ \bar{m} \end{bmatrix} \tag{12}$$

For the isotropic network the following equation applies:

$$\begin{bmatrix} \bar{i}_d \\ \bar{i}_q \end{bmatrix} = \begin{bmatrix} \bar{a} & -\bar{b} \\ \bar{b} & \bar{a} \end{bmatrix} \begin{bmatrix} \bar{u}_d \\ \bar{u}_q \end{bmatrix} = \bar{\mathbf{Y}} \begin{bmatrix} \bar{u}_d \\ \bar{u}_q \end{bmatrix} \tag{13}$$

References p. 78

The matrix $\bar{\mathbf{Y}}$ includes the symmetric element $\bar{a}$ and the antisymmetric element $\bar{b}$. The elements $\bar{a}$ and $\bar{b}$ are real for frequency independent networks and complex for frequency dependent networks. With eqns. (11), (12) and (13) there are five equations for the six variables $\bar{i}_d$, $\bar{i}_q$, $\bar{u}_d$, $\bar{u}_q$, $\bar{m}$ and $\bar{\delta}$. From the first two lines of eqn. (12) one obtains, with eqn. (13),

$$\left\{\bar{\mathbf{Y}} - \begin{bmatrix} \bar{F}_{11} & \bar{F}_{12} \\ \bar{F}_{21} & \bar{F}_{22} \end{bmatrix}\right\} \begin{bmatrix} \bar{u}_d \\ \bar{u}_q \end{bmatrix} = \begin{bmatrix} \bar{F}_{13} \\ \bar{F}_{23} \end{bmatrix} \bar{m} \tag{14}$$

By introducing $\bar{u}_d$ and $\bar{u}_q$ into the third line of eqn. (12) the solution for $\bar{\delta}$ becomes:

$$\bar{\delta} = \left\{ \begin{bmatrix} \bar{F}_{31} \\ \bar{F}_{32} \end{bmatrix}^T \left(\bar{\mathbf{Y}} - \begin{bmatrix} \bar{F}_{11} & \bar{F}_{12} \\ \bar{F}_{21} & \bar{F}_{22} \end{bmatrix} \right)^{-1} \begin{bmatrix} \bar{F}_{13} \\ \bar{F}_{23} \end{bmatrix} + \bar{F}_{33} \right\} \bar{m} \tag{15}$$

i.e.

$$\bar{\delta} = \bar{D}\bar{m} \tag{16}$$

where $\bar{D}$ is the inverse completed synchronizing coefficient.

In the following the generalization to n machines is discussed. The connection of several generators to one network is illustrated in Fig. 6.

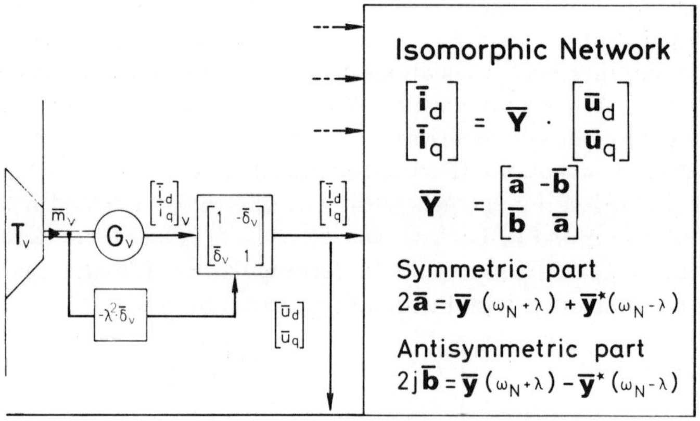

Fig. 6. Network in d,q-components oscillating with angular frequency $\lambda < \omega_N$.

For the νth machine the transformation matrix $\boldsymbol{\Xi}_\nu^*$, which transforms a vector $\mathbf{g}_{abc}$ expressed in the three-phase system into a vector $\mathbf{g}_{0dq}$ expressed in the two-axis system is given by eqn. (17):

$$\mathbf{g}_{abc} = \boldsymbol{\Xi}_\nu \mathbf{g}_{0dq} \tag{17}$$

* $\boldsymbol{\Xi}_\nu$ is the Park transformation matrix.

or explicitly

$$\begin{bmatrix} g_a \\ g_b \\ g_c \end{bmatrix} = \begin{bmatrix} 1 & \cos \xi_\nu & -\sin \xi_\nu \\ 1 & \cos(\xi_\nu - 2\pi/3) & -\sin(\xi_\nu - 2\pi/3) \\ 1 & \cos(\xi_\nu + 2\pi/3) & -\sin(\xi_\nu + 2\pi/3) \end{bmatrix} \begin{bmatrix} g_0 \\ g_d \\ g_q \end{bmatrix} \tag{18}$$

The matrix $\mathbf{\Xi}_\nu$ may be factorized provided use is made of the relationship

$$\xi_\nu = \omega_N t + \delta_\nu \tag{19}$$

Then in matrix notation $\mathbf{\Xi}_\nu$ is given by

$$\mathbf{\Xi}_\nu = \mathbf{\Omega}\mathbf{\Gamma}_\nu \tag{20}$$

or explicitly in component representation by

$$\begin{bmatrix} 1 & \cos \xi_\nu & -\sin \xi_\nu \\ 1 & \cos(\xi_\nu - 2\pi/3) & -\sin(\xi_\nu - 2\pi/3) \\ 1 & \cos(\xi_\nu + 2\pi/3) & -\sin(\xi_\nu + 2\pi/3) \end{bmatrix} =$$

$$\begin{bmatrix} 1 & \cos \omega_N t & -\sin \omega_N t \\ 1 & \cos(\omega_N t - 2\pi/3) & -\sin(\omega_N t - 2\pi/3) \\ 1 & \cos(\omega_N t + 2\pi/3) & -\sin(\omega_N t + 2\pi/3) \end{bmatrix} \begin{bmatrix} 1 & 0 & 0 \\ 0 & \cos \delta_\nu & -\sin \delta_\nu \\ 0 & \sin \delta_\nu & \cos \delta_\nu \end{bmatrix} \tag{21}$$

In this factorization the matrix $\mathbf{\Omega}$ is common to all n machines while the matrix $\mathbf{\Gamma}_\nu$ ($\nu = 1, 2, \ldots, n$) is machine dependent. It must be kept in mind that the arguments δ_ν appearing in $\mathbf{\Gamma}_\nu$ are small. This is a clear advantage in view of the numerical accuracy. The second and third rows and columns of $\mathbf{\Gamma}_\nu$ imply a torsion of the d,q-coordinates of the νth machine against the d,q-coordinates of the network by small angles δ_ν.

Figure 6 shows that the external variables of the νth generator are the shaft torque $\bar{m}_\nu$ and the network voltages $\bar{u}_d$ and $\bar{u}_q$. The internal variables $\bar{i}_d$ and $\bar{i}_q$ are governed by the induction equation. The rotor angle oscillations $\bar{\delta}_\nu$ are determined through the acceleration eqn. (4). The matrix $\mathbf{\Gamma}_\nu$ is further used for transforming the d,q-coordinates from the νth machine to the network. Due to the fact that $\delta_\nu \ll 1$ the matrix $\mathbf{\Gamma}_\nu$ in eqn. (21) may be linearized as follows:

$$\mathbf{\Gamma}_\nu \doteqdot \begin{bmatrix} 1 & 0 & 0 \\ 0 & 1 & -\delta_\nu \\ 0 & \delta_\nu & 1 \end{bmatrix} \tag{22}$$

For the small range of the interference frequencies $\lambda/\omega_N < 0.10$ the frequency dependence of the network admittances contained in $\bar{a}$ and $\bar{b}$ may be neglected. Thus, the matrix elements a and b in eqn. (13) are real. They can be computed from the nodal admittance matrix $\bar{\mathbf{Y}}$ as shown in Fig. 6.

For a network with n generators eqn. (15) still applies if the parameters $\bar{F}_{ij}$ are replaced by diagonal matrices $\bar{\mathbf{F}}_{ij}$ and the elements $\bar{a}$, $\bar{b}$ and $\bar{D}$ by quadratic matrices

References p. 78

$\bar{\mathbf{a}}$, $\bar{\mathbf{b}}$ and $\bar{\mathbf{D}}$. The six variables $\bar{i}_d$, $\bar{i}_q$, $\bar{u}_d$, $\bar{u}_q$, $\bar{m}$ and $\bar{\delta}$ have to be replaced by six column vectors each of dimension n. The elements $\bar{D}_{\nu\nu}$ on the main diagonal of the "coupling coefficient matrix" $\bar{\mathbf{D}}$ describe the rotor angle oscillations according to eqn. (16) provided all elements in the torque vector $\mathbf{m}$ except the element $\bar{m}_\nu$ are set equal to zero. Therefore the desired inverse interference function of the νth generator is

$$\frac{\bar{m}_\nu}{\bar{\delta}_\nu} = \frac{1}{\bar{D}_{\nu\nu}} \tag{23}$$

The stability criterion presented in this paper may be applied to eqn. (23). The computation of the quadratic matrix $\bar{\mathbf{D}}$ by a digital computer program is described in ref. 10. For every value of λ the required computing time t in seconds on an IBM 7040 may be estimated with the following empirical relation:

$$t = 0.5564 + 0.1975n^2 + 0.0496n^3 \tag{24}$$

where n is the number of rotating machines.

Figure 7(a) shows an example in which the 220 kV system consists of 30 nodes and 10 power plants. It includes a link to the 380 kV grid at node 1. Speed governors and voltage regulators are out of action. Figure 7(b) shows the locus plots of the inverse interference functions $1/\bar{D}_{\nu\nu}$ for the power plants $\nu = 3$, 7 and 10. All machines are electromechanically stable because the loci intersect the positive ordinate for $0.01 \leq \lambda/\omega_N \leq 0.10$.

5. SPEED GOVERNOR LAYOUT

The frequency response of the controlled system is obtained by dividing the ratio $\bar{m}_\nu/\bar{\delta}_\nu$ by $\bar{p} = \mathrm{j}\lambda$:

$$\frac{\bar{m}_\nu}{\bar{p}\bar{\delta}_\nu} = \frac{1}{\mathrm{j}\lambda\bar{D}_{\nu\nu}} \tag{25}$$

Interpreting $\bar{p}$ as the differentiation operator this gives the speed variation $p\delta_\nu = \dot{\delta}_\nu$ of the νth machine. An adequate speed governor may be designed and laid out according to the well-known methods of control design.

Figure 5 serves again as a typical example. The case $R_v = 0.35\,Z_N$ is unstable without a speed governor. Hence the function $\bar{m}_\nu/\mathrm{j}\lambda\bar{\delta}_\nu$ is approximated by a quadratic polynomial in the vicinity of the origin and a proportional control is then designed. The broken curve in Fig. 5 now intersects the positive ordinate and the system is therefore electromechanically stable. A disadvantage of the chosen example is the fact that such high resistances R_v hardly occur in networks. Moreover, the instantaneously operating speed governors that would be required are not available. However, major simplifications best demonstrate the basic properties of the stability analysis presented in this paper. Realistic cases such as are reported in refs. 4 and 5 may be treated by the same method. By setting the speed governor of the μth turbine the element $\bar{D}_{\nu\nu}(\mathrm{j}\lambda)$ with $\nu \neq \mu$ undergoes some variation. This effect is in general so small that no additional iterations are required.

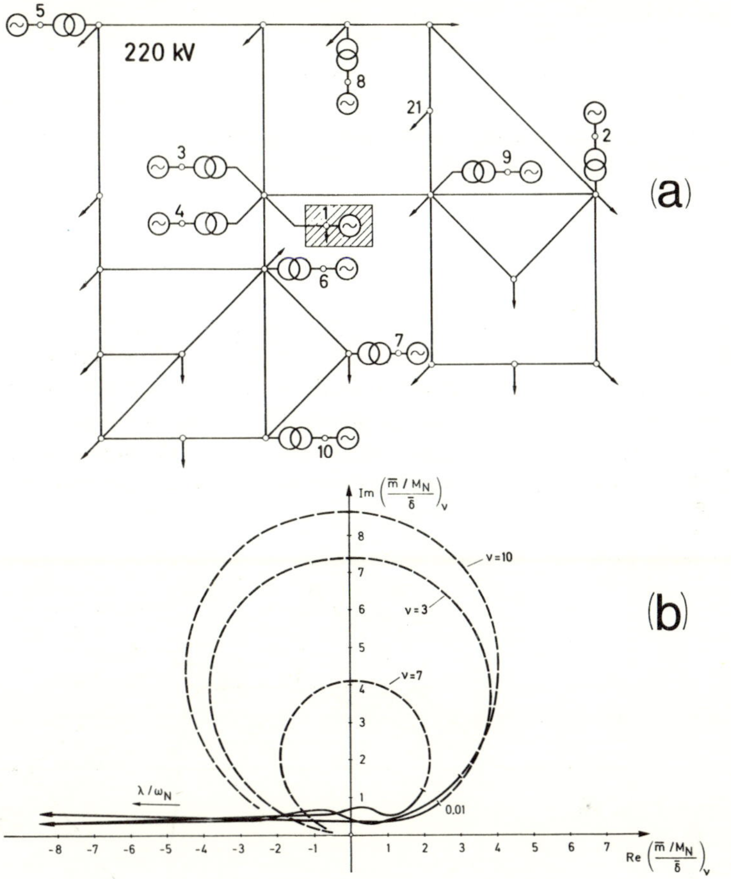

Fig. 7. (a) Circuit diagram of a ten-machine system with 30 nodes.
(b) Inverse interference function $(\bar{m}/M_N)/\bar{\delta}$ for generator $\nu = 3$, 7 and 10.

6. VOLTAGE REGULATOR LAYOUT

The input variables of the voltage regulator of the νth generator are given by the terminal voltage magnitude $\bar{u}$ and the real and imaginary currents $\bar{i}_r$ and $\bar{i}_i$ of the network. In the case of small oscillations the output quantity is the excitation voltage oscillation $\bar{u}_f$:

$$\bar{u}_f = \bar{c}_1\bar{u} + \bar{c}_2\bar{i}_r + \bar{c}_3\bar{i}_i \tag{26}$$

The currents $\bar{i}_r$ and $\bar{i}_i$ depend on the voltage magnitude $\bar{u}$ because of the coupling provided by the network. The coupling admittances are denoted by $\bar{Y}_r$ and $\bar{Y}_i$. If the effects of the voltage oscillations of all the other generators on the νth generator

References p. 78

are neglected, then

$$\bar{u}_f = (\bar{c}_1 + \bar{c}_2\bar{Y}_r + \bar{c}_3\bar{Y}_i)\bar{u} \tag{27}$$

An adequate voltage regulator may be designed and laid out according to well-known control theory using the frequency response $\bar{u}_f/\bar{u}$ and the matrix $\bar{\mathbf{A}}$ from eqn. (11).

7. FURTHER EXAMPLES OF APPLICATIONS

Figure 8(a) shows the cyclic active and reactive load program of a synchronous cyclotron installed in node 21 of the network shown in Fig. 7(a). This example is also contained in ref. 10. A Fourier analysis is used to represent the real and reactive load consumed by the cyclotron. This analysis ranges from $k = 0$ to $k = 30$ for $\lambda/\omega_N = k/165$. The associated variations of the active power $P_9(t)$ and voltage $U_9(t)$ at the nearest power station in node 9 are illustrated in Fig. 8(b).

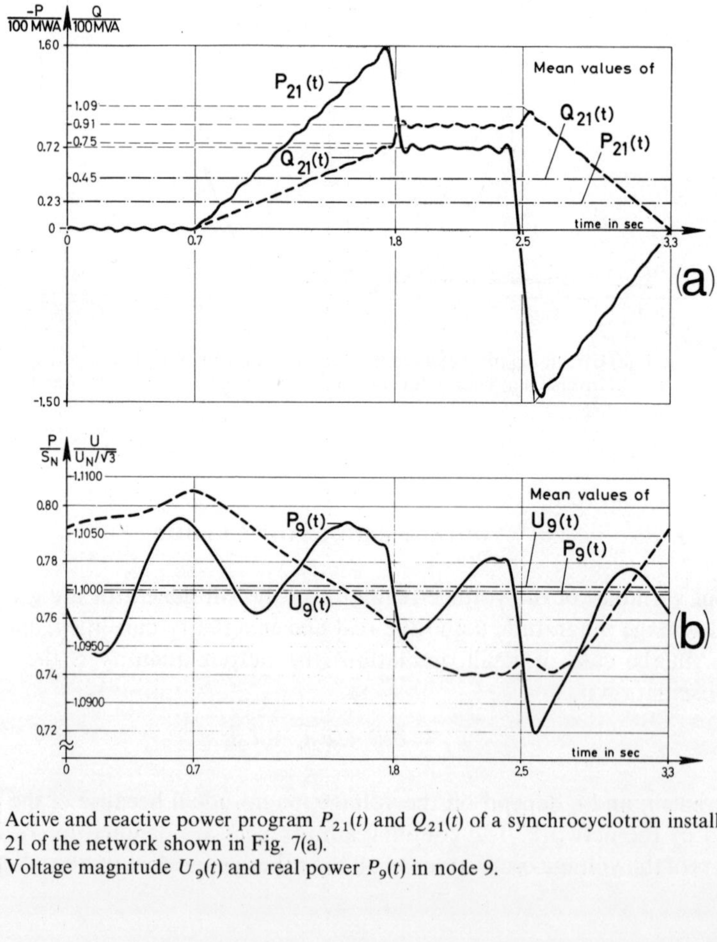

Fig. 8. (a) Active and reactive power program $P_{21}(t)$ and $Q_{21}(t)$ of a synchrocyclotron installed at node 21 of the network shown in Fig. 7(a).
(b) Voltage magnitude $U_9(t)$ and real power $P_9(t)$ in node 9.

Speed governor and voltage regulator are out of action. The reactive power program $Q_{21}(t)$ shown in Fig. 8(a) of the cyclotron is noticeable from the voltage variations $U_9(t)$ shown in Fig. 8(b). On the other hand there is no similarity between the active power program of the cyclotron and the power response $P_9(t)$ of the generator in node 9.

For the calculation a computer program may be used as described in ref. 10. Because high interference angular frequencies λ are also considered in the Fourier analysis, it is necessary to take the frequency dependence of the network admittances into account. The elements of the matrix $\bar{\mathbf{Y}}$ in Fig. 6 and eqn. (13) become complex.

The procedure described in the paper should in principle also be suitable for analysing non-cyclic electromechanical processes in networks. Experiences on this aspect are not yet available.

ACKNOWLEDGEMENT

The authors are indebted to dipl.Ing. K. F. Eichhorn for his valuable assistance.

NOTATION

λ	angular frequency
ω_N	steady state angular frequency
$\bar{\boldsymbol{\delta}}$	n-dimensional complex vector of rotor angle oscillations based on a two-pole machine
$\bar{\dot{\boldsymbol{\delta}}}$	n-dimensional complex vector of angular velocity oscillations based on a two-pole machine
$\bar{\mathbf{m}}$	n-dimensional complex vector of shaft torque oscillations
n	number of generators in multi-machine system
$\bar{\mathbf{D}}(\mathrm{j}\lambda)$	complex transfer matrix (coupling coefficient matrix)
$\boldsymbol{\Delta}_o$	n-dimensional vector of rotor angles under normal operating conditions
$\mathbf{M}_o$	n-dimensional vector of shaft torques under normal operating conditions
J	inertia
M_N	rated torque
M_E	electric torque
M_A	driving torque
M_R	additional driving torque controlled by the governor
P	active power
Q	reactive power
$\bar{y}_{\mathrm{T}}$	complex input admittance in node T of network of Fig. 2(a)
K, C	capacitance
L	inductance

References p. 78

G	real conductance
Z_N	rated impedance of generator
$\bar{u}_d, \bar{u}_q$	direct and quadrature axis voltage
$\bar{u}, \bar{\gamma}$	voltage magnitude and phase in polar coordinates
$\bar{i}_d, \bar{i}_q$	direct and quadrature axis currents
$\bar{i}_r, \bar{i}_i$	real and imaginary parts of complex current
$\bar{\mathbf{A}}, \bar{\mathbf{F}}$	transformation matrices
$\bar{\mathbf{Y}}$	network admittance matrix
$\bar{a}, \bar{b}$	elements of $\bar{\mathbf{Y}}$
$\mathbf{g}_{abc}$	arbitrary vector in the three-phase coordinate system (*a,b,c*-system)
$\mathbf{g}_{0dq}$	arbitrary vector in the 0,*d,q*-coordinate system
$\mathbf{\Xi}_\nu$	Park transformation matrix from 0,*d,q*- to *a,b,c*-coordinate system for νth machine
$\mathbf{\Omega}$	machine independent factor of $\mathbf{\Xi}_\nu$
$\mathbf{\Gamma}_\nu$	machine dependent factor of $\mathbf{\Xi}_\nu$
$\bar{p} = \sigma + j\lambda$	complex variable
u_f	excitation voltage

REFERENCES

1 A. EL-SERAFI, Selbsterregte Schwingungen der Synchronmaschine unter dem Einfluß von ohmschen Reihenwiderständen und von Reihenkondensatoren, *Elektrotech. Z.*, **A87** (1966) 17–24.

2 A. EL-SERAFI, Synchronisierungs- und Dämpfungskoeffizieten der Synchronmaschine und ihre Anwendung auf Stabilitätsuntersuchungen, *Scientia Electrica,* **XVI** (1970) 1–13.

3 A. W. KRON AND H. W. LORENZEN, Die selbsterregten Pendelungen von Drehstrom-Asynchronmaschinen, *Elektrotech. Z.*, **90** (1969) 200–205.

4 S. LALANDER, L. GUSTAFSON AND B. MATTSON, Seventeen years' experience of service reliability in the Swedish 400 kV transmission system, *CIGRE,* **23** (1970) 31–02.

5 Å. ÖLWEGÅRD, Improvement of system stability in interconnected power systems, *CIGRE*, **23** (1970) 32–17.

6 R. JÖTTEN AND R. FOERST, Modell-plant tests on behaviour of a HVDC-transmission during AC and DC system disturbances, *CIGRE*, **22** (1968) 43–08.

7 E. UHLMANN, AC network stabilisation by DC-links, *CIGRE*, **23** (1970) 32–01.

8 W. FREY, Beweis einer Verallgemeinerung des Stabilitätskriterium von Nyquist und Leonhard, *Brown Boveri Mitt.*, **33** (1946) 59–65.

9 R. HARTENSTEIN, Oscillations impressed upon diverse synchronous machines coupled to pulsating-torque prime movers in a common supply system, *Proc. Power System Computation Conf.*, Vol. I, 1969, Paper DC9 and Supplement.

10 R. HARTENSTEIN, Erzwungene Schwingungen von Drehstrommaschinen mit ungleichförmigem Antriebsmoment, die über ein beliebiges Netz miteinander verbunden sind, *Darmstädter Dissertation*, (D 17) 1970.

11 G. HOSEMANN, Drehstromnetze in Zwei-Achsen-Komponenten, *Elektrotech. Z.*, **92** (1971) 618–622, 746.

12 V. CRASTAN, Zur Rheorie der Selbsterregung von Synchronmaschinen, *Brown Boveri Mitt.*, **49** (1962) 80–94.

Discussion

J. P. BOWLES (*Hydro Quebec Institute of Research, Varennes, Canada*)
Did you attempt to represent within your system the effects of excitation stabilizing, or modifications of it, to help with the stability of the system?

H. Glavitsch (*Brown, Boveri & Co. Ltd., Baden, Switzerland*)
My question also pertains to the effect of excitation systems. In our experience the slip or acceleration signal is very effective in stabilizing the system. Today the tendency exists to include this signal in the excitation system.

E. Ferrari (*ENEL, Milan, Italy*)
I would like to outline the possibility of improving the damping of electromechanical oscillations by means of a suitable structure of excitation control, as mentioned by Mr. Glavitsch.

For small variations, we too obtained good results by the use of angle derivative feedbacks. The damping action of these feedbacks in the simplified case of an alternator connected to an infinite bus through a series reactance can easily be visualized. However, without further analysis the results cannot be extended to more general models of the network. For instance, in a more general case, the speed and electrical power can no longer be considered as the first and second derivatives of the angle; furthermore, the results of a linearized analysis may not be applicable to oscillations between weakly connected areas. As the additional feedbacks in excitation control could give a solution for such a difficult problem by relatively simple means, the results of theoretical or experimental work in this specific field are of the highest interest.

G. Hosemann
For avoiding or reducing electromechanical oscillations by means of control action you require quick power shifts. They can be realized by an alteration of the torque through the speed governor. Another possibility, according to the simplified eqns. (6) and (7), is to increase or decrease the angle Δ between $\bar{E}$ and $\bar{U}$ by excitation. The success of this procedure depends on $\partial\Delta/\partial E$, which is unfortunately zero for the unloaded generator in both directions d and q. If the rotors were equipped with two orthogonal excitation systems, as proposed in the *Cigré Report 11/01* (1970) for other reasons, you could influence E_q or E_d so that $\partial\Delta/\partial E$ would be satisfactory under all load conditions. Special investigations referring to these different possibilities have not yet been carried out.

H. Glavitsch (*Brown, Boveri & Co. Ltd., Baden, Switzerland*)
The real merit of the contribution by Professor Hosemann and Dr. Hartenstein is that you can study multi-machine systems without any extra effort, even if you have complex control systems behind your machines. The astonishing thing is that the entire solution can be found in the frequency domain.

W. B. Jervis (*Merz and McLellan, Newcastle upon Tyne, England*)
The authors appear to have developed a refined means of quantitatively assessing the electromechanical oscillations for interconnected multi-machines. I understand that the studies are based on very simplified machine representation, and I would like to ask the authors if they have verified their calculations from results obtained from actual network measurements and, if so, what degree of correlation has been obtained.

G. HOSEMANN

As you may see from refs. 9 and 10, our first effort was to investigate forced oscillations of interconnected machines caused, for example, by piston engines or by periodical load changes. The frequency range gave us the opportunity to consider also the influence of many other parameters, *e.g.* frequency dependent transfer functions, governors and regulators, solid iron in rotors etc. By the obtained matrix $\overline{\mathbf{D}}$ we were also able to study self-excited oscillations based on the same extended machine representation according to the complete two-axis theory. Therefore the results are expected to be as exact for small oscillations as the well-known Park equations. The very simplified model of Fig. 2 and Table 1 was not used for evidence but for discussing the complicated results of, for example, Fig. 7. The authors couldn't think of any better way to explain such curves than by plausible simple models particularly suited for understanding the method.

A Review of Minimization Techniques with Reference to Power System Engineering

A. BRAMELLER AND K. L. LO

Department of Electrical Engineering and Electronics,
The University of Manchester Institute of Science and Technology, Manchester, England

SUMMARY

Current practice for on-line control approximates the non-linear relations in a power system to a linear programming problem. The approximated equations are then solved by well-established and efficient linear programming techniques. In recent years there has been a steady growing interest towards more accurate representation of the system. This demands an efficient non-linear solution technique.

The paper reviews some of the more applicable non-linear optimization methods. It begins by formulating the optimization problem in the power system field as a general convex programming problem, using simple matrix notations and common engineering terminologies.

The next section gives a critical review of some practical non-linear techniques without detailed mathematical manipulation. It then follows with two algorithms reviewed in detail.

The final section presents an alternative practical approach to the solution of large-scale optimization problems. The proposed method takes advantage of the topology of the network with particular form of decomposition and, together with sparsity techniques, offers economical storage and computer time. The method is especially suitable for a multi-computer environment.

The linear formulation to the problem is given in Appendix 1.

1. INTRODUCTION

The optimization problem in a power system can arise in connection with the design and the operation of the system. In both cases the main objective is to meet the demand subject to operational limits of the system parameters within some degree of security at the minimum cost. The optimal operation can be achieved by optimal active and reactive power flow.

The nature of the problem is such that it can be defined by non-linear relations. Current practice for on-line control approximates these relations to linear equations and uses well-established linear programming (LP) techniques. They have the advantage of simplicity and assurance of a solution. The main disadvantage lies in the d.c. approximation. Other more accurate linear approximations are possible but these generally involve more variables and longer computing time. For linear programming an effective simplex method has been developed; furthermore, there is no lack of review papers or references on this field. This is equally true for integer and mixed integer programming.

Currently, a great deal of effort is placed on the application of non-linear optimization for on-line control. This has the advantage of more accurate representation and the possibility of dealing with larger problems. This paper is concerned with the review of the more well-known minimization techniques and of a practical non-linear optimization procedure particularly suitable for large-scale optimization. Complicated mathematical details are avoided as much as possible but the principle and general approach to each application are expressed in the simplest form.

2. GENERAL FORMULATION OF THE PROBLEM

In optimal control the objective is to minimize a scalar objective function, such as system losses or instantaneous operating cost, subject to constraints.

In order to achieve this optimality, parameters in the system are divided into control, dependent and fixed parameters; *e.g.* transformer taps are treated as control variables while voltage magnitudes and angles at P, Q nodes are treated as dependent variables. Constant parameters are those which are specified.

For optimal active and reactive power flow the objective function f is the sum of instantaneous operating cost over all controllable power generation. Using the linearized approximation of the power plant characteristic, f can be expressed as follows:

$$f(\mathbf{P}_\theta^G) = \mathbf{A}_\theta^T \mathbf{P}_\theta^G + \mathbf{C}_\theta^T \mathbf{E}_\theta \tag{1}$$

where $\mathbf{P}_\theta^G$ = vector representing the total active power outputs in the system
$\mathbf{A}_\theta$ = vector representing the cost coefficients of $\mathbf{P}_\theta^G$
$\mathbf{C}_\theta$ = vector representing the constants in the cost characteristics
$\mathbf{E}_\theta$ = summation vector containing unity elements

Alternatively, the objective function can be a loss function in the minimization of active power loss in the system when the active power dispatch has already been decided. Using the usual notations the total complex loss in a system is expressed as*

$$S_{\text{loss}} = \sum_k V_k \sum_i Y_{ki}^* V_i^*$$

or, in matrix form,

$$S_{\text{loss}} = \mathbf{V}_\alpha^T \mathbf{Y}_{\alpha\alpha}^* \mathbf{V}_\alpha^* \tag{2}$$

* Complex conjugate quantities are denoted by an asterisk: V^*.

and the active power loss is the real part of the above equation where $\mathbf{V}_\alpha$ is a vector for nodal voltages and $\mathbf{Y}_{\alpha\alpha}$ is the admittance matrix.

2.1 Constraints

The constraints that any feasible solution must satisfy can be divided mainly into two types: the network equations which govern the current, voltage and power flow relationships; and the operating limits of the system parameters such as voltage limit, current ratings of the line, active and reactive power limits. These are usually imposed because of engineering design standards.

The network flows must satisfy both Kirchhoff's and Ohm's laws. In its simplest form this means that the power injected into a node must be equal to that extracted from the node. Thus both the summations of real and reactive powers at a node must be zero:

$$\mathbf{P}_\alpha^L + \mathbf{P}_\alpha^N - \mathbf{P}_\alpha^G = \mathbf{0} \tag{3}$$

and

$$\mathbf{Q}_\alpha^L + \mathbf{Q}_\alpha^N - \mathbf{Q}_\alpha^G = \mathbf{0} \tag{4}$$

where $\mathbf{P}_\alpha^L + \mathrm{j}\mathbf{Q}_\alpha^L$ = vectors representing the total load demands at various nodes
$\mathbf{P}_\alpha^N + \mathrm{j}\mathbf{Q}_\alpha^N$ = vectors representing the total net power transfer at various nodes
$\mathbf{P}_\alpha^G + \mathrm{j}\mathbf{Q}_\alpha^G$ = vectors representing the total power generated at various nodes.

With an assumed set of values for $\mathbf{P}_\alpha^G$ and expressing the transferred powers in terms of the busbar voltage magnitudes and angles, an independent solution of eqns. (3) and (4) results in an ordinary load flow. This is an important feature as it will be seen later that standard load flow algorithms are incorporated in many conventional optimization techniques.

The final solution from the optimization algorithm must also satisfy, in addition to the network equations, the engineering design and operational limits. These limitations are usually expressed in the form of inequalities. To begin with, the active generations of the generators are restricted to lie within given minimum and maximum limits which are determined by the permissible extremes of boiler operating conditions, *i.e.* at the optimum solution $\mathbf{P}_\theta^G$ must fall within the following inequality:

$$\mathbf{P}_\theta^{G,\min} \leq \mathbf{P}_\theta^G \leq \mathbf{P}_\theta^{G,\max} \tag{5}$$

where $\mathbf{P}_\theta^{G,\min}$ and $\mathbf{P}_\theta^{G,\max}$ are the vectors for lower and upper limits of the generators.

Similarly, the reactive power limits of the generators can be expressed

$$\mathbf{Q}_\theta^{G,\min} \leq \mathbf{Q}_\theta^G \leq \mathbf{Q}_\theta^{G,\max} \tag{6}$$

In order to satisfy statutory legal requirements and design limitations the voltage magnitudes are restricted to limits

$$\mathbf{V}_\alpha^{\min} \leq \mathbf{V}_\alpha \leq \mathbf{V}_\alpha^{\max} \tag{7}$$

References pp. 98–99

Limits on the transfer of power through transmission lines can be achieved either by limiting the angular displacement between the busbars concerned or by the thermal rating of the line. However, this can be replaced by an equivalent upper limit on the line loss

$$-\mathbf{P}_{\beta}^{B,\max} \leq \mathbf{P}_{\beta}^{B} \leq \mathbf{P}_{\beta}^{B,\max} \tag{8}$$

where $\mathbf{P}_{\beta}^{B}$ is a vector representing the transmission losses in the various lines and $\mathbf{P}_{\beta}^{B,\max}$ is the equivalent upper limit on the line losses.

Taps of in-phase tap transformers must also lie within the range

$$\mathbf{T}_{\gamma}^{\min} \leq \mathbf{T}_{\gamma} \leq \mathbf{T}_{\gamma}^{\max} \tag{9}$$

Similarly, phase-shifting transformers introduce limits also:

$$\mathbf{R}_{\phi}^{\min} \leq \mathbf{R}_{\phi} \leq \mathbf{R}_{\phi}^{\max} \tag{10}$$

where $\mathbf{T}_{\gamma}$ is a vector giving the tap positions of the transformers and $\mathbf{R}_{\phi}$ a vector giving the angular positions of phase shifters. The superscripts "min" and "max" denote the lower and upper limits.

In addition to the above limits, network security limits might be superimposed. These are the unplanned circuit outages and thus cause violations of the limits. The types of limits imposed would depend on the degree of intended circuit security. The introduction of such limits would certainly further complicate the optimization problem and so far has not been properly resolved. This presents an additional difficulty for application of non-linear programming for on-line control.

3. NON-LINEAR OPTIMIZATION

In the field of non-linear optimization numerous methods are available and in general each method is specifically suited to a particular problem. These methods are mainly divided into non-derivative and gradient methods. However, in the power system engineering field non-derivative approaches have not been widely applied because of the computational inefficiency when used for a large number of variables.

In the gradient approach the derivative of the objective function is taken and this gradient vector is used to update the variables concerned. During the updating process it is necessary to perform a single dimensional search such that the optimum correction is used. This is a process which is widely employed.

In the power system field most of the methods used incorporate standard load flow solution between intermediate steps so that the solution is feasible at each iteration for the network constraints. Furthermore, during the updating process it is necessary to ensure that no violation of the limits is encountered.

Carpentier[1] in 1962 formulated the general problem of minimizing the instantaneous operating cost of a power system subject to both network constraints and inequality limits.

In the following year, Smith and Tong[2] put forward an algorithm for minimizing power transmission losses. In this approach the total system losses are assumed to be a quadratic function of the busbar voltage magnitude. Thus by assigning different voltage magnitude values to a generator busbar, while all other voltages are temporarily held constant, the transmission losses are calculated. Using these points, the losses are approximated to a quadratic function. The optimum voltage magnitude of the generator busbar corresponding to the minimum loss is evaluated. Treating each generator busbar in turn, a set of optimum voltage magnitudes can be obtained and after performing a load flow solution the whole procedure is repeated until convergence is obtained.

The whole algorithm is simple. However, for large systems the number of iterations will be high. Furthermore, the extent of the effect of loss minimization is only on branches which are connected to reactive power sources. Consequently, the optimum point obtained is only partial.

Dopazo *et al.*[3] presented another algorithm which optimized both the active as well as the reactive power allocation. This algorithm is actually divided into two separate sections: the first section optimizes the active power distribution, while the second section optimizes the reactive power distribution. The program thus alternates between them until the total cost of fuel input to the system remains unchanged or within a specified tolerance.

In the economic schedule of active power distribution optimization is obtained by an equal incremental cost solution. The minimum and maximum active power limits of the generators are imposed.

In the reactive power allocation the technique employed is a simple steepest-descent gradient step. This gradient vector is the derivative of the active power loss with respect to the reactive power at each reactive source. Reactive power limits and voltage limits are imposed.

The method is neat and simple. Computational efficiency as applied to practical systems is not known since at the time of publication the method was untried on a large practical system.

Peschon *et al.*[4] summarized Carpentier's formulation of the Lagrangian–Kuhn–Tucker conditions and pointed out the major difficulty and complication involved in implementing a computer program for resolving the optimization equations. They proposed a solution algorithm for optimum control of reactive power flow and the method has been successfully tried on a 500-node system at BPA*. The method is equally applicable to the optimization of active power allocation but other techniques have been recently published for solving problems of a similar size.

The method published by El-Abiad and Jaimes[5] also uses Lagrangian multipliers but has close similarity to that of Dommel and Tinney[6], whose method will be described in greater detail in the next section.

The following two sections review the more promising practical methods. One published by Shen and Laughton[7] deals rigorously with all the system constraints

* Bonneville Power Administration, Portland, Oregon, U.S.A.

References pp. 98–99

and the other, published by Dommel and Tinney[6], offers a more practical approach to the problem.

3.1 *First approach*[7, 8]

This approach, which incorporates Lagrangian and Kuhn–Tucker dual variables, is presented by Shen and Laughton[7], The objective function used is the cost characteristic of all the generators and the constraints are the network constraints and the inequality listed in §2.

The optimization conditions provided by the theorem of Kuhn and Tucker can be visualized in the following way. If f represents the cost function and $\mathbf{G} = \mathbf{0}$ the network constraints, $\mathbf{H} \leq \mathbf{0}$ are the inequality limits then the Lagrangian function is

$$L = f + \boldsymbol{\lambda}^T\mathbf{G} + \mathbf{u}^T\mathbf{H} \tag{11}$$

where

$$\boldsymbol{\lambda} \text{ is arbitrary} \qquad \mathbf{u} \geq \mathbf{0} \tag{12}$$

and

$$\mathbf{u}^T\mathbf{H} = 0 \tag{13}$$

The variables $\boldsymbol{\lambda}$ and $\mathbf{u}$ are known as dual variables associated with the equality and inequality constraints respectively. Condition (13) implies that $u_i = 0$ if the constraint h_i is not reached.

Thus, using eqn. (11) the complete Lagrangian function for minimization incorporating all the constraints in §2 will be as follows:

$$\begin{aligned} L = {} & \mathbf{A}_\theta^T\mathbf{P}_\theta^G + \mathbf{C}_\theta^T\mathbf{E}_\theta + (\boldsymbol{\lambda}_\alpha^P)^T(\mathbf{P}_\alpha^L + \mathbf{P}_\alpha^N - \mathbf{P}_\alpha^G) \\ & + (\boldsymbol{\lambda}_\alpha^Q)^T(\mathbf{Q}_\alpha^L + \mathbf{Q}_\alpha^N - \mathbf{Q}_\alpha^G) + (\mathbf{u}_\theta^{PL})^T(\mathbf{P}_\theta^{G,\min} - \mathbf{P}_\theta^G) \\ & + (\mathbf{u}_\theta^{PU})^T(\mathbf{P}_\theta^G - \mathbf{P}_\theta^{G,\max}) + (\mathbf{u}_\theta^{QL})^T(\mathbf{Q}_\theta^{G,\min} - \mathbf{Q}_\theta^G) \\ & + (\mathbf{u}_\theta^{QU})^T(\mathbf{Q}_\theta^G - \mathbf{Q}_\theta^{G,\max}) + (\mathbf{u}_\alpha^{VL})^T(\mathbf{V}_\alpha^{\min} - \mathbf{V}_\alpha) \\ & + (\mathbf{u}_\alpha^{VU})^T(\mathbf{V}_\alpha - \mathbf{V}_\alpha^{\max}) + (\mathbf{u}_\beta^{BU})^T(\mathbf{P}_\beta^B - \mathbf{P}_\beta^{\max}) \\ & + (\mathbf{u}_\gamma^{TL})^T(\mathbf{T}_\gamma^{\min} - \mathbf{T}_\gamma) + (\mathbf{u}_\gamma^{TU})^T(\mathbf{T}_\gamma - \mathbf{T}_\gamma^{\max}) \\ & + (\mathbf{u}_\phi^{SL})^T(\mathbf{R}_\phi^{\min} - \mathbf{R}_\phi) + (\mathbf{u}_\phi^{SU})^T(\mathbf{R}_\phi - \mathbf{R}_\phi^{\max}) \end{aligned} \tag{14}$$

where the superscripts of $\boldsymbol{\lambda}$ and $\mathbf{u}$ denote the system parameters associated with the dual variables.

It can be seen that this function gives a complete representation of the system constraints. The conditions for the optimum are that the partial derivatives of L with respect to its variables must be zero: $\partial L/\partial x_i = 0$. The strategy for solving problems such as the above is essentially to use an iterative process. This necessitates dividing the variables under the categories of dependent and control variables. It can be a physical division, *e.g.* the transformer tap can be a control variable while the voltage is a dependent one, or an arbitrary division.

The success of the method depends on how all these derivatives are grouped and solved. The method adopted by Shen and Laughton[7], which they term a variational method, is based on the solution of non-linear simultaneous equations by the Newton–Raphson method. It can be divided into four main steps and in each step some variables are kept constant while others are adjusted.

(1) Adjust active power generation, voltage phase angles (transformer settings and voltage magnitudes are kept constant).

(2) Adjust angular positions of phase shifters, voltage phase angles (voltage magnitudes and auto-transformer settings are kept constant).

(3) Adjust voltage magnitudes, reactive power generations (transformer settings and voltage phase angles are kept constant).

(4) Adjust ratios of auto-transformers, voltage magnitudes (voltage phase angles and phase shifter settings are kept constant).

This method gives a rigorous mathematical solution to the problem and system parameters are represented as fully as possible. At the time of publication the method was tried on a medium-sized system. For practical on-line application it may be desirable to dispense with some of these complex representations in order to achieve a simpler algorithm for solution.

3.2 Second approach[6, 9]

This method published by Dommel and Tinney[6] offers a significant improvement in the methods for optimization of power systems. The whole formulation builds on a Newton–Raphson power flow program and can be used for optimal control of both real and reactive power dispatch.

As stated in §2, solving eqns. (3) and (4) independently represents an ordinary load flow solution. This solution can be achieved by the well-known method of Newton–Raphson and the Jacobian matrix used during the process of solution is highly sparse. Matrix factorization technique can be incorporated so that only the non-zero elements are stored. This approach not only decreases the storage requirement but also increases computational efficiency.

The optimization problem developed depends on the assumption that initially the inequality limits are not incorporated. Thus if the objective function is represented by f with the network constraints as $\mathbf{G} = \mathbf{0}$, the Lagrangian formulation without inequality limits is

$$L = f + \boldsymbol{\lambda}^T \mathbf{G} \tag{15}$$

where $\boldsymbol{\lambda}$ is known as the Lagrangian multipliers. The objective function f may be the cost characteristic of the generator or the system losses. The choice will depend on the type of optimization required.

If $\mathbf{x}$ and $\mathbf{y}$ represent the dependent and control variables then for the optimum the following conditions need to be satisfied:

$$\frac{\partial L}{\partial \mathbf{x}} = \frac{\partial f}{\partial \mathbf{x}} + \frac{\partial \mathbf{G}^T}{\partial \mathbf{x}} \boldsymbol{\lambda} = \mathbf{0} \tag{16}$$

References pp. 98–99

$$\frac{\partial L}{\partial \mathbf{y}} = \frac{\partial f}{\partial \mathbf{y}} + \frac{\partial \mathbf{G}^T}{\partial \mathbf{y}} \boldsymbol{\lambda} = \mathbf{0} \tag{17}$$

$$\frac{\partial L}{\partial \boldsymbol{\lambda}} = \mathbf{G} = \mathbf{0} \tag{18}$$

It is worth noting that $\partial \mathbf{G}/\partial \mathbf{x}$ is the Jacobian matrix in a load flow solution and the satisfaction of eqn. (18) is simply an ordinary load flow. Thus by solving the three equations iteratively the optimum can be obtained, using eqn. (17) for updating the control variables. The iterative steps are as follows:

$$\mathbf{G} = \mathbf{0} \qquad \text{(load flow solution)}$$

$$\boldsymbol{\lambda} = -\left(\frac{\partial \mathbf{G}^T}{\partial \mathbf{x}}\right)^{-1} \frac{\partial f}{\partial \mathbf{x}} \qquad \text{(Lagrangian multipliers)}$$

$$\frac{\partial L}{\partial \mathbf{y}} = \frac{\partial f}{\partial \mathbf{y}} + \frac{\partial \mathbf{G}^T}{\partial \mathbf{y}} \boldsymbol{\lambda} \qquad \text{(gradient vector)}$$

$$\mathbf{y}^{\text{new}} = \mathbf{y}^{\text{old}} + k \frac{\partial L}{\partial \mathbf{y}} \qquad \text{(update variables)}$$

where k is the step length.

A more careful observation reveals that eqn. (17) holds only if no inequality limit is imposed. Thus, at the optimum where some of the variables are on their limits $\partial L/\partial \mathbf{y} \neq \mathbf{0}$.

The handling of inequality limits for control variables offers no difficulty as they can be set and held to their limits upon violation. When the dependent variables have violated their limits the penalty method is used to force the variables back to their limits. The reasons for choosing this approach are that these limits are not usually hard limits and that penalty terms can easily be incorporated.

This method offers a neat and efficient algorithm for solution. The main factors which need to be carefully computed are the length of the step sizes and the value of the weighting factor for the penalty functions.

Practical experience indicates that for certain classes of problems the method offers good convergence but there are other problems which are physically or mathematically ill-conditioned where the method becomes less effective.

4. MULTILEVEL COMPUTER CONTROL—DIAKOPTICAL APPROACH[10]

When the size and complexity of the system increases it will be difficult to solve the complete system in one optimization procedure. A possible approach to overcome the problem proposed by Brameller and Hamam[10] is to divide the large system into smaller ones by removing certain interconnections. Each isolated subsystem can be optimized in its own right without reference to its neighbouring system. From the solutions of the subdivisions and using the removed interconnections, the over-all solution to the problem may be obtained.

The over-all procedure is an iterative one and the schematic outline of the method is shown in Fig. 1.

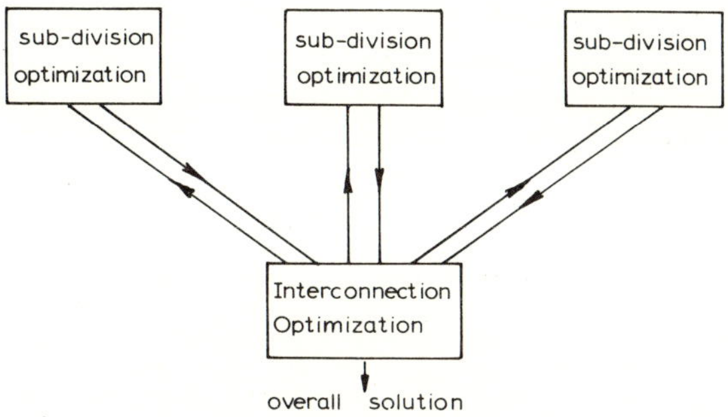

Fig. 1. Multilevel optimization.

4.1 *Principles of diakoptics*[10–12]

In order to understand the underlying theory of diakoptics consider the linear network shown in Fig. 2. Treating the branches AD and CE as interconnections the network is split into two subdivisions as shown in Fig. 3.

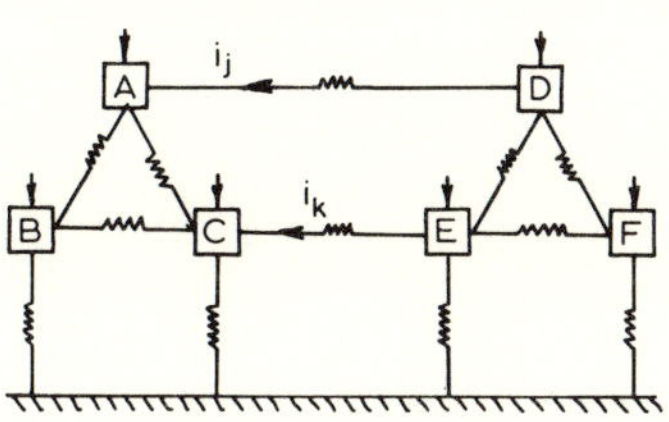

Fig. 2. Original network.

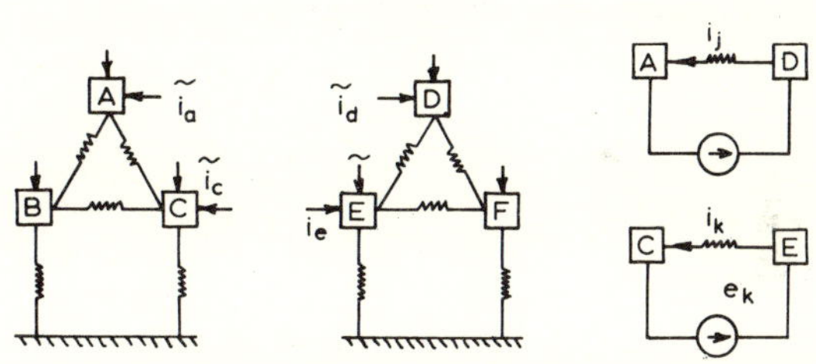

Fig. 3. Subdivided network.

In order to retain the physical identity of the subdivided system, unknown hypothetical voltages and currents have been introduced and the equations can now be expressed as follows:

$$\left[\begin{array}{c|c} \mathbf{Y}_{11} & 0 \\ \hline 0 & \mathbf{Y}_{22} \end{array}\right]\left[\begin{array}{c} \mathbf{V}_1 \\ \hline \mathbf{V}_2 \end{array}\right] = \left[\begin{array}{c} \mathbf{I}_1 \\ \hline \mathbf{I}_2 \end{array}\right] + \left[\begin{array}{c} \tilde{\mathbf{i}}_1 \\ \hline \tilde{\mathbf{i}}_2 \end{array}\right] \tag{19}$$

or in general matrix form

$$\mathbf{Y}_{\alpha\alpha}\mathbf{V}_\alpha = \mathbf{I}_\alpha + \tilde{\mathbf{i}}_\alpha \tag{20}$$

where $\mathbf{I}_\alpha$ = nodal currents
$\tilde{\mathbf{i}}_\alpha$ = nodal currents from interconnections
$\mathbf{V}_\alpha$ = nodal voltages

References pp. 98–99

The equations for the removed interconnection are

$$\mathbf{Z}_{\psi\psi}\mathbf{i}_{\psi} = \mathbf{e}_{\psi} \tag{21}$$

where $\mathbf{Z}_{\psi\psi}$ = impedance matrix
$\mathbf{i}_{\psi}$ = current in interconnections
$\mathbf{e}_{\psi}$ = hypothetical voltages

The relationship between $\tilde{\mathbf{i}}_{\alpha}$ and $\mathbf{i}_{\psi}$ is

$$\tilde{\mathbf{i}}_{\alpha} = \mathbf{C}_{\alpha\psi}\mathbf{i}_{\psi} \tag{22}$$

where $\mathbf{C}_{\alpha\psi}$ is a topological (incident) matrix. Expressing eqn. (22) in full yields

$$\begin{bmatrix} \tilde{i}_a \\ \tilde{i}_b \\ \tilde{i}_c \\ \hline \tilde{i}_d \\ \tilde{i}_e \\ \tilde{i}_f \end{bmatrix} = \left[\begin{array}{c|c} 1 & 0 \\ 0 & 0 \\ 0 & 1 \\ \hline -1 & 0 \\ 0 & -1 \\ 0 & 0 \end{array}\right] \begin{bmatrix} i_j \\ i_k \end{bmatrix}$$

Similarly,

$$\mathbf{e}_{\psi} = -\mathbf{C}_{\psi\alpha}^T\mathbf{V}_{\alpha} \tag{23}$$

Combining eqns. (20)–(24) gives

$$\mathbf{Y}_{\alpha\alpha}\mathbf{V}_{\alpha} - \mathbf{C}_{\alpha\psi}\mathbf{i}_{\psi} = \mathbf{I}_{\alpha} \tag{24}$$

$$\mathbf{C}_{\psi\alpha}^T\mathbf{V}_{\alpha} + \mathbf{Z}_{\psi\psi}\mathbf{i}_{\psi} = \mathbf{0} \tag{25}$$

The above equations can be solved for $\mathbf{V}_{\alpha}$ retaining the isolated structure of the subdivision.

4.2 Diakoptical optimization[17]

For a general non-linear network eqns. (24) and (25) can be expressed as follows:

$$\mathbf{G}_{\alpha} - \mathbf{C}_{\alpha\psi}\mathbf{i}_{\psi} = \mathbf{0} \tag{26}$$

$$\mathbf{C}_{\psi\alpha}^T\mathbf{x}_{\alpha} + \boldsymbol{\phi}_{\psi} = \mathbf{0} \tag{27}$$

where $\mathbf{G}_{\alpha}$ = a function of $\mathbf{x}_{\alpha}$ and $\mathbf{y}_{\beta}$
$\mathbf{x}_{\alpha}$ = dependent variables (*e.g.* voltages)
$\mathbf{y}_{\alpha}$ = control variables (*e.g.* transformer taps)
$\boldsymbol{\phi}_{\psi}$ = a function of $\mathbf{i}_{\psi}$

Using the Lagrangian formulation of Dommel and Tinney eqn. (15) can be expressed in diakoptical form as

$$L = f + \boldsymbol{\lambda}_{\alpha}^T(\mathbf{G}_{\alpha} - \mathbf{C}_{\alpha\psi}\mathbf{i}_{\psi}) + \boldsymbol{\lambda}_{\psi}^T(\mathbf{C}_{\psi\alpha}^T\mathbf{x}_{\alpha} + \boldsymbol{\phi}_{\psi}) \tag{28}$$

where $\boldsymbol{\lambda}_{\alpha}$ are Lagrangian multipliers associated with isolated network equations and $\boldsymbol{\lambda}_{\psi}$ are Lagrangian multipliers associated with interconnections.

Differentiating L with respect to the variables $\mathbf{y}_\beta$, $\mathbf{x}_\alpha$, $\mathbf{i}_\psi$, $\boldsymbol{\lambda}_\alpha$ and $\boldsymbol{\lambda}_\psi$ gives the equations corresponding to eqns. (16), (17) and (18).

$$\frac{\partial L}{\partial \mathbf{y}_\beta} = \frac{\partial f}{\partial \mathbf{y}_\beta} + \frac{\partial \mathbf{G}_\alpha^T}{\partial \mathbf{y}_\beta}\boldsymbol{\lambda}_\alpha = \mathbf{0} \tag{29}$$

$$\frac{\partial L}{\partial \mathbf{x}_\alpha} = \frac{\partial f}{\partial \mathbf{x}_\alpha} + \frac{\partial \mathbf{G}_\alpha^T}{\partial \mathbf{x}_\alpha}\boldsymbol{\lambda}_\alpha + \mathbf{C}_{\alpha\psi}\boldsymbol{\lambda}_\psi = \mathbf{0} \tag{30}$$

$$\frac{\partial L}{\partial \mathbf{i}_\psi} = -\mathbf{C}_{\psi\alpha}^T\boldsymbol{\lambda}_\alpha + \frac{\partial \boldsymbol{\phi}_\psi^T}{\partial \mathbf{i}_\psi}\boldsymbol{\lambda}_\psi = \mathbf{0} \tag{31}$$

$$\frac{\partial L}{\partial \boldsymbol{\lambda}_\alpha} = \mathbf{G}_\alpha - \mathbf{C}_{\alpha\psi}\mathbf{i}_\psi = \mathbf{0} \tag{32}$$

$$\frac{\partial L}{\partial \boldsymbol{\lambda}_\psi} = \mathbf{C}_{\psi\alpha}^T\mathbf{x}_\alpha + \boldsymbol{\phi}_\psi = \mathbf{0} \tag{33}$$

It is worth noting that the last two equations are the network constraints which can be solved by the Newton–Raphson method as follows:

$$\mathbf{J}_{\alpha\alpha}\Delta\mathbf{x}_\alpha - \mathbf{C}_{\alpha\psi}\Delta\mathbf{i}_\psi = \boldsymbol{\varepsilon}_\alpha \tag{34}$$

$$\mathbf{C}_{\psi\alpha}^T\Delta\mathbf{x}_\alpha + \mathbf{J}_{\psi\psi}\Delta\mathbf{i}_\psi = \boldsymbol{\varepsilon}_\psi \tag{35}$$

where $\mathbf{J}_{\alpha\alpha} = \partial\mathbf{G}_\alpha/\partial\mathbf{x}_\alpha$, $\mathbf{J}_{\psi\psi} = \partial\boldsymbol{\phi}_\psi/\partial\mathbf{i}_\psi$, $\boldsymbol{\varepsilon}_\alpha$ = residual of eqn. (32) and $\boldsymbol{\varepsilon}_\psi$ = residual of eqn. (33).

From eqn. (34)

$$\Delta\mathbf{x}_\alpha = \mathbf{J}_{\alpha\alpha}^{-1}(\mathbf{C}_{\psi\alpha}\Delta\mathbf{i}_\psi + \boldsymbol{\varepsilon}_\alpha) \tag{36}$$

Substituting into eqn. (35) yields

$$\Delta\mathbf{i}_\psi = \tilde{\mathbf{J}}_{\psi\psi}^{-1}(\boldsymbol{\varepsilon}_\psi - \mathbf{C}_{\psi\alpha}^T\mathbf{J}_{\alpha\alpha}^{-1}\boldsymbol{\varepsilon}_\alpha) \tag{37}$$

where

$$\tilde{\mathbf{J}}_{\psi\psi} = \mathbf{C}_{\psi\alpha}^T\mathbf{J}_{\alpha\alpha}^{-1}\mathbf{C}_{\alpha\psi} + \mathbf{J}_{\psi\psi} \tag{38}$$

and

$$\mathbf{J}_{\alpha\alpha} = \begin{bmatrix} \mathbf{J}_{11} & & & \\ & \mathbf{J}_{22} & & \\ & & \ddots & \\ & & & \mathbf{J}_{nn} \end{bmatrix}$$

The Lagrangian multipliers can be obtained from eqns. (30) and (31) where

$$\boldsymbol{\lambda}_\alpha = -\left(\frac{\partial \mathbf{G}_\alpha^T}{\partial \mathbf{x}_\alpha}\right)^{-1}\left(\frac{\partial f}{\partial \mathbf{x}_\alpha} + \mathbf{C}_{\alpha\psi}\boldsymbol{\lambda}_\psi\right) \tag{39}$$

Substituting into eqn. (31) yields

$$\boldsymbol{\lambda}_\psi = -\left[\frac{\partial \boldsymbol{\phi}_\psi^T}{\partial \mathbf{i}_\psi} + \mathbf{C}_{\psi\alpha}^T\left(\frac{\partial \mathbf{G}_\alpha^T}{\partial \mathbf{x}_\alpha}\right)^{-1}\mathbf{C}_{\alpha\psi}\right]^{-1}\left[\mathbf{C}_{\psi\alpha}^T\left(\frac{\partial \mathbf{G}_\alpha^T}{\partial \mathbf{x}_\alpha}\right)^{-1}\frac{\partial f}{\partial \mathbf{x}_\alpha}\right] \tag{40}$$

Thus the gradient vector which is used to update the independent variables can be obtained from eqn. (29). This completes one iteration of the algorithm. A closer examination of the term $\partial \mathbf{G}_\alpha/\partial \mathbf{x}_\alpha$ would reveal that it is a block diagonal matrix. Each block represents a subdivision and thus the control computer would assemble all these "local blocks" to evaluate $\boldsymbol{\lambda}_\psi$.

It can easily be verified that one step of the proposed algorithm will produce the same result as one step of solving the complete system as a whole. It can also be seen that an alternative scheme is to optimize each subdivision to some degree of accuracy before the optimization on the interconnection is performed.

5. CONCLUSION

For on-line control a speedy and reliable solution is essential and for this reason linear approximation and linear programming are widely used. The application of non-linear optimization is of interest and the authors believe that it is feasible to consider its application for on-line control.

For large integrated networks some form of decomposition along the line proposed would be essential. Although a great deal of effort has been devoted to investigating the practical application of non-linear programming to on-line control, further researches are desirable to achieve a degree of efficiency and reliability comparable with the linear techniques.

ACKNOWLEDGEMENTS

The authors wish to express their gratitude to Professor L. M. Wedepohl for providing the facilities in the Power Systems Laboratory at the University of Manchester Institute of Science and Technology and to their colleagues for stimulating and interesting discussions.

NOTATION

P active power
Q reactive power
$\bar{V}$ voltages
$\bar{\mathbf{Y}}$ admittance matrix
T transformer tap positions

R angular shifts of phase shifters
L Lagrangian function
$\boldsymbol{\lambda}$ Lagrangian multipliers
$\mathbf{u}$ Kuhn–Tucker multipliers
$\mathbf{G}$ network constraints
$\mathbf{H}$ inequality limits
I nodal currents
i current in interconnections
e hypothetical voltages
$\overline{\mathbf{Z}}$ impedance matrix
$\mathbf{C}$ topological incident matrix
$\mathbf{J}$ Jacobian matrix
f objective function
$\mathbf{x}$ dependent variables
$\mathbf{y}$ control variables

Superscripts
G generation quantities
T transposition of matrices
L load quantities
N net quantities
B branch quantities
$\sim$ quantities associated with interconnections

Subscripts
θ number of generators
α number of nodes in the system
β number of branches
ϕ number of phase shifters
γ number of in-phase tap transformers
ψ number of interconnections

In order to distinguish between various matrices each vector or matrix is printed in bold type with its dimensions indicated by the corresponding subscripts. Complex quantities are denoted by a bar, *e.g.* $\overline{V}$. Complex conjugate is denoted by an asterisk, *e.g.* $\overline{V}^*$.

APPENDIX 1

In the linear formulation of the problem the objective function of eqn. (1) can be used. The main concern now lies with the linearization of the network equations. The complex power at a node can be expressed as

$$P_i^N + jQ_i^N = \overline{V}_i\overline{I}_i^* \tag{41}$$

References pp. 98–99

Expressing the above equation in terms of the real and imaginary part of the admittance matrix yields

$$P_i^N + jQ_i^N = V_i(\cos\delta_i + j\sin\delta_i)\left\{\sum_k V_k[(G_{ik}\cos\delta_k - B_{ik}\sin\delta_k) - j(B_{ik}\cos\delta_k + G_{ik}\sin\delta_k)]\right\} \tag{42}$$

where

$$\overline{Y}_{ik} = G_{ik} + jB_{ik}$$

Multiplying and rearranging the terms gives

$$P_i^N + jQ_i^N = \sum_k V_iV_kG_{ik}\cos(\delta_i - \delta_k) + \sum_k V_iV_kB_{ik}\sin(\delta_i - \delta_k) + j\sum_k V_iV_kG_{ik}\sin(\delta_i - \delta_k) - j\sum_k V_iV_kB_{ik}\cos(\delta_i - \delta_k) \tag{43}$$

It is now necessary to make certain assumptions so that the above equation can be linearized. The assumptions are:

(1) reactive power is not taken into consideration;
(2) voltage magnitudes are assumed equal to unity;
(3) $\cos(\delta_i - \delta_j) \approx 1$
$\sin(\delta_i - \delta_j) \approx \delta_i - \delta_j$
(4) resistances of lines are ignored so that there are no losses in the system; and
(5) shunt elements in the system are ignored.

Equation (43) is thus simplified into the form

$$P_i^N \doteqdot -\sum_k B_{ik}\delta_k$$

or in matrix form

$$\mathbf{P}_\alpha^N \doteqdot -\mathbf{B}_{\alpha\alpha}\boldsymbol{\delta}_\alpha \tag{44}$$

Thus using eqn. (44) an inequality linear constraint of the following form is established

$$\mathbf{P}_\alpha^{\min} - \mathbf{P}_\alpha^L \leq -\mathbf{B}_{\alpha\alpha}^l\boldsymbol{\delta}_\alpha \leq \mathbf{P}_\alpha^{\max} - \mathbf{P}_\alpha^L \tag{45}$$

where $\mathbf{B}_{\alpha\alpha}^l$ is the reduced admittance matrix with the slack busbar angle set equal to zero.

It is also necessary to ensure that lines are not overloaded. Using the same derivation as for eqn. (42), the real power in the line between node i and j is

$$P_i^{BF} = G_{ij}[V_i^2 - V_iV_j\cos(\delta_i - \delta_j)] + B_{ij}[V_iV_j\sin(\delta_i - \delta_j)] \tag{46}$$

Incorporating the assumptions listed previously it becomes

$$P_i^{BF} = -B_{ij}(\delta_i - \delta_j) \tag{47}$$

Thus the constraint which needs to be satisfied is

$$-\mathbf{P}_\beta^{BFM} \leq -\mathbf{U}_{\beta\beta}\mathbf{C}_{\beta\alpha}\boldsymbol{\delta}_\alpha \leq \mathbf{P}_\beta^{BFM} \tag{48}$$

where $\mathbf{P}_{\beta}^{BFM}$ = vector giving the maximum flow limit in a line
$\mathbf{U}_{\beta\beta}$ = diagonal matrix which gives the susceptance of the lines
$\mathbf{C}_{\beta\alpha}$ = connection matrix

In addition to the above constraints it is also necessary to consider line outages. Furthermore, when a line is out, generation distribution must be such that powers at generator nodes remain unchanged:

$$\mathbf{P}_{\alpha}^{N} = -\mathbf{B}_{\alpha\alpha}\boldsymbol{\delta}_{\alpha} = -\mathbf{B}_{\alpha\alpha}^{m}\boldsymbol{\delta}_{\alpha}^{m}$$

where $\mathbf{B}_{\alpha\alpha}^{m}$ is the modified susceptance matrix and $\boldsymbol{\delta}_{\alpha}^{m}$ are the modified phase angles. It follows that line outage constraints will be similar in form to eqn. (48), *i.e.*

$$-\mathbf{P}_{\beta}^{BFM} \leq -\mathbf{U}_{\beta\beta}^{m}\mathbf{C}_{\beta\alpha}\boldsymbol{\delta}_{\alpha}^{m} \leq \mathbf{P}_{\beta}^{BFM} \quad (49)$$

It is worth noting that if there is no line outage no such constraint exists. This constraint is only added one at a time.

The above formulation is a linear model for active power optimization. There are numerous standard commercial programs available to solve such linear programming problems[13–16], the most widely used and the most suitable being the dual-linear programming methods. Since there are papers and books published on this topic the authors will not follow the actual algorithm of solution any further.

APPENDIX 2

To illustrate the diakoptical procedure this section shows one complete cycle applied to a simplified electrical system shown in Fig. 4.

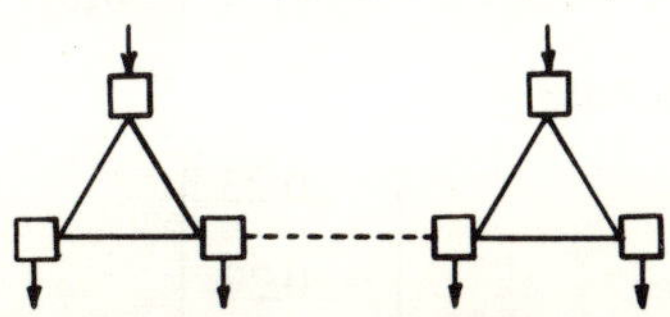

Fig. 4. Numerical example.

For this problem, node 6 is the reference (slack) busbar with $\theta_6 = 0$,

$$y = [\theta_5]$$

$$\mathbf{x} = \begin{bmatrix} \theta_1 \\ \theta_2 \\ \hline \theta_3 \\ \theta_4 \end{bmatrix}$$

References pp. 98–99

$$\mathbf{C}_{\alpha\psi} = \begin{bmatrix} 0 \\ -1 \\ --- \\ 1 \\ 0 \end{bmatrix} = \begin{bmatrix} \mathbf{C}_{1\psi} \\ --- \\ \mathbf{C}_{2\psi} \end{bmatrix}$$

$$i_\psi = i_j$$

$$\phi_\psi = Z_{\psi\psi} i_\psi = 0.2 i_j$$

$$G_1 = 1 - 0.5\theta_1^2 + 10.5\theta_1 - 5\theta_2$$

$$G_2 = 0.8 - 5\theta_1 - 0.4\theta_2^2 + 10.3\theta_2$$

$$G_3 = 1 - 0.5\theta_3^2 + 10.2\theta_3 - 5\theta_4 - 5\theta_5$$

$$G_4 = 1.5 - 5\theta_3 - 0.75\theta_4^2 + 10.3\theta_4 - 5\theta_5$$

$$\mathbf{G} = \begin{bmatrix} G_1 \\ G_2 \\ --- \\ G_3 \\ G_4 \end{bmatrix}$$

$$\begin{aligned} F = & -(\theta_1 + \theta_2) + 0.625(\theta_1 + \theta_2)^2 \\ & + 1.5[(\theta_5 - \theta_3)(1 + 0.5\theta_5\theta_3) + (\theta_5 - \theta_4)(1 + 0.5\theta_4\theta_5)] \\ & + 0.25[(\theta_5 - \theta_3)(1 + 0.5\theta_5\theta_3) + (\theta_5 - \theta_4)(1 + 0.5\theta_4\theta_5)]^2 \end{aligned}$$

Starting with an initial estimate

$$\mathbf{x} = \begin{bmatrix} -0.23 \\ -0.27 \\ ----- \\ -0.40 \\ -0.48 \end{bmatrix}$$

$$y = 0$$

then from eqns. (32) and (33)

$$\boldsymbol{\varepsilon} = \begin{bmatrix} -0.091 \\ -0.160 \\ ----- \\ -0.160 \\ -0.317 \end{bmatrix} \qquad \varepsilon_\psi = [0.01]$$

$$\mathbf{J}_{11} = \begin{bmatrix} 10.73 & -5 \\ -5 & 10.52 \end{bmatrix} \qquad \mathbf{J}_{22} = \begin{bmatrix} 10.60 & -5 \\ -5 & 11.02 \end{bmatrix}$$

$$J_{\psi\psi} = 0.2$$

Hence substituting in eqn. (38) gives

$$\tilde{J}_{\psi\psi} = [0.4422]$$

and Δi_ψ is then calculated using eqn. (37) giving

$$\Delta i_\psi = [-0.0493]$$

Consequently, the corrections for **x** can be calculated from eqn. (36):

$$\Delta\mathbf{x} = \begin{bmatrix} \Delta\theta_1 \\ \Delta\theta_2 \\ \hdashline \Delta\theta_3 \\ \Delta\theta_4 \end{bmatrix} = \begin{bmatrix} 0.0228 \\ 0.0307 \\ \hdashline 0.0306 \\ 0.0426 \end{bmatrix}$$

Having found the corrections $\Delta\mathbf{x}$, the next step is to calculate the gradients:

$$\partial F/\partial\theta_1 = -1.5581$$
$$\partial F/\partial\theta_2 = -1.5581$$
$$\partial F/\partial\theta_3 = -1.6828$$
$$\partial F/\partial\theta_4 = -1.6886$$
$$\partial F/\partial\theta_5 = +3.6485$$

λ_ψ can then be calculated from eqn. (40):

$$\lambda_\psi = [0.0337]$$

and $\boldsymbol{\lambda}_\alpha$ from eqn. (39):

$$\boldsymbol{\lambda}_\alpha = \begin{bmatrix} 0.2771 \\ 0.2812 \\ \hdashline 0.2898 \\ 0.2847 \end{bmatrix}$$

Finally, the gradient $\partial L/\partial y$ is calculated from eqn. (29) giving

$$\partial L/\partial y = [0.3280]$$

For comparison consider the whole problem in one part (second approach):

$$G_1 = 1 - 0.5\theta_1^2 + 10.5\theta_1 - 5\theta_2$$
$$G_2 = 0.8 - 5\theta_1 - 0.4\theta_2^2 + 15.3\theta_2 - 5\theta_3$$

References pp. 98–99

$$G_3 = 1 - 5\theta_2 - 0.5\theta_3^2 + 15.2\theta_3 - 5\theta_4 - 5\theta_5$$

$$G_4 = 1.5 - 5\theta_3 - 0.75\theta_4^2 + 10.3\theta_4 - 5\theta_5$$

$$\frac{\partial \mathbf{G}}{\partial \mathbf{x}} = \mathbf{J}_{\alpha\alpha} = \begin{bmatrix} 10.73 & -5 & 0 & 0 \\ -5 & 15.52 & -5 & 0 \\ 0 & -5 & 15.6 & -5 \\ 0 & 0 & -5 & 11.02 \end{bmatrix}$$

and

$$\Delta \mathbf{x} = \left(\frac{\partial \mathbf{G}}{\partial \mathbf{x}}\right)^{-1} \mathbf{G} = \begin{bmatrix} 0.0228 \\ 0.0307 \\ ---- \\ 0.0306 \\ 0.0426 \end{bmatrix}$$

which is the same as that calculated using diakoptics. The Lagrangian multipliers calculated from eqn. (16)

$$\boldsymbol{\lambda} = \begin{bmatrix} 0.2771 \\ 0.2812 \\ 0.2898 \\ 0.2847 \end{bmatrix}$$

are also the same as those calculated by the diakoptical method.

REFERENCES

1 J. Carpenter, Contribution à l'étude du dispatching économique, *Bull. Soc. Trans. Elect.*, **8** (1962) A31–44.

2 H. M. Smith and S. Y. Tong, Minimizing power transmission losses by reactive-volt-ampere control, *IEEE Trans. Power Apparatus Syst.*, **PAS-82** (1963) 542–544.

3 J. F. Dopazo, O. A. Klitin, G. W. Stagg and M. Watson, An optimization technique for real and reactive power allocation, *Proc. IEEE*, **55** (1967) 1877–1885.

4 J. Peschon, D. S. Piercy, W. F. Tinney, O. J. Treit and M. Cuenod, Optimum control of reactive power flow, *IEEE Trans. Apparatus Syst.*, **PAS-87** (1968) 40–48.

5 A. H. El-Abiad and F. J. Jaimes, A method for optimum scheduling of power and voltage magnitude, *IEEE Trans. Power Apparatus Syst.*, **PAS-88** (1969) 413–422.

6 H. W. Dommel and W. F. Tinney, Optimal power flow solution, *IEEE Trans. Power Apparatus Syst.*, **PAS-87** (1968) 1866–1874.

7 C. M. Shen and M. A. Laughton, Determination of optimum power system operating conditions under constraints, *Proc. IEE*, **116** (1969) 225–239.

8 M. A. Laughton, Optimal power system operation—analytical formulation, *Proc. Optimal Power System Operation, U.M.I.S.T., September, 1969.*

9 H. W. Dommel, Constrained optimal control of real and reactive power dispatch, *Proc. Optimal Power System Operation, U.M.I.S.T., September, 1969.*

10 A. Brameller and Y. M. Hamam, Application of diakoptics to non-linear optimization, *Proc. Conference on Numerical Methods for Non-Linear Optimization, Dundee, July, 1971.*

11 A. BRAMELLER, M. N. JOHN AND M. R. SCOTT, *Practical Diakoptics for Electrical Networks*, Chapman and Hall, London, 1969.
12 G. KRON, *Diakoptics*, Macdonald, London, 1963.
13 G. HADLEY, *Linear Programming*, Addison-Wesley, Reading, Mass., 1962.
14 J. C. KALTENBACH AND L. P. HAJDU, Optimal corrective rescheduling for power system security, Paper presented at IEEE Summer Power Conference, July, 1970.
15 M. L. BAUGHMAN AND F. C. SCHWEPPE, Contingency evaluation: real power flows from a linear model, Paper presented at IEEE Summer Power Conference, July, 1970.
16 C. M. SHEN AND M. A. LAUGHTON, Power system load scheduling with security constraints using dual linear programming, *Proc. IEE*, **117** (1970) No. 11.
17 Y. M. HAMAM, Optimization of large systems, *Ph.D. thesis*, University of Manchester Institute of Science and Technology, Manchester, 1972.

Discussion

H. A. NOUR ELDIN (*Eidg. Technische Hochschule, Zürich, Switzerland*)
I would like to ask the authors the following questions.

(1) Which method do they recommend for the one-dimensional optimization? My special interest is the unimodality of the problem and how to reach the global optimum when the one-dimensional optimization problem is not unimodal. How sensitive is the solution of the non-linear optimization problem to the optimization in one dimension?

(2) Is there any reference on the convexity of the optimization problem in power systems? Can the authors give figures and references on the mode and rate of convergence in practical problems?

(3) The authors emphasize decomposition of large systems by neglecting some coupling between subsystems. Do they suggest a measure (or measures) for this coupling?

A. BRAMELLER
(1) At the present moment it appears that the most promising method for the one-dimensional optimization is the quadratic approximation. Of course, this will not always be the case; it really depends on the type of problem that you have in mind. When the problem is not unimodal the whole thing becomes complicated. We don't think you will be able to get to the global optimum without repeating the optimization procedure from different starting points. However, even this does not guarantee a global optimum solution.

(2) One of the fundamental approaches to the optimization problem in power systems is to assume its convexity. It will further complicate the problem if this is not done. The mode and rate of convergence differ from system to system. It is difficult to generalize.

(3) The coupling effect between subsystems is not neglected in our algorithm for decomposition. It is already taken into account in the torn branches.

S. K. MITTER (*Massachusetts Institute of Technology, Cambridge, Mass., U.S.A.*)
I want to add to Dr. Nour Eldin's comment. The one-dimensional search in unconstrained minimization problems is delicate and crucial, especially if more sophisticated techniques such as conjugate direction techniques are used. It is

also very difficult to perform when the cost function is augmented by adding penalty functions. Would the authors comment on this?

A. BRAMELLER
When penalty functions are added to the cost function the contours to the problem are distorted. One-dimensional search procedure usually fails to give a sensible step length. Under such circumstances an alternative approach is to return to small arbitrary step sizes.

M. A. LAUGHTON (*Queen Mary College, London, England*)
We have evaluated some of the non-linear programming methods with several test problems. The power system problem is less complicated then these test problems.

With regard to the one-dimensional search problem I think one can prove that the Fibonacci search is the optimal method. The necessary algorithmic performance can be specified by the following requirements: a comprehensive model and a secure solution which must be acceptable under all single-line outage conditions. The effect when all lines are removed is to generate a new system. If you have the objective function and you start removing lines it means that extra blocks of constraints are added. The only technique which will solve this sort of problem is a dual linear programming model.

A. BRAMELLER
The method that Dr. Laughton has just demonstrated is the sequential unconstrained minimization technique. The problem used for testing the method is a small one with few variables. We also have some experience of the method and find that it is not suitable for large problems. If one observes closely one can easily see that each constraint associated with a weighting constant is included in the objective function. All these additions will distort the contours of the original problem, resulting in a more difficult problem for optimization.

As far as security checking is concerned, the dual linear programming approach is certainly appropriate. The main reason is that security constraints can be added one at a time. At the present moment, to incorporate security constraint in non-linear programming is certainly not easy.

F. ARIATTI (*ENEL, Rome, Italy*)
I am referring to our system consisting of 250 busses, 400 transmission lines and 60 thermal units, where the problem of optimal active power generation has been studied. An iterative method is used where in each step a feasible solution is obtained. No convergence problem was encountered and it was found very advantageous to use the second order derivatives (*i.e.* the Hessian matrix) calculated in an approximate way. This reduced the computation time considerably[1].

A more recent version by D. Grohmann includes security constraints so that no line will be overloaded when a single other line is lost. These additional constraints are incorporated by linear programming methods within the framework of the non-linear optimization approach.

1 F. ARIATTI, D. GROHMANN AND D. VENTURINI, A method for economic load dispatching in a thermal power system, *3rd Power System Computation Conference, Rome, 1969.*

A. BRAMELLER
The method that you have used may suit your system very well. We notice that you mention an iterative method, with security constraint incorporated by linear programming. This approach is not unconventional. However, without knowing the full details of your method, it is difficult for us to comment, but in general a method suitable for one system may not be ideal for another.

H. J. KOGLIN (*Technische Hochschule, Darmstadt, Germany*)
Is there any algorithm dividing the whole system into subsystems in an optimal way? If it exists, is it comparatively efficient with the human eyes watching the graph of the whole system?

A. BRAMELLER
There is an algorithm which can divide the system into subsystems in an optimal way. At the present time this is very inefficient and manual tearing is preferred.

J. P. WAHA (*INTERCOM, S.A., Brussels, Belgium*)
Every optimization algorithm uses, directly or implicitly, a load flow calculation at each iteration step. When one uses decomposition techniques for load flow calculations one may proceed in two different directions:

(1) using a geographical decomposition as proposed by Dr. Brameller, and
(2) using a decomposition based on tree and loops of the associated graph.

The second decomposition technique is interesting in the following aspects. Having solved the problem for the "tree" part by an iterative scheme and having done the same for the loops, we may obtain the exact load flow solution for the complete network by pure linear manipulations of the partial results without any other iterative computation.

In the first decomposition technique, called "bloc iteration" technique, having solved the load flow for each subnetwork by an iterative scheme, there is still a lot of iterative computation to be done in order to solve the complete system by matching the boundary conditions. This means that there will be a high transfer of data between decentralized and centralized computers, as proposed by Dr. Brameller. Unfortunately, the second method proposed here is obviously not feasible for decentralized dispatching.

A. BRAMELLER
There are many different criteria for tearing a system. For on-line control the object is to minimize core store and time of calculations. A certain amount of data transfer which is associated with diakoptics can generally be tolerated, particularly when it is carried out in parallel with the arithmetical calculations.

S. DESPOTOVIĆ (*Nikola Tesla Institute, Belgrade, Yugoslavia*)
What is the limit of decomposition techniques and when is it better to use a different method to solve the optimization problem? I would like to know whether the authors have some experience with the solution of large scale problems on small computers using sparsity techniques, and how they compare with decomposition techniques.

A. BRAMELLER

Because of the complications involved in writing a program for diakoptical solution, sparsity techniques are advisable whenever possible. For small- and medium-sized problems it is our experience that these are perfectly adequate. Modern computer practice is to use a multiprogramming environment. When this includes accountancy and other data processing requirements, adequate core store and access time may prove difficult, as has been our experience. To overcome this difficulty some form of decomposition such as diakoptics is the only way of obtaining a solution without upsetting the accountants.

The Economic Load Dispatch—State of the Problem and Developments at Electricité de France

J. G. SIROUX, J. POUGET AND A. MERLIN

Electricité de France, Clamart, France

SUMMARY

A great amount of work has already been carried out on the economic load dispatch (ELD) problem. After experimenting with different approaches for several years, ideas are clarifying as to the best ways of matching

(1) the actual needs: the network operations one must carry out;

(2) the methods: the mathematical algorithms; and

(3) the computer constraints: size and duration of programs.

In the present paper we investigate what different formulations of the problem can be made depending on the kind of difficulties encountered by the dispatcher. The present state and trends in solving the ELD problem as seen today at Electricité de France (EdF) are also given.

1. GENERAL REMARKS ABOUT THE STATEMENT OF THE ELD PROBLEM

Considering the various decisions taken by dispatchers it has been found advantageous (for about 12 years) to use the following optimization tool:

> a computer program minimizing the instantaneous operating cost of a generating system, taking into account the constraints of the transmission system.

From this general definition of the ELD problem the following general remarks can be made.

(1) The formulation is an instantaneous optimization; however, many of the decisions to be taken will be related to the long-term behaviour of the system.

(2) The formulation starts from an economic viewpoint—essentially to save fuel costs. In a complementary way another viewpoint, the security aspect, is also considered; this is very important and requires much additional work.

(3) The network constraints may be written in many ways according to the actual problem.

The security aspect has been a main consideration from the beginning. It has now become necessary to include it in the computer programs used by dispatchers.

In previous years, the dispatcher's programs have been developed starting from economic considerations without security constraints. These can be more or less compulsory according to the different kinds of networks operating in the world. For some networks security is even more important than economy.

Concerning the third point is should be noted that the way to state the network constraints differs when real-time solutions are needed and when the dispatcher is optimizing over longer periods.

The results issued from an ELD program can be used either

(1) directly for a given instant in real time or in forecast-scheduling; or

(2) as elements of a more intricate computation (such as unit commitment, the amount of spinning reserves, etc.).

At first the ELD was stated with simplifications in the network equations (the *B* coefficients method in ref. 1). More accurate formulations were given in ref. 2 but it appeared that to solve the complete instantaneous problem was a complicated matter. It was difficult and also quite insufficient, since the actual requirement was to use the results for long-term optimization. Some compromises had therefore to be made among approximations in the statement of the problem, such as the computer time and the characteristics of the network. The relative importance of these aspects depends on the expected use of the ELD results.

Thus, the numerical traps, the computer constraints and the increase in difficulty with the number of busses were underestimated at first. It can now be said that no general method has wholly solved the problem for large systems with up to 400 nodes, including all the network constraints. Moreover, we think that this would not be useful. The dispatcher's requirements are concerned (in order of decreasing importance) with such items as unit commitment, hydro-operation, spinning reserve and all other power margins meeting the random fluctuations. Once more, the methods for ELD must be different according to the kind of decision (real-time or forecast-scheduling) and according to the number of nodes and of line constraints.

We now state the different approximations on the network model.

1.1 Interesting limit cases of the ELD problem

In what we call a one-point model, *i.e.* a model without any transmission system, the ELD problem is easily solved by taking the thermal groups in their order of merit (with increasing incremental costs). The ELD can be defined as that solution which gives the best balance between this generation planning and the shift from this policy through the inclusion of the network laws. With actual systems, two cases may occur with this merit order remaining the optimum. Neither case has flow constraints as an important factor. They appear

(1) when the thermal costs are widely separated even if the differential losses $\partial\pi/\partial P$ are important; and

(2) when the differential losses are small whatever the spectrum of thermal costs may be.

In these cases it is not useful to make computations with the network equations.

Some simpler computations can be of interest, *e.g.* looking for intersection points in thermal cost curves.

Assuming that these two cases are excluded the reasons for a computer solution of ELD differ according to the following problems. Firstly, in connection with real-time decisions, the computations will be of interest only if the conditions are sufficiently different from the forecast ones. Here, there is a link with the study of random variations occurring in operating conditions (thermal plants outages, demand, hydro-availability, exchanges with neighbouring systems). If the circumstances happen to be too far from the expected ones, flow constraints may appear, and they compel the dispatcher to resort to running a new computation. Secondly, in connection with forecast-scheduling decisions, the reasons for computing an ELD are related to the evolution of the operating network topology, changes in committing peak-power groups, and here again, when flow constraints appear, it is desirable to use a computer program.

1.2 The general power flow equations

The following notation will be used at node n:

N	number of nodes
$P_n + jQ_n$	generated power
$V_n \exp(j\theta_n)$	voltage
$C_n + jD_n$	power demand

At the branch joining nodes i and k:

T_{ik}	active power flow
$Y_{ik} \exp(j\xi_{ik})$	element of the admittance matrix
π_{ik}	active losses

The flow equations express the power balance at each node n for a steady state:

$$P_n - C_n = V_n \sum_{i=1}^{N} Y_{in} V_i \cos(\theta_n - \theta_i - \xi_{in}) \tag{1}$$

$$Q_n - D_n = V_n \sum_{i=1}^{N} Y_{in} V_i \sin(\theta_n - \theta_i - \xi_{in}) \tag{2}$$

In vector form these non-linear equations can be written as:

$$\mathbf{P} - \mathbf{C} = \boldsymbol{\varphi}(\boldsymbol{\theta}, \mathbf{V}) \tag{3}$$

$$\mathbf{Q} - \mathbf{D} = \boldsymbol{\psi}(\boldsymbol{\theta}, \mathbf{V}) \tag{4}$$

The Jacobian $\mathbf{J}$ expresses the sensitivity of the unknowns $\mathbf{P}$ and $\mathbf{Q}$ with respect to $\boldsymbol{\theta}$ and $\mathbf{V}$:

$$\mathbf{J} = \begin{bmatrix} \boldsymbol{\varphi}_\theta^T & \boldsymbol{\varphi}_V^T \\ \boldsymbol{\psi}_\theta^T & \boldsymbol{\psi}_V^T \end{bmatrix} \tag{5}$$

References p. 112

The Jacobian matrix $\mathbf{J}$ has the following properties:

(1) $\mathbf{J}$ is a singular matrix for all values of $\boldsymbol{\theta}$ and $\mathbf{V}$. Thus we have:

$$\boldsymbol{\lambda}\boldsymbol{\varphi}_\theta^T + \boldsymbol{\mu}\boldsymbol{\psi}_\theta^T = 0 \tag{6}$$

$$\boldsymbol{\lambda}\boldsymbol{\varphi}_V^T + \boldsymbol{\mu}\boldsymbol{\psi}_V^T = 0 \tag{7}$$

with $[\boldsymbol{\lambda}, \boldsymbol{\mu}] \neq \mathbf{0}$.

(2) $\mathbf{J}$ is a sparse matrix. Only 1–5 % of its elements have non-zero values. This property will be used in the computing methods.

(3) The elements of $\boldsymbol{\varphi}_V$ and $\boldsymbol{\psi}_\theta$ are small compared with the elements of $\boldsymbol{\varphi}_\theta$ and $\boldsymbol{\psi}_V$. Thus, the first-order approximation to the non-linear eqns. (3) and (4) becomes:

$$\Delta\mathbf{P} = \boldsymbol{\varphi}_\theta^T \Delta\boldsymbol{\theta} \tag{8}$$

$$\Delta\mathbf{Q} = \boldsymbol{\psi}_V^T \Delta\mathbf{V} \tag{9}$$

The active power is mainly related to phase angles; the reactive power depends mainly on the voltage magnitudes.

(4) If eqns. (3) and (4) are linearized around a feasible solution $(\mathbf{P}_0, \boldsymbol{\theta}_0)$ and if the voltage magnitudes are assumed constant at each bus, the following formula holds:

$$\mathbf{P} - \mathbf{P}_0 = \boldsymbol{\varphi}_{\theta_0}^T(\boldsymbol{\theta} - \boldsymbol{\theta}_0) \tag{10}$$

Linearized around $\boldsymbol{\theta}_0 = \mathbf{0}$ this gives the *d.c. approximation.* In this case the matrix $\boldsymbol{\varphi}_0^T$ is symmetric.

1.3 The network constraints

Compared with a one-point model, the network requires additional equations or inequalities:

the mesh laws,

the losses $\boldsymbol{\pi}$,

limits for the flows: $-\mathbf{T}_{max} \leq \mathbf{T} \leq \mathbf{T}_{max}$,

limits for the voltage magnitudes at the consuming nodes: $\mathbf{V}_{min} \leq \mathbf{V} \leq \mathbf{V}_{max}$.

The generation system also has its constraints:

$$\mathbf{P}_{min} \leq \mathbf{P} \leq \mathbf{P}_{max} \tag{11}$$

$$\mathbf{Q}_{min} \leq \mathbf{Q} \leq \mathbf{Q}_{max} \tag{12}$$

The voltage magnitudes are fixed at the generating nodes.

The security aspect has not been included so far. The existing lines and groups are assumed to be fully reliable and their failure cannot influence the optimization.

1.4 The objective function

The economic aim is to minimize the operating cost:

$$\min \sum_{n=1}^{N} L_n(P_n) \tag{13}$$

where L_n is a piecewise linear cost. Instead of attempting the solution of the entire problem with all the constraints mentioned above, the question of whether its results would be of any value should be answered.

If no constraints for the voltage magnitudes appear, the complete solution has only an economic interest, and this is of very little importance in this case. If, on the other hand, voltage constraints appear, they can in general be solved without resuming the entire computation. This allows separation into two optimization steps. Such a possibility also arises as a result of the remarks about the Jacobian. Thus the minimization consists of the following two steps:

(1) minimization of the operating cost, including only the active power constraints and keeping the voltage magnitudes fixed;

(2) the use of reactive generation at each node; the objective function

$$\min \sum_{n=1}^{N} |Q_n| \tag{14}$$

is an example.

In the second step, the reactive power and voltage inequalities hold and the voltage phase angles $\boldsymbol{\theta}$ are fixed. Other objective criteria could be taken, as long as the aim of avoiding excessive reactive loads for the groups (especially during the off-hours) and great differences in voltages (which minimize reactive flows) is satisfied.

After the two optimization steps are completed it can happen that a further iteration is necessary. In such a case, simultaneous active and reactive constraints are occurring (which is quite uncommon in our network when active constraints are already satisfied as a result of a previous computation).

2. USE OF THE ELD IN REAL-TIME DECISIONS

With the support of these considerations the actual problems in which the ELD can be used will now be briefly examined. Two main classes can be distinguished:

(1) the real-time operations; and

(2) the forecast and scheduling computations.

When the network is operating in a steady state, two questions may be of interest for the dispatcher at any instant of time to help him in his real-time decisions:

(1) Is the present state a safe state with respect to the security of supply?

(2) If it is safe, is it the most economic among all safe states?

Safety of the network operation means essentially that, if there is any line trip or sudden outage of any generator, it is possible to have another load flow in which the network constraints hold. The ELD greatly helps in answering these questions,

References p. 112

but for this purpose it is not always necessary to use sophisticated computer programs.

2.1 ELD without flow constraints and with fixed voltage magnitudes

In cases where no problem appears with overloaded transmission lines or with voltages outside their boundaries, a program without flow constraints and with fixed voltage magnitudes might be of interest. It only remains to compare the expected savings on the fuel cost with the computing expenses according to the number of times such an ELD program is run.

(1) If the conditions are near the two former limit cases, mentioned in §1.1, an ELD program will provide nothing better than the merit order planning.

(2) If the incremental costs are quite closely related (which is the general case for huge new thermal plants) when the differential losses are not small, ELD can be used. The following figures hold for the EdF network:

(a) the differential losses $\partial\pi/\partial P$ are about 0.1 or 0.2;

(b) due to the ELD program described in this paper the saving of the total operation cost ranges from 0.5 to 1 %.

The point of matching the precision of data and the complexity of computations must be considered. With a teleprocessing system the dispatcher can have the accurate description of the network and its flows without additional computing devices. Thus, economical computations make these devices relatively less expensive. Moreover, such teleprocessing is an important requirement for a network with as many constraints as there are in the EdF network.

2.2 ELD with flow constraints and fixed voltage magnitudes

Even with a well-conditioned network, circumstances occur when some dangerous limits are reached concerning the security of supply. This may be the case when several transmission lines get overloaded. For the EdF transmission system, three limits are taken into account to estimate this overloading:

(1) a steady state capacity for each line;

(2) a temporary overload capacity; and

(3) a quick overload capacity.

The first limit must not be reached in a steady state. Although this limit should physically be expressed in ampères, in this approximation of ELD it is translated into watts. The second limit may be reached if the duration does not exceed 20 min. Generally, it must hold in any case of single contingency (one line only being tripped). The third limit (about 1.6 times the first limit) gives a line trip after 15 sec.

In these cases (more than in the previous case, §2.1) it is compulsory to describe the system accurately in order to detect flow constraints as soon as they occur. To obtain significant improvements, methods such as the "injection method" and the "transit method" have been developed[2]. The latest method, called the MAYA model, will be explained further (see also refs. 3 and 4).

2.3 ELD with variable voltage magnitudes

An ELD program with variable voltage magnitudes will be of interest when reactive or voltage problems arise. Reasons for separating the active and reactive optimizations have already been mentioned. They have even more importance in real time, when reactive generation is generally performed according to the voltage levels and the levels are often regulated by committing new plants (or compensators).

It must be questioned whether it is useful to include the reactive flows in the ELD. If its limits are not reached, the reactive energy only plays a role through its influence on the active losses $\boldsymbol{\pi}$, and we have seen that it is a very weak effect. Practically, reactive flows could only be important because of constraints (limits for $\mathbf{Q}$ and $\mathbf{V}$ being reached). Actually, voltage levels are determined by the tap-changers and there are other means of satisfying the reactive constraints: committing new local plants, altering the network topology, etc.

It was decided in EdF to design a voltage control similar to the so-called secondary control (power-frequency automatic control). Voltage will be regulated in each area with national coordination. As mentioned before, the object of this coordination should be the minimization of the generation or absorption of reactive energy. Computer constraints also indicate a tendency towards the subdivision of voltage control according to areas. An ELD program without reactive flows can be run on a smaller computer. After the active flows have been optimized it is easy to check what happens to the reactive energy, and this check can also be done with a small conventional program. Finally, a program with reactive optimization is only run when constraints appear. In such situations, the regulation of different area voltage levels is done. This is not a heavy load for the dispatching computers.

3. USE OF THE ELD IN SCHEDULING COMPUTATIONS

Up to now more use of the ELD has been made in another domain: the scheduling of generation planning one day or one week in advance. The scope of such an ELD program is bounded by the connection with other programs which optimize on a longer range: several months or one year. The length of the planning period must include the hydro-generation because of the regulating role of water MWh. After such a chain of programs[5] has been run, long-range decisions can be made, especially for the amount of hydro-energy available and for the thermal maintenance scheduling.

The short-range decisions, for which the ELD is the main subroutine, consist of the thermal unit commitment and the hydro-power scheduling. They can be included directly in an ELD program with an incremental cost derived from long-range considerations. The choice of this cost is still an open problem.

(1) If a decision has to be taken for one week, the hydraulic cost should balance starting-up and other thermal costs, and random fluctuations should be taken into consideration for the load demand.

References p. 112

(2) If the decision is only for one day (tomorrow's allocation) the hydraulic cost should compensate only incremental costs and this benefit is more certain. The random variations on the demand load curve are less important.

The main role of the ELD in these short-range decisions is to consider the network. On the other hand, we think that in long-range decisions the network limitations are rather small and, because the introduction of the network involves great intricacies, we prefer to check after a one-point model whether there are such limitations and what to change so that they hold.

The ELD used with different conditions, such as loads, amount of hydro-power in run-of-the-river plants, and network topologies, shows the predictable behaviour of the system. Of most importance here are the ELD results known as the dual values associated with the optimum. They can be understood as the cost of energy per node. The ELD thus becomes the main subroutine, especially in the phase of search for the thermal unit commitment. The huge number of combinations with different hours and groups requires a simplified formulation for the ELD in this case. The d.c. approximation seems sufficient here. The possibility of including flow constraints remains available.

The computer time saved in this manner can be used for a more accurate search among the vast number of feasible schedules. The branch and bound technique has here proved to be very efficient. Improvements of more than 0.5% have been obtained in the operating cost[6].

4. PRESENT TRENDS IN THE COMPUTING METHODS

Following the survey of the different uses of ELD according to the pending problems, the recent progress in methods will now be summarized. ELD is a non-linear optimization problem and several methods have been tried for its solution:

(1) penalty methods,
(2) gradient projection,
(3) linearizing,
(4) Lagrangian methods and
(5) direct search.

Our experience indicates that the three main features in numerical progress are:

(1) introducing a Newton method, instead of computing node after node;
(2) using the safe basis of linear programming and linearizing equations; and
(3) taking advantage of the sparse matrix properties in linear programming techniques.

The first point tends to be conventional now. The second and third points seem rich in developments for the future. These features are included in the ELD program called MAYA.

4.1 The MAYA algorithm

The program solves an ELD with flow constraints and fixed voltage magnitudes.

As mentioned earlier, the objective function is a piecewise linear sum. The main steps in the algorithm are as follows.

4.1.1 Search for a feasible solution

(a) Computing the optimal solution without losses and with flow constraints. This is done by the dual simplex method starting from the merit order solution.
(b) Search for a solution with losses and flow constraints. To this end the load flow equations are solved by Newton's method starting from solution (a).

4.1.2 Improving a feasible solution

(a) The previous feasible state is used to linearize the flow equations around it. The simplex method gives the optimum for this linear problem.
(b) Newton's method supplies a new feasible solution. If this last solution is a better one, the program returns to step 4.1.1. Otherwise, a new solution is looked for between the last solution and the best obtained previously (dichotomies are made).

4.2 Reducing the core memory storage

It has been stated that the Jacobian **J** is a sparse matrix. This sparsity is used in Newton's method by an LU factorization of **J**. An ordering scheme of the matrix lines according to their number of non-zero elements gives the least number of new non-zero terms.

Moreover, at each step of the algorithm only a few lines have reached their capacity limits. This is used to reduce the size of the simplex array. Only the previously overloaded lines have their constraints introduced in the simplex tableau. Examples of computing times are given here for a case where ten lines satisfy the flow constraints.

Computer	*Number of nodes*	*Duration (sec)*	*Core memory words (decimal)*
CDC 6600	400	15	35 000
CDC 6600	200	5	19 000
CII-Iris 50	100	15	11 000

The following comments may be made about this ELD program.

(1) The form of constraints (inequalities with energy flows) provides the possibility of including more intricate rules of security, such as safe generation plan, even if any three thermal groups and any two transmission lines are out.

(2) No convergence trouble has occurred with this algorithm; linear programming is a sound basis without too many numerical traps.

(3) It can be used from a system planning standpoint in such problems as the balancing of operating costs with investment costs, where the line capacities directly describe the amount of the invested network.

References p. 112

5. CONCLUSIONS

The main trends in current work on ELD problems can be summarized as follows.

(1) A better definition is being made of the situations in which the ELD is a powerful aid. Different approximations of the problem are being developed taking into account both the requirements of the situation and the available computing power. The same dispatcher can now use different ELD programs according to his problems.

(2) A better evaluation of operation savings is being made as a result of combining economical computations within a wider control system.

(3) Mathematical progress is being made.

(4) The initial mainly economic framework is being modified, with emphasis being placed more and more on the security aspect.

REFERENCES

1 L. KIRCHMAYER, *Economic Control of Interconnected Systems,* Wiley, New York, 1959.

2 J. G. SIROUX AND J. CARPENTIER, L'optimisation de la production à EDF, *Rev. Gén. Elec.,* March, 1963.

3 J. C. DODU AND A. MERLIN, Méthode de résolution du dispatching économique dans l'approximation du courant continu (modèle MAYA), *EDF Bull. Etudes et Rech.,* **B3** (1970).

4 J. C. DODU, P. MARTIN, A. MERLIN AND J. POUGET, An optimal formulation and solution of short-range operating problems for a power system with flow constraints, *Proc. IEEE,* special issue on computers and design, to be published in 1972.

5 J. G. SIROUX, *EDF modern dispatching,* EDF publication, June, 1970.

6 P. MARTIN AND A. MERLIN, Méthode combinatoire de répartition à court terme d'un ensemble de production thermique hydraulique, *Rev. Gén. Elec.,* October, 1970.

7 G. DAUPHIN, D. FEINGOLD AND D. SPOHN, Méthodes d'optimisation de la production des groupes d'un réseau électrique, *EDF Bull. Etudes et Rech.,* **1** (1966).

8 W. F. TINNEY AND C. E. HART, Power flow solution by Newton's method, *IEEE Trans.,* **PAS-86** (1967) 1449–1460.

9 L. P. HADJU, J. PESCHON, W. F. TINNEY AND D. S. PIERCY, Optimum load-shedding policy for power systems, *IEEE Trans.,* **PAS-87** (1968) 40.

Discussion

K. L. LO (*University of Manchester, Manchester, England*)

The use of the d.c. approximation combined with the Newton–Raphson iteration process around the feasible point has been implemented by us for a gas dispatch problem. We would confirm from our experience that this method gives speedy convergence, with practical accuracy comparable with that obtained by non-linear programming methods.

The d.c. approximation, by neglecting the off-diagonal elements of the Jacobian matrix, has effectively partitioned the problem into two independent systems. This is a form of the diakoptical approach which we are proposing in our paper and provides an alternative criterion for tearing the system by taking advantage of the engineering knowledge of the problem. For a full diakoptical treatment the neglected terms could form the intersection system, which could be incorporated in the iterative cycle to improve the speed of convergence and accuracy of solution.

We understand that branch and bound methods have been used in your unit commitment problem. Can you elaborate on the approach that you have used?

J. POUGET
The branch and bound method, designed by Merlin to solve the unit commitment problem, has already been explained in the UNO (Economical Commission for Europe) Symposium at Varna (25–27 May 1970). It is coming into current use now at EdF. It will also be included in a paper to be published in *IEEE Trans.* (see ref. 4).

H. GLAVITSCH (*Brown, Boveri & Co. Ltd., Baden, Switzerland*)
You mention the small CII-IRIS 50 computer. Do you plan to use a multi-processor configuration and diakoptics for security and optimization?

J. POUGET
Not exactly. The whole computation, especially for security, will be processed on a central computer. This program will work for a 200 or 400 node transmission system. Some of its results could possibly be sent to regional centres which, using smaller computers (the IRIS-50 being only an example), could make a few optimization decisions more precise (in the hydro-power field).

K. J. ÅSTRÖM (*Lund Institute of Technology, Lund, Sweden*)
I found the figures on the economic incentive for doing the optimization very interesting. Could you elaborate upon these figures and the particular properties of your system that influence them?

J. POUGET
The economic interest of optimizing the operation of the EdF network is evident to us. This is so because we have compared the cost of computation and the savings due to an improvement of 0.5 or 1.0 % on the fuel cost. It must be added that in the EdF transmission system the flow constraints are often being reached and we think that only an ELD with flow constraints brings improvements, by comparison with the merit order solution. This is the case for MAYA. Moreover, the differential losses also have the same influence in the EdF situation.

Load-Frequency Control with Particular Emphasis on Thermal Power Stations

H. GLAVITSCH AND F. D. GALIANA

Brown Boveri Group Research Center, Baden, Switzerland

SUMMARY

Conventional load-frequency control, as well as more recent concepts employing modern control theory for stochastic and deterministic disturbances, works with a single setting of the control parameters.

With the present concept an attempt is made to separate the system conditions and disturbances into classes. Within the class the load-frequency control problem is solved with known techniques, considering the possible implementation of the concept in a real system with particular emphasis on thermal power stations. The separation is performed by observing the load changes and the changes in the tie-line power which drive the system. It is actually done by a detection scheme, which makes use of a load model and state estimation.

Since, in a given area, only partial information of the state of the system is available and various phenomena due to the outside world have to be modelled by uncertainties, an observer–estimator is employed to extract the relevant state information. From its outputs the controls are computed using a linear feedback law derived from cost functions which are assigned to the various classes. The formulation of these cost functions is specifically geared towards the properties of thermal power stations. Little control action is emphasized in the case of small disturbances, whereas system security is in the foreground when the disturbances are larger and an active control is needed.

Simulations of the observer–estimator and the optimal control together with the power system prove the validity of the models and the feasibility of the suggested control strategies.

1. INTRODUCTION

Load-frequency control (LFC) in electric power systems represents the first realization of a higher level control system. It has made the operation of interconnected systems possible and today it is still the basis of any advanced concept for the guidance of large systems. A peculiarity of LFC lies in the fact that each partner in the interconnection has equal rights and possibilities, being limited only by the installed power in the area and the capability of the tie-lines; thus it is not a centralized control system when the total interconnection is considered.

Economic dispatch control and security control schemes demand the LFC system as a prerequisite. In the historical development, LCF systems were installed because of stability problems and the need for better control of the MWh exchange, etc. In Europe the relative ease of controlling hydraulic power stations has contributed very much to the success of LFC. In the United States LFC has gained in importance with the growth of interconnected systems mainly supplied by thermal power stations.

In all systems the percentage of thermal power stations is at present increasing and the control requirements are becoming more demanding. More and more thermal units will have to take over the task of area control. In opposition to these demands, thermal units have the disadvantage that they respond much more slowly to the control signal than hydraulic units. This inferior response is due to a number of things, such as boiler control, limitations derived from mechanical and thermal stresses, limitations in the boiler pressure variations, steam storage in the reheater, etc.

Another consideration is the economic operation of the units. In this respect the prevailing tendency is to operate the units at the valve position limiter to achieve the best efficiency. Maintaining this policy for a large percentage of units does not allow a satisfactory control of the area. Including a large number of units in the control scheme will therefore require the smoothing of the control signal when the system is disturbed by small random load variations.

It is the concern of this paper to investigate the possibilities of a better control of an area exclusively supplied by thermal power stations. This configuration was chosen for the sake of clarity, but as far as real systems are concerned it is not too unrealistic if we look some years ahead.

In this development a concept for LFC will be suggested which is supposed to cover many facets of area control by a unified approach, rather than to present just a singular solution to one particular aspect of LFC. Modern control theory will therefore be emphasized.

The concept to be presented would be of little value if it were not feasible. Hence, the possibilities for a realization on a process computer will be dealt with.

2. REVIEW OF CONVENTIONAL LFC

Conventional LFC is based upon tie-line bias control (see refs. 1 and 2) where each area tends to reduce the area control error (ACE)

$$ACE = \Delta P_{tie} + K\Delta f \tag{1}$$

to zero. When this aim is achieved the system frequency equals the desired value and the interchange schedule is met. The area bias K determines the amount of interaction during a disturbance in the neighbouring areas. For K equal to the area-frequency response characteristic, the system is non-interactive for slow transients.

The control law is proportional–integral:

$$u(t) = CP \times ACE + \frac{1}{TN}\int ACE\,\mathrm{d}t \tag{2}$$

where the integral part takes care that the ACE becomes zero in the steady state. The gains CP and $1/TN$ determine the speed of response. Typical values are

$$CP = 0.1\text{–}1.0 \text{ and } TN = 10\text{–}30\text{ s}$$

When CP is at the upper limit the system results in a fast response for large disturbances, but also tries to make the mechanical power follow small disturbances in a quasi-steady state. The latter is less desirable for economic reasons.

Other quantities to be controlled are the inadvertent interchange $\int P_{tie}\,\mathrm{d}t$ and the time error $\int(1 - f/f_0)\mathrm{d}t$. Usually it is not a difficult problem to keep these deviations small and even the operator can take care of them by manual intervention if his actions are coordinated with those of the other areas.

Practical experience has shown that a fixed setting of the three parameters K, CP and TN cannot cover the needs for all operating conditions, in particular for areas with a large percentage of thermal power stations, as is apparent from the large number of schemes proposed in this problem area (see refs. 3–6).

To give an idea of the operation of a conventional LFC system an example is shown in Fig. 1. Here, a step increase in load is assumed in the base area by an amount of 4% of the spinning power. In the figure, deviations in frequency, tie-line power, control signal and mechanical power are shown. According to tie-line bias control with $CP = 0.2$ and $TN = 30$ s the control signal builds up quite slowly and the controlled variables return to their designated values after a considerable

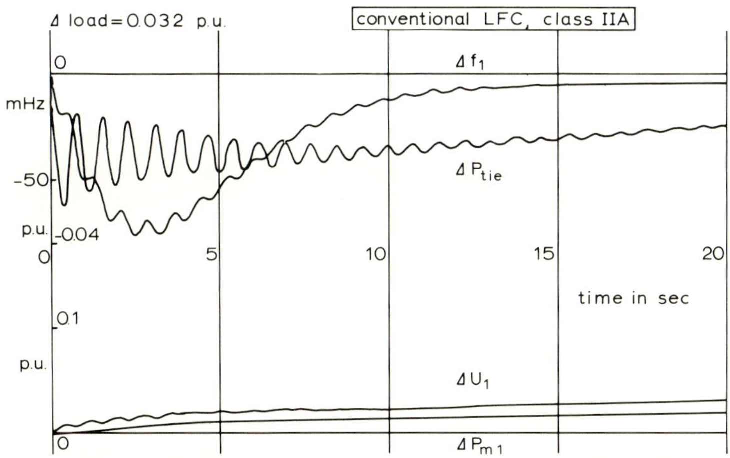

Fig. 1. Conventional LFC, class IIA; load disturbance in base area of 4% of rated power (0.8 p.u.). Δf_1 is the frequency deviation from 50 Hz in mHz, ΔP_{tie} is the tie-line power deviation in p.u. defined positive out of base area, Δu_1 is the LFC signal deviation in p.u. while ΔP_{m1} is the mechanical power deviation in p.u. The subscript 1 refers to the base area, area 1.

References p. 139

delay. It should be noted that the ramp-like increase in the control signal is due to the integrator in the controller. The proportional action is not very significant.

Later in this paper a comparable case will be shown where the proportional action will be accentuated to improve the performance.

3. THERMAL UNITS AS CONTROLLING UNITS IN LFC

Before the design of an advanced LFC system is presented, the properties of typical thermal units of interest to LFC are described. Referring to Fig. 2, two portions in the block diagram may be distinguished. One is a signal processing portion consisting of the speed governor, and the other is represented by the turbine and the generator.

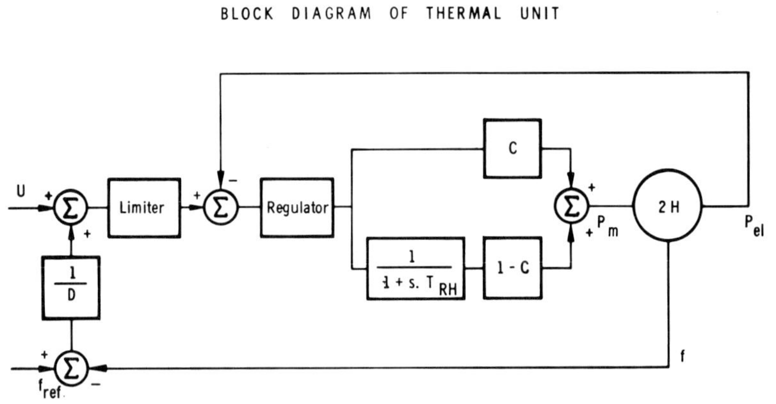

Fig. 2. Block diagram of thermal unit. C is that fraction of the mechanical power output P_{m1} which is due to the high-pressure portion of the turbine; $(1 - C)$ is provided by the reheater ($C = 0.3$). $2H$ is the inertia constant of the generator in p.u.; $2H = 10$ to 20 s. P_{el} is the electrical power demand. D is the droop of the unit, $f_{ref} = 50$ Hz. The limiter is described in the appendix, and the governor is of the proportional plus integral type.

At the input of the speed governor, the summing junctions combine the frequency deviation, the LFC signal and the electrical power output. The limiter, which stands for quite a complex device[7], affects the control action if mechanical and thermal stresses tend to become excessive, if the boiler pressure reaches a lower limit, the step change or the gradient change in the demand signal is too large, etc. The output of the speed governor is converted to valve positions which determine the thermal power input to the turbine. The mechanical power at the turbine shaft is made up of several components produced in the high, medium and low pressure cylinders.

In Fig. 2 just one reheater is shown, which is characterized by a time constant T_{RH} of the order of 5–10 s. The generation of the electrical output is further delayed because the motion of the rotor is characterized by the inertia constant H, which is also of the order of 4–6 s.

The properties of the boilers are incorporated in this model only in the limiter, where the maximum step change of the order of 3–10% of the rating and the gradient change of the order of 1–3% of the rating per minute are derived from boiler characteristics. The permissible step change depends upon the storage capability and is a very important feature of the thermal unit as far as LFC is concerned.

Due to the reheater and the small permissible gradient change, the response of the electrical output to a large step in the LFC signal is quite slow, as can be seen from Fig. 3. Here, the limited value of the LFC signal was used as an input. It

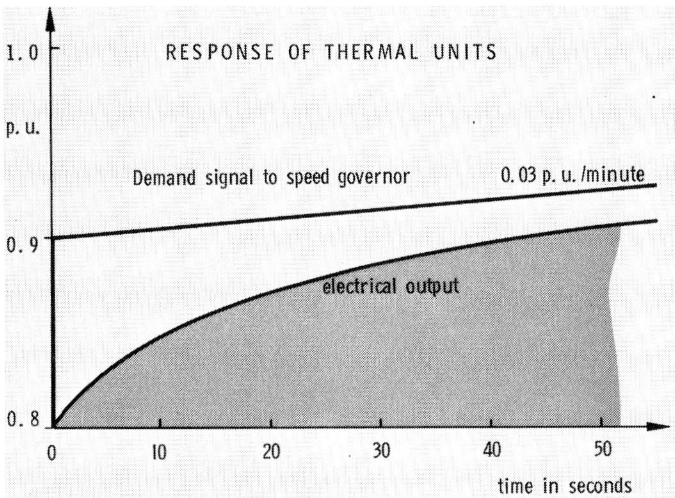

Fig. 3. Limiter and maximum electrical output. The top curve shows the maximum output permitted by the limiter while the lower curve shows the corresponding power generated.

takes about 1 min for the electrical output to approach the desired value. From this result it should be quite clear that it is not possible to affect the power swings appearing on a tie-line as in Fig. 1 at a frequency of 0.5–2.0 Hz by the action of the speed governors. This holds even for small disturbances when the limits are not reached.

In spite of all these limitations LFC in an area exclusively supplied by thermal units has to change the electrical output of these units sufficiently fast for deviations in frequency and tie-line power to stay within given bounds.

In the quasi-steady state operating condition, where load variations are small and random, the speed of response is of no concern; it is rather desirable to affect the electrical output as little as possible to achieve the best economic result. This is the area where error adaptive control[8] has been applied so far.

The slow response of thermal units has quite an important bearing on the security considerations for the area or the interconnected system. The response in Fig. 3 shows that the spinning reserve, as calculated by the difference between the momentary loads and the ratings of the units, is not effective within a period of

References p. 139

50–200 s or even longer after initiating the increase. This means that the assessment of spinning reserve has to take the individual responses of the units into account.

4. THE STRUCTURE OF LFC SYSTEMS

Within an area the LFC system takes the place of a system at the upper level, whereas the units with their speed governor form the lower level. The structure of this two-level system is shown in Fig. 4. On the lower level the units respond to the

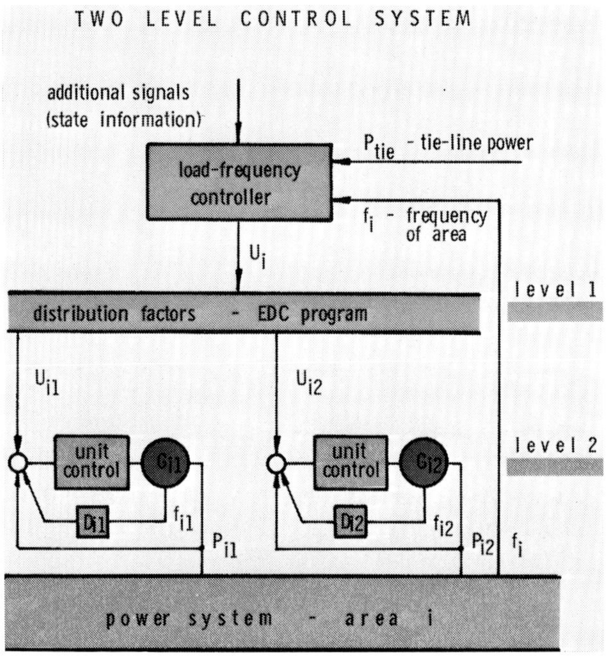

Fig. 4. Two-level control system. Generator G_{i1} in area i has LFC input U_{i1} and electrical power P_{i1}. D_{i1} is the droop of machine 1 while f_{i1} is the instantaneous frequency of machine 1. f_i is the average frequency of area i.

local frequency deviations and to deviations in the unit output. In the case of a loss of generation or of an increase in load, the lower level is able to make up for the unbalance in load according to the static characteristic only. There is no way of controlling the area net interchange directly from the lower level.

On the upper level the area frequency and the area net interchange are monitored. Additional signals may be used for control which will supplement the state information. The control, *i.e.* the LFC signal sent out to the various units, is modified in the simple form by participation or distribution factors and in a more advanced form by an economic dispatch program.

With respect to reliability this system has some outstanding features as is the case for hierarchical structures. Either one of the levels may fail and the system will still be operative, although not as perfect. The loss of the upper level will result in steady state errors but this drawback is quite acceptable for the period of the disturbance.

5. LFC REQUIREMENTS

Since the advanced control system for LFC is to be treated here, the tasks and aims of this system should be examined. Special attention has to be paid to the peculiarities of thermal power stations. The normal control functions, such as keeping deviations of frequency, tie-line power and inadvertent interchange to zero in the steady state, have to be guaranteed in any case, and are of no concern.

In the case of large disturbances when control action is definitely needed, it is recommended, for reasons of security and economics, that control is activated in the area where the disturbance took place. Some support by the other areas will always be given by the mere existence of interconnections.

LFC demands some well-defined control action within the area. If this control action is supported by all the thermal units the individual contribution is small and the slowness of the response of the units is less pronounced.

A consequence of this requirement is another one which aims at generating a smooth control signal. It is known that a constant power demand signal will allow the unit to operate at maximum efficiency. For this reason many units are at present operated at the valve position limiter. To achieve a similar effect when all units are taking part in control, additional measures are required to avoid unnecessary control action in the operating state, where the load variations are random and their amplitudes are small. The statistics of frequency variations show that this state prevails most of the time and frequency deviations larger than 100 mHz occur at intervals larger than 30 min.

Hence, most of the time little control action is required—just sufficient to follow small trends in the load curve. The resulting deviations in frequency and tie-line power in this operating period have to be considered permissible since no value with respect to either security or economics can be attached to them.

Another important aspect which may be treated within the framework of LFC is the allocation of spinning reserve. This question arises when very large disturbances are considered. It is partly a control problem since control action is limited, as explained in connection with the response of thermal units. The electrical output within the area in response to a very large demand signal should be taken as the base for the determination of the spinning power, and should be coordinated with the support of power available from other areas, thereby observing limits of power transfer and frequency deviations.

To achieve an appropriate on-line control performance this allocation of spinning reserve has to be done ahead of time, thereby adapting the generation of the load demand and associated contingencies. The allocation program will have to determine the control strategy besides the required set of generators. The strategy will be realized in on-line control.

References p. 139

6. PRESENT TRENDS IN LFC

With the development of modern control theory, several concepts for LFC have been investigated[3–6] which go beyond the simple tie-line bias control. The fundamental approach is the use of more extended mathematical models. In retrospect, the *ACE* can also be considered as a model of the area. The recent concepts, however, involve the representation of the dynamics of the area, or even of the complete system.

With the availability of models, other concepts of modern control theory are being employed such as state estimation and optimal control. As far as optimal control is concerned the linear regulator, which employs constant feedback gains, is used most frequently. This is also the case in conventional LFC, but there the complete state of the system is not fed back.

Some investigators suggest models which describe the complete interconnected system. The use of such a model, however, cannot be realized at the moment since the state information of the complete system will hardly be available at the location of the area controller.

In addition to the structures which aim at the control of deterministic signals and disturbances, there are schemes under investigation which employ stochastic control concepts, *e.g.* minimization of some expected value of an integral quadratic error criterion[9]. Usually this results in the design of a Wiener or Kalman filter which is of value for the control of small random disturbances or the control of a system corrupted by noise.

Finally, a very simple procedure may be mentioned. This is the filtering of a known frequency, such as the frequency of the power swings. If a filter with such a purpose is used, the control signal is smoothed but the response may also be slowed down, since the properties of the model are not taken into account.

The present trends may be summarized by saying that more information is used in designing controls. Both deterministic and stochastic concepts are being employed.

7. ADAPTATION OF CONTROLS

The different objectives of control in stationary operation and in the disturbed state of the power system suggest the employment of different strategies for the different disturbances. In stationary operation some control will be needed as the system will be able to follow slowly varying load trends, but essentially it is desirable to keep the control signal smooth.

During a large disturbance, specifically if it appears in the base area, a faster control strategy is needed. This consideration leads to two classes of disturbances where the classification is based on the magnitude of the load change. A further distinction may be made if the load change takes place in the base area or in other areas. There may be occasions when some control action is needed for disturbances in the other areas.

A further class is the one in which maintaining frequency and tie-line power depends heavily on the spinning reserve. For this class the spinning reserve has to be predetermined in an allocation process together with the control strategy. The control is stored and called upon in the actual event.

In Table 1 the classification of disturbances is summarized, together with the objective of control. There are three classes with a further subdivision in class II. Class II is the main area of application for tie-line bias control; class III is related to system security as explained above.

TABLE 1

CLASSIFICATION OF DISTURBANCES

Class	*Magnitude of disturbances*	*Location*	*Control action*	*Remark*
I	$\leq 2\%$	Base area	Little	Economic operation
IIA	$< 5\%$	Base area	Required	Security
IIB	$< 5\%$	Other areas	Required	Security
III	$\geq 5\%$	Base area	Required	Security spinning reserve

Having distinguished the various disturbances it is obvious to adapt the control strategies to the various classes. Thus the control objectives are the following.

Class I — Maintenance of a smooth control signal. The system should be able to follow trends in the load.

Class IIA — Frequency and tie-line power deviations should be as small as possible and should return to zero within a reasonable time. Limited control action is to be used. This objective will be realized by employing a quadratic error criterion.

Class IIB — Here the objective is similar to the one in class IIA with the emphasis on the small frequency deviations. The base area is led to support the troubled area.

Class III — The spinning reserve is allocated in such a way that the expected operating cost is a minimum for a given risk of failure.

In the following, an attempt will be made to devise a control strategy which is valid for classes I and II in which just the gains have to be adapted to the individual class.

8. POWER SYSTEM MODELS FOR LFC STUDIES

To arrive at a more efficient LFC control scheme, one must start with a more detailed model between the control action and the significant variables, frequency and tie-line power. A compromise must be reached, however, so as not to require the availability of a mathematical model which from the practical point of view would be unrealistic. This model should make use of system parameters and

References p. 139

observable quantities readily available in present systems or within the foreseeable future.

In order to make a reliable and honest evaluation of the capabilities of such a scheme, a relatively complete model of a real power system has been realized, which is henceforth called the "real world" (see Appendix). It must, however, be emphasized that when designing LFC control strategies only partial knowledge, such as might be available in a real system, is assumed about the "real world".

8.1 Reduced models

The purpose of a reduced model is to permit the design of control schemes in a judicious manner without having to consider all the intricacies of the "real world". In conventional LFC a reduced model is used which is simply the natural steady state frequency–tie-line power characteristics of each area. From this graph the *ACE* control concept is derived which yields non-interacting steady state LFC action.

In this study a reduced model has been considered which describes transient behaviour in the range from 1 s to the steady state with considerable accuracy. By this it is meant that the behaviour of the reduced model is close to that of the "real world" for the same disturbances. Of considerable significance is the fact that the steady state behaviour of the model and the "real world" are identical.

There are various ways to proceed but only the main approach used here is described, which is based on engineering judgment.

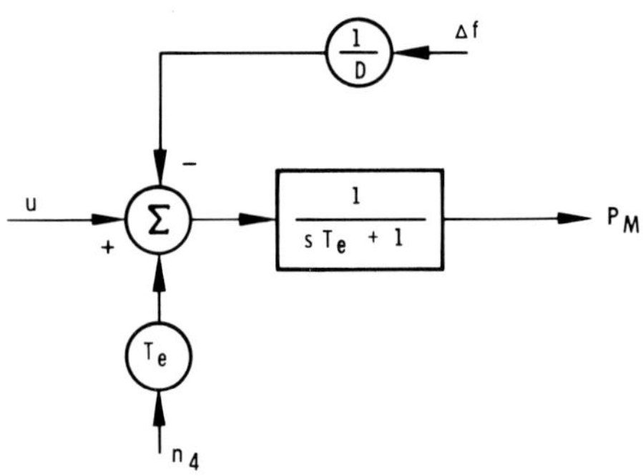

Fig. 5. Reduced model of thermal unit. n_4 is an uncertain process, T_e is the equivalent unit time constant and D is the droop of the unit.

Referring to Fig. 3, the dominant time constant between the output of the limiter and P_{m1} is approximately 10 s. This, together with the fact that in steady state the control input u must equal P_{m1}, led to the reduced model of Fig. 5. All variables are in p.u. based on the rated power of the base area.

The term n_4 is an uncertain process which models the effect of all the terms that are not considered in this reduced system. The introduction of uncertainty in reduced models is an important concept which should be emphasized. It recognizes the fact that the model is not perfect and that it can only describe the real world with a degree of inherent uncertainty. It also has the appealing facet that if one is cunning enough to describe this uncertainty by, for example, a random process with known covariance or by a bounded process, then the reduced model will not only yield an estimate of the system variables, but will also yield a measure of how accurate this estimate is. The term n_4, and the equivalent 10 s time constant T_e, form then the basis of our reduced model for mechanical power P_{m1}.

In equation form one can write

$$\dot{P}_m = \frac{1}{T_e}\left(u - \frac{1}{D}\Delta f - P_m\right) + n_4 \tag{3}$$

8.2 Complete reduced two-area model

Referring to Fig. 5 and the Appendix the complete reduced model can be written as a set of first-order differential equations:

$$\dot{P}_{m1} = \frac{1}{T_e}\left(u - \frac{1}{D_1}\Delta f_1 - P_{m1}\right) + n_4 \tag{4}$$

$$\Delta\dot{f}_1 = \frac{1}{2H_1}(P_{m1} - P_{E1}) \tag{5}$$

where

$$P_{E1} = P_{tie} + P_{L1} \tag{6}$$

and

$$P_{tie} = k_0(\delta_{g1} - \delta_{g2}) - k_1 P_{L1} + k_2 P_{L2} \tag{7}$$

Also

$$\dot{\delta}_{g1} = \Delta f_1 \tag{8}$$

and

$$\dot{\delta}_{g2} = \Delta f_2 \tag{9}$$

while

$$\Delta\dot{f}_2 = \frac{1}{2H_2}(P_{m2} - P_{E2}) \tag{10}$$

where

$$P_{E2} = -P_{tie} + P_{L2} \tag{11}$$

Note that k_0, k_1 and k_2 are defined by eqns. (A9), (A10) and (A11) in the Appendix.

Furthermore, it is assumed that the loads P_{L1} and P_{L2} are of the form

$$P_{L1} = \hat{P}_{L1} + n_1 \tag{12}$$

$$P_{L2} = \hat{P}_{L2} + n_2 \tag{13}$$

References p. 139

where n_1 and n_2 are zero mean white random processes, whereas $\hat{P}_{L1}$ and $\hat{P}_{L2}$ are constant (perhaps slowly varying) mean load levels. In the case of a disturbance, $\hat{P}_{L1}$ and $\hat{P}_{L2}$ may have a discontinuity, but the new mean values then remain approximately constant according to the relations

$$\dot{\hat{P}}_{L1}(t) = w_6(t) \tag{14}$$

$$\dot{\hat{P}}_{L2}(t) = w_7(t) \tag{15}$$

where w_6 and w_7 are zero mean white random processes.

Introducing the definitions

$$y_1 = \Delta f_1 \tag{16}$$

$$y_2 = \Delta f_2 \tag{17}$$

$$y_3 = k_0(\delta_{g1} - \delta_{g2}) - k_1\hat{P}_{L1} + k_2\hat{P}_{L2} \tag{18}$$

$$y_4 = P_{m1} - \hat{P}_{L1} \tag{19}$$

$$y_5 = P_{m2} - \hat{P}_{L2} \tag{20}$$

$$y_6 = \hat{P}_{L1} \tag{21}$$

$$y_7 = \hat{P}_{L2} \tag{22}$$

$$\tilde{u} = u - \hat{P}_{L1} \tag{23}$$

and defining the dynamics of y_5 by

$$\dot{y}_5 = w_5 \tag{24}$$

where w_5 is a zero mean white random process, then if

$$\mathbf{y} = (y_1, y_2, y_3, \ldots, y_7)^T \tag{25}$$

it follows that

$$\dot{\mathbf{y}} = \mathbf{F}\mathbf{y} + \mathbf{B}u + \mathbf{w} \tag{26}$$

where $\mathbf{F}$ is the 7×7 system matrix and $\mathbf{B}$ the 7×1 input matrix corresponding to eqns. (4)–(15) according to definitions (16)–(24). The system uncertainty term $\mathbf{w}$ can easily be shown to be

$$\mathbf{w} = \left(\frac{n_1}{2H_1}, \frac{n_2}{2H_2}, -k_1 n_1 + k_2 n_2, n_4 - w_6, w_5, w_6, w_7\right)^T \tag{27}$$

A number of comments must be made about this reduced model. The significance of n_1, n_2, n_4, w_6 and w_7 has already been described, but a new uncertainty term, w_5, has been introduced to describe the dynamics of y_5 or $P_{m2} - \hat{P}_{L2}$. This is consistent with the philosophy that each area does not have complete information, such as turbine dynamics and control action, about other areas. Instead, y_5 has been modelled as a slowly time-varying uncertain process—a not unreasonable assumption. The remaining constants in the reduced model are all within reach of the base area. H_1 and H_2 are the areas' inertia constants, $1/D$ is the equivalent droop of the base area, T_e is the equivalent time constant of the base area turbine dynamics, and k_0, k_1 and k_2 are transmission constants which can be measured or estimated.

In addition, a number of measurements can be made on-line which permit the intelligent use of the above model. These can be frequency, tie-line flows and own load level.

It should be noted that, due to the assumed linear nature of the model, each state y_i is assumed to represent a deviation from its equilibrium point in per unit.

8.3 Modelling of uncertainty in reduced model

The reduced model of eqn. (26) has a number of variables which are uncertain and which describe one's inability or unwillingness to describe the system exactly.

Reasonable results were obtained by assuming that the variables w_i, $i = 1, \ldots, 7$, are zero mean white random processes with

$$\begin{aligned} E[w_i(t)w_j(\tau)] &= Q_i(t-\tau) \qquad & i = j \\ &= 0 \qquad & i \neq j \end{aligned} \tag{28}$$

The actual values of Q_i were picked by engineering judgment, but more sophisticated identification methods[10] could be used. Typical values were, in p.u. quantities,

$$\begin{aligned} Q_1 &= 10^{-8} \\ Q_2 &= 10^{-8} \\ Q_4 &= 10^{-6} \\ Q_5 &= 10^{-5} \\ Q_6 &= 10^{-6} \\ Q_7 &= 10^{-5} \end{aligned} \tag{29}$$

This is consistent with the fact that there is more uncertainty in the behaviour of the interconnected areas than in the base area.

9. ADVANCED LFC STRATEGIES

The area control strategy used in this study is of the form

$$u(t) = \sum_{i=1}^{n} G_i \hat{x}_i(t) + G_I \int_0^t ACE(\tau)\mathrm{d}\tau \tag{30}$$

As in conventional LFC this has a linear feedback character, but in addition to frequency and tie-line power (the components of ACE) other system states are fed back also.

References p. 139

The last term in eqn. (30) is also present in conventional LFC schemes and its purpose is to eliminate steady state errors as well as to track long-term trends. The purpose of the proportional term, $\sum_{i=1}^{n} G_i \hat{x}_i$, is to improve dynamic performance through its more rapid response to changes in system variables.

The variables $\hat{x}_i$ are the filtered values of those system states which are deemed significant for load-frequency control. Thus, in conventional LFC $\hat{x}_i$ would be the single filtered value of *ACE*, such as is given by the error adaptive control computer EACC[8]. The feedback gains G_i and G_I are adjustable constants and their choice, together with the filtered states $\hat{x}_i$, define the over-all control strategy.

It is felt that the main guide-lines in designing improved and practical LFC schemes are as follows, in approximate order of importance.

(a) Compatibility with present schemes and available system information. This is partly accomplished through eqn. (3) and the reduced system model of §8.2.

(b) Adaptation of the control action to the different classes of disturbances. This is accomplished by automatically readjusting the gains G_i and G_I to one of the various predetermined values according to the classes of disturbances described in §7, *i.e.* class I, class IIA and class IIB.

(c) A systematic method for the rapid detection of the disturbance class. This is discussed in §9.1.

(d) A systematic method of filtering the system variables being fed back to avoid "chasing" small random disturbances (see §9.1).

(e) A systematic method of designing the feedback gains G_i and G_I (see §9.2).

(f) The feedback of more than one system variable. It is felt, however, that the use of a very large number of system variables is not very realistic.

9.1 Filtering, estimation and class detection

As discussed in the previous section, these topics form an important part of improved LFC schemes.

Filtering yields smooth system variable estimates, thus reducing the "chasing" of small random disturbances.

Estimation yields reliable approximations of those important system variables which are not directly measurable.

Detection provides a quantitative systematic method of differentiating between the classes of disturbances.

In order to accomplish these objectives in a practical and systematic manner, use is made of Kalman filtering techniques[11] together with the uncertain model of eqn. (26), §8.2, and a set of system variable observations. The model dynamics provide *a priori* system behaviour information, while the observations provide sampled information about the present state of the system. The filter weighs the two information sources according to the degree of confidence one has in each source. Detection of anomalous behaviour is accomplished by comparing the variances of the errors between observed and expected system variables.

Here it was assumed that the following system variables were observed:

$$z_1 = f_1 + v_1 \tag{31}$$

$$z_2 = f_2 + v_2 \tag{32}$$

$$z_3 = P_{tie} + v_3 \tag{33}$$

$$z_4 = P_{L1} + v_4 \tag{34}$$

where $v_i, i = 1, \ldots, 4$, are zero mean white random observation errors with

$$\begin{aligned} \mathrm{E}[v_i(t)v_j(\tau)] &= R_i(t - \tau) \qquad & i = j \\ &= 0 & i \neq j \end{aligned} \tag{35}$$

which can be easily written in vector form as

$$\mathbf{z} = \mathbf{H}\mathbf{y} + \mathbf{v} \tag{36}$$

where $\mathbf{H}$ is the 4×7 observation matrix defined by eqns. (31)–(34) and the definitions of the state, $\mathbf{y}$. The term $\mathbf{v}$ can be shown to be

$$\mathbf{v}^T = (v_1, v_2, v_3 - k_1 n_1 + k_2 n_2, v_4 + n_1) \tag{37}$$

The filtered state $\hat{\mathbf{y}}$ is then given by

$$\dot{\hat{\mathbf{y}}} = \mathbf{F}\hat{\mathbf{y}} + \mathbf{B}\tilde{u} + \mathbf{K}(\mathbf{z} - \mathbf{H}\hat{\mathbf{y}}) \tag{38}$$

where $\mathbf{K}$ is the Kalman gain[11] corresponding to the system of eqn. (26), the observations (36) and the system and observation uncertainty covariance matrices. The vector $\hat{\mathbf{y}}$ is thus a reliable, filtered estimate of all the states of the system as defined by eqns. (16)–(22).

The detection of anomalies is done by the on-line analysis of the residuals $\mathbf{e}$, *i.e.*

$$\mathbf{e}(t) = \mathbf{z}(t) - \mathbf{H}\hat{\mathbf{y}}(t) \tag{39}$$

Each residual e_i has an expected variance. Monitoring of the third and fourth residuals for persistent deviations from one or two standard deviations could be used as a quantitative test for the rapid and reliable detection of class IIA or IIB disturbances, while the frequency residuals can be used to detect distant faults, slower disturbances or failure of the model. If the disturbance is a load change, the tracking nature of the model will follow and identify the new load level.

9.2 Design of control gains

In this section a systematic method of designing the control gains of eqn. (30) is discussed.

Although the seven order model described in the previous section is important for the function of systematic data monitoring, filtering estimation and anomaly detection, it is readily seen that not all states are controllable from the base area alone, and some are simply uncontrollable. These are clearly $y_5 = P_{m2} - \hat{P}_{L2}$, $y_6 = \hat{P}_{L1}$ and $y_7 = \hat{P}_{L2}$, the mechanical power in other areas and the mean load levels respectively. It can readily be shown that the states $y_1 = \Delta f_1, y_2 = \Delta f_2$,

References p. 139

$$y_3 = k_0(\delta_{g1} - \delta_{g2}) - k_1\hat{P}_{L1} + k_2\hat{P}_{L2} = \hat{P}_{tie},$$

where $\hat{P}_{tie}$ is the mean value of P_{tie}, and $y_4 = P_{m1} - \hat{P}_{L1}$ are controllable*, *i.e.* after a disturbance, the LFC control u can be manipulated to yield zero deviations in steady state.

If $\mathbf{x}$ is defined as

$$\mathbf{x} \triangleq (y_1, y_2, y_3, y_4)^T \tag{40}$$

then from eqns. (4)–(15) one gets

$$\dot{\mathbf{x}} = \mathbf{A}\mathbf{x} + \mathbf{b}\tilde{u} + \tilde{\mathbf{w}} \tag{41}$$

where $\tilde{u} = u - \hat{P}_{L1}$ and

$$\mathbf{A} = \begin{bmatrix} 0 & 0 & -1/2H_1 & 1/2H_1 \\ 0 & 0 & 1/2H_2 & 0 \\ k_0 & -k_0 & 0 & 0 \\ -1/T_eD & 0 & 0 & -1/T_e \end{bmatrix} \tag{42}$$

$$\mathbf{b} = \begin{bmatrix} 0 \\ 0 \\ 0 \\ 1/T_e \end{bmatrix} \tag{43}$$

$$\mathbf{w} = \begin{bmatrix} w_1 \\ w_2 \\ w_3 \\ w_4 \end{bmatrix} + \begin{bmatrix} 0 \\ y_5/2H_2 \\ 0 \\ 0 \end{bmatrix} \tag{44}$$

where $w_i, i = 1, \ldots, 4$, are zero mean white random processes defined by eqns. (27) and (28).

The vector $\mathbf{w}$ is not in general white because of the effect of $y_5 = P_{m2} - \hat{P}_{L2}$. For small disturbances in interconnected areas, $\hat{P}_{L2} = 0$ so that $P_{m2} = 0$ and $y_5 = 0$. For larger disturbances in P_{L2}, $y_5 \neq 0$, but it is a zero mean oscillation. In this study the term y_5 was nevertheless always assumed to be part of the system noise $\mathbf{w}$ and this gave reasonable results in designing *control* schemes. For *estimation* purposes it was necessary, however, to describe the dynamics of y_5 by eqn. (24) in order to improve the estimation of the other state variables.

To control $\mathbf{x}$, an integral quadratic criterion was used. The variable $\tilde{u}$ was chosen to be that control which minimized $\mathrm{E}[J]$ where

$$J = \int_0^\infty (\mathbf{x}^T(t)\mathbf{S}\mathbf{x}(t) + \tilde{u}^2(t)T)\mathrm{d}t \tag{45}$$

* Controllable in mean value under small random disturbances.

and where $\mathbf{S}$ and T are weighting matrices picked according to the class of disturbance. Under the given assumptions the optimum $\tilde{u}$ is

$$\tilde{u} = \mathbf{G}^T \hat{\mathbf{x}} \tag{46}$$

where $\mathbf{G}^T$ is the optimal control gain for system (41) with zero disturbance which minimizes J, while $\hat{\mathbf{x}}$ is the best linear estimate of $\mathbf{x}$, such as that given by the filter of eqn. (38).

In terms of filtered physical variables one has

$$u = G_1 \hat{f}_1 + G_2 \hat{f}_2 + G_3 \hat{P}_{tie} + G_4(\hat{P}_{m1} - \hat{P}_{L1}) + \hat{P}_{L1} \tag{47}$$

In practice an integral of ACE term was arbitrarily added to eqn. (47) with a 30 s time constant. This term could also be included as a new state to eqn. (41) and appropriately weighted in eqn. (45) through the matrix $\mathbf{S}$.

The advantages of this approach are as follows.

(i) Through $\mathbf{S}$ and T one can put more or less emphasis on the various variables of the system in a simple and logical manner.

(ii) The control gains G_i always result in a stable feedback system. This permits use of larger gains with more confidence.

(iii) The control strategy is linear and time invariant.

(iv) The gains can be readily precalculated by solving the steady state Riccati equation[12].

The primary disadvantage is that a system model must be available; however, this model is quite simple and its parameters readily accessible. In the simulations it controlled a much more complex "real world" quite well.

9.3 Adaptation of gains

The control gains were varied according to the classes of disturbances. Their calculation was done off-line and used values of $\mathbf{S}$ and T approximately as follows.

Class I

$$\mathbf{S} = \begin{bmatrix} 1 & 0 & 0 & 0 \\ 0 & 4 & 0 & 0 \\ 0 & 0 & 25 \times 10^{-4} & 0 \\ 0 & 0 & 0 & 0 \end{bmatrix} \tag{48}$$

$$T = 50 \times 10^{-4} \tag{49}$$

Class IIA

$\mathbf{S}$ is as before but

$$T = 25 \times 10^{-4} \tag{50}$$

References p. 139

Class IIB

S is as before, but T can be decreased further if additional support to interconnected areas is desired, so that

$$5 \times 10^{-4} \leq T \leq 25 \times 10^{-4} \tag{51}$$

It should be noted that decreasing T results in more freedom for u, and hence larger gains G_i. In general, this eventually results in control signals which reach limiter bounds and are no longer optimum.

In Fig. 6 the concepts of observation, estimation, detection and control are illustrated. OED stands for *o*bserver, *e*stimator and *d*etector.

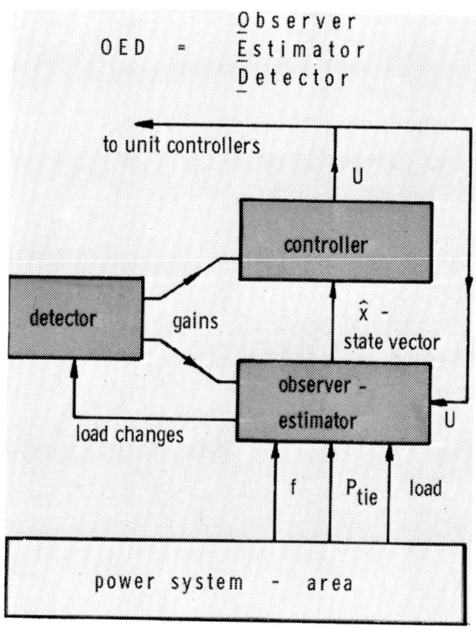

Fig. 6. Block diagram description of over-all OED scheme.

10. RESULTS

The "real world" together with the reduced model and an optimal control scheme have been simulated on a digital computer. Representative disturbances for the classes I, IIA and IIB have been considered.

In order to illustrate the possible improvements obtained by the proposed advanced LFC concept, a comparison is made for each case with a conventional LFC system. The control parameters, *i.e.* $CP = 0.5$, $TN = 10$ s, have been kept the same for all cases, which corresponds to a real system.

10.1 Class I

In this class the power system is disturbed by random load changes only. In the simulation a sequence of steps changing every second is assumed. The magnitude of the steps is uniformly distributed over the range $\pm 1.0\%$ of the steady state generation of the base area. The mean value of the disturbance is zero. In the other area no disturbance is assumed.

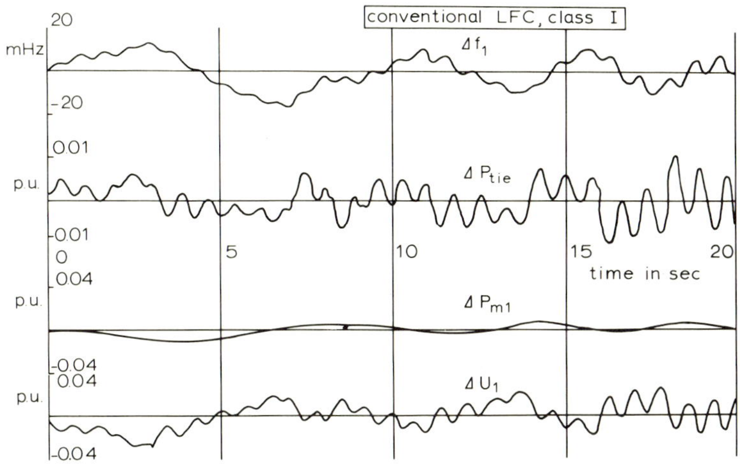

Fig. 7. Conventional LFC, class I, small random disturbance.

How a conventional LFC system responds to this sequence is shown in Fig. 7. Frequency and tie-line power deviations stay within a certain band which is related to the range of the disturbance. The deviations of the control signal

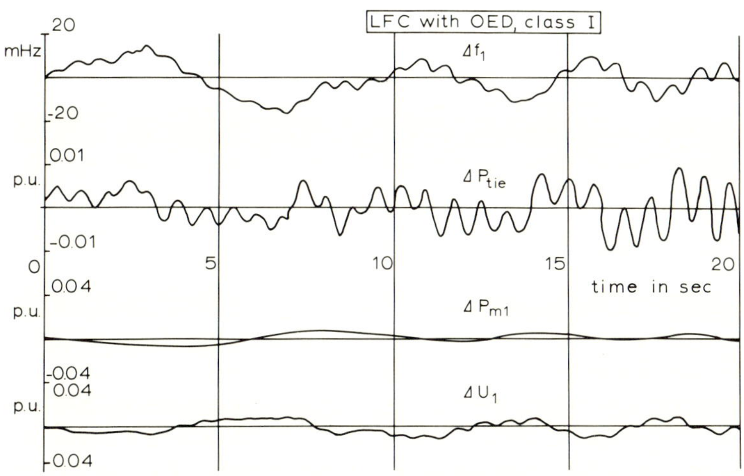

Fig. 8. OED control, class I, small random disturbance.

References p. 139

amount to approximately 3% of the steady state value. They reflect the deviations in frequency and tie-line power quite well, being due to the proportional gain *CP*.

The performance of the estimator together with the optimal control can be seen in Fig. 8 where the same sequence of disturbances is assumed. The deviations of the control signal are much smoother and the mechanical power deviations do not go beyond 1% of the steady state generation. It must be noted that some portion of this deviation is due to the action of the speed governors at the lower control level. Hence, the point is reached where further reduction of the LFC action will not reduce the mechanical power deviations.

The outstanding result of this comparison is that frequency and tie-line power are hardly affected by the difference in control action.

It may be added that the setting of gains, as used in Fig. 8, was able to follow a load increase in the form of a ramp of 3% of rated generation per minute without any deterioration of frequency and tie-line power control.

10.2 Class IIA

A step change in load of 4% of rated generation in the base area was considered in this class. With conventional LFC the control signal and the mechanical power show a ramp-like increase as illustrated in Fig. 1; the maximum frequency deviation reaches 87 mHz.

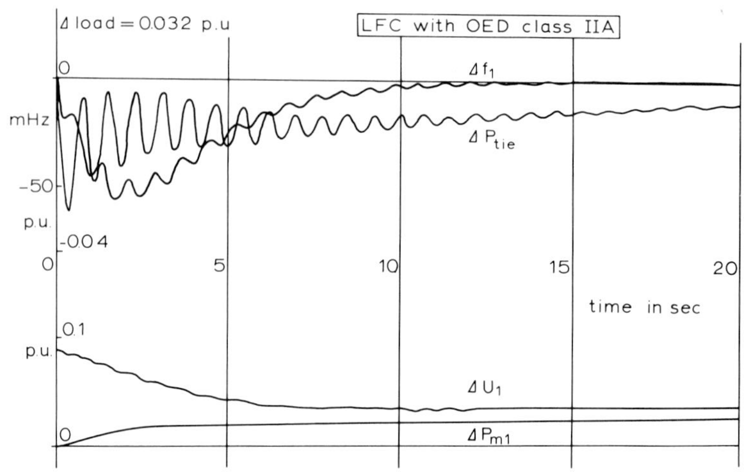

Fig. 9. OED control, class IIA, load disturbance of +4% of rated power in the base area.

The advanced control scheme shows a very desirable jump in control signal with the appearance of the load change, as can be seen in Fig. 9. Afterwards it returns smoothly to the new steady state value. The mechanical power reaches its new value within 5 s. As a result the frequency deviation drops just to 57 mHz and returns to zero after 15 s. The tie-line power deviations are considerably smaller than with conventional LFC.

In this class the advanced control scheme shows a very pronounced improvement over conventional LFC.

10.3 *Class IIB*

According to the classifications in §7, a disturbance appears outside the base area and it may be desirable to support the system by control action in the base area. Conventional LFC does not show any support, as seen in Fig. 10. This is quite correct since the adjustment was made to be non-interactive.

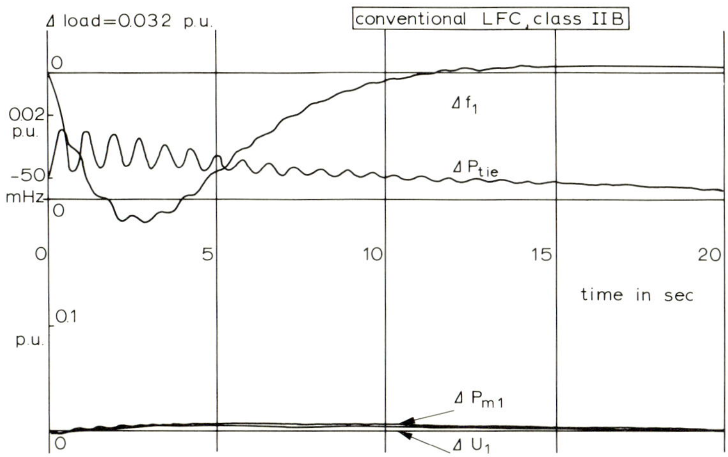

Fig. 10. Conventional LFC, class IIB, load disturbance of 4% of rated power in the other area.

In the advanced control scheme the objective function was chosen such that the control signal in the base area responds to the external disturbance. This is quite noticeable in Fig. 11, where the control signal deviation causes a positive deviation in mechanical power. The support given by the base area reflects itself also in the tie-line power. Thus, the frequency deviation is reduced from 70 mHz to 57 mHz. The tie-line power deviation returns to zero faster than with conventional LFC.

The deviation in mechanical power returns to zero within the time period considered, *i.e.* 20 s. In this way, it is demonstrated that the output of thermal units over a short period can be used quite effectively in the control. This output can easily be drawn from the stored energy of the boiler without resorting to additional measures in the control of the boiler.

The result of the conventional LFC also shows that tie-line power is drawn from the base area without any control action, which proves that true non-interaction cannot be realized. The definite support of the other area, on the other hand, costs very little extra effort in the base area and has a very pronounced effect.

Finally, the implementation of the OED concept should be briefly mentioned. If a process computer is envisaged for LFC with OED, there is just a program extension due to the estimation and detection part which requires slightly more storage than the corresponding conventional LFC program. The observed

References p. 139

quantities are readily available from the base area except the over-all system frequency which was required here to yield information about the interconnected system.

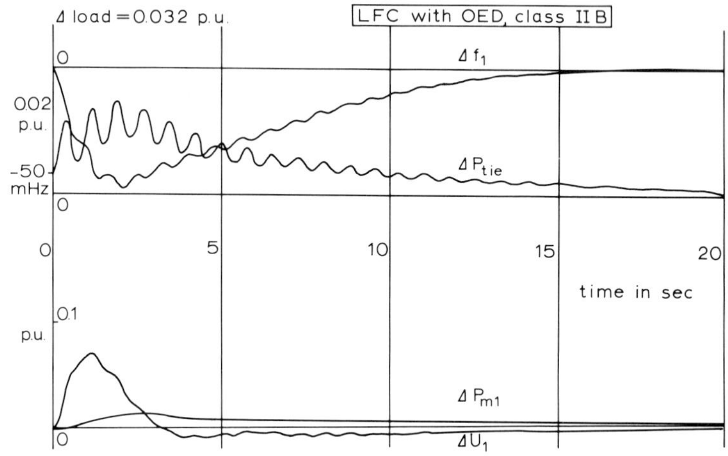

Fig. 11. OED control, class IIB, load disturbance of 4% of rated power in the other area.

Emphasis is put on the compatibility of OED with conventional LFC which was proved by the simulations.

11. CONCLUSIONS

In this paper various extensions to the present methods of LFC were presented which, in the authors' opinion, lead to more reliable control, more efficient use of control action under a wide range of conditions and improved dynamic performance. The schemes proposed have the attractive feature that they are practical, as the models used require parameter values and measurements generally available or within reach of each area. In addition, the estimation, detection and control portion of the scheme can easily be implemented in an on-line computer.

The main points of this approach are: the use of a reduced probabilistic model to describe system dynamics; the use of an on-line estimator with the multiple purpose of estimating states not directly measured and filtering observation and system noise; plus the use of an on-line detector for the reliable and rapid differentiation of the type of disturbance. The disturbances were classified as small and random with slow trends (class I) and larger and deterministic (class IIA in the base area and class IIB in interconnected areas).

The choice of the control gains was made systematically by the off-line *a priori* solution of a Riccati equation corresponding to the minimization of an integral quadratic performance criterion of frequency and tie-line power deviations, which are weighted differently according to the type of disturbance.

The results based on simulated data support the claims made, but additional tests and extensions are needed.

It is felt that the proposed scheme, although subject to some limitations, provides practical improvements to conventional LFC through the classification of disturbances and control adaptation to these classes, the systematic estimation and filtering of system variables, and the rapid and reliable detection of system changes.

APPENDIX

The model used to describe the "real world" consists of two generators interconnected by a lossless network, as shown in Fig. 12.

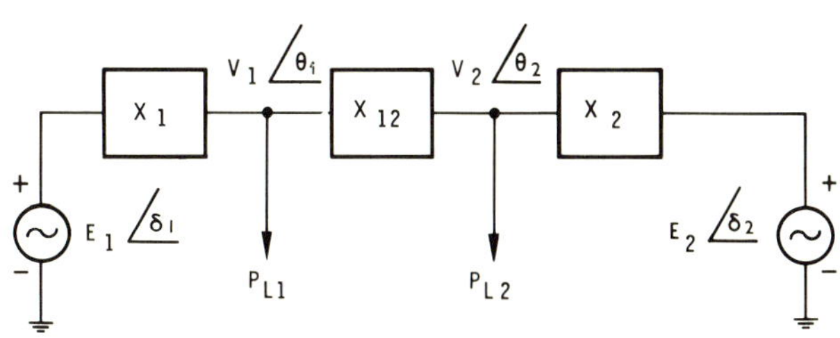

Fig. 12. Network structure of two-area, two-generator power system.

The loads P_{L1} and P_{L2} are modelled as real power sinks. The network reactances had the per unit values

$$X_1 = 0.2 \tag{A1}$$

$$X_{12} = 0.4 \tag{A2}$$

$$X_2 = 0.1 \tag{A3}$$

A d.c. approximation is used where it is assumed that, in per unit,

$$E_1 = V_1 = V_2 = E_2 = 1 \tag{A4}$$

and that, in radians,

$$\delta_1 - \theta_1 \ll 1 \tag{A5}$$

$$\theta_1 - \theta_2 \ll 1 \tag{A6}$$

$$\theta_2 - \delta_2 \ll 1 \tag{A7}$$

These assumptions are quite valid for the study of small to medium load-frequency transients. One can then readily express the tie-line power flowing across X_{12} as

$$P_{tie} = k_0(\delta_1 - \delta_2) - k_1 P_{L1} + k_2 P_{L2} \tag{A8}$$

References p. 139

where

$$k_0 = \frac{2\pi f_0}{X_1 + X_{12} + X_2} \quad \text{(A9)}$$

$$k_1 = \frac{X_1}{X_1 + X_{12} + X_2} \quad \text{(A10)}$$

$$k_2 = \frac{X_2}{X_1 + X_{12} + X_2} \quad \text{(A11)}$$

while f_0 is the reference frequency, or 50 Hz. The electrical power at generator 1 is

$$P_{E1} = P_{L1} + P_{tie} \quad \text{(A12)}$$

The phase angles δ_1 and δ_2 are determined by the turbine–generator dynamics, as defined in Fig. 2, and the relation

$$\delta_1 - \delta_2 = \int_0^t (f_1(\tau) - f_2(\tau))\mathrm{d}\tau \quad \text{(A13)}$$

We note that only the difference in phase angles is needed to define tie-line power.

Referring to Fig. 2 of the thermal unit used in this model, consider the following blocks.

(i) *Limiter*. This is a non-linear device which limits the instantaneous control signal change to 10 % and to a growth rate thereafter of 3 %/min. These are boiler-imposed limitations. Under small disturbances one is always below these limits.

(ii) *Regulator*. This consists of electric and hydraulic portions of the form

$$CP\left(1 + \frac{1}{sTN}\right)\frac{\exp(-sT_d)}{(sT_h + 1)} \quad \text{(A14)}$$

(iii) *Generator*. This is the usual energy balance equation plus a damping term, $d\Delta f$:

$$\Delta\dot{f} = \frac{1}{2H}(P_m - P_E - d\Delta f) \quad \text{(A15)}$$

$$\Delta f = f - f_0 \quad \text{(A16)}$$

The following parameter values were used in the turbine generator models:

$D_1 = 0.05$	$TN = 20$ s	$C = 0.3$
$D_2 = 0.025$	$T_d = 0.25$ s	$H_1 = 5$ s
$T_e = 2$ s	$T_h = 0.8$ s	$H_2 = 10$ s
$CP = 0.5$	$T_{RH} = 10$ s	$d = 5$

Subscripts 1 and 2 refer to areas 1 and 2 respectively; otherwise both areas have the same parameter values.

REFERENCES

1 N. Cohn, *Control of Generation and Power Flow on Interconnected Systems*, Wiley, New York, 1966.

2 A. de Quervain and W. Frey, Methods of controlling interconnected power systems, *Brown Boveri Rev.*, **44** (1957) 472–487.

3 O. I. Elgerd and C. E. Fosha, Optimum megawatt-frequency control of multiarea electric energy systems, *IEEE Trans.*, **PAS-89** (1970) 556–563.

4 C. E. Fosha and O. I. Elgerd, The megawatt-frequency control problem: a new approach via optimal control theory, *IEEE Trans.*, **PAS-89** (1970) 563–577.

5 R. K. Cavin, M. C. Budge and P. Rasmussen, An optimal linear systems approach to load-frequency control, *IEEE Trans. Paper 71 TP 52-PWR*, presented at IEEE Winter Power Meeting, New York, January 31–February 5, 1971.

6 J. D. Glover, Load frequency control of electric power systems, *Report 26*, Electric Power Systems Engineering Laboratory, Massachusetts Institute of Technology, Feb. 1971.

7 E. Andres, Load frequency control by steam turbines with electro-hydraulic governors, *Brown Boveri Rev.*, **51** (1964) 180–185.

8 C. W. Ross, Error adaptive control computer for interconnected power systems, *IEEE Trans.*, **PAS-85** (1966) 742–749.

9 J. S. Meditch, *Stochastic Optimal Linear Estimation and Control*, McGraw-Hill, New York, 1969, Ch. 10.

10 K. J. Åström and T. Bohlin, Numerical identification of linear dynamic systems from normal operating records, *Proc. 2nd IFAC Symposium on Self-Adaptive Systems, 1965*, pp. 76–111.

11 R. E. Kalman, A new approach to linear filtering and prediction problems, *Trans. ASME, J. Basic Eng., Series D*, **82** (1960) 35–45.

12 M. Athans and P. L. Falb, *Optimal Control*, McGraw-Hill, New York, 1966, Ch. 9.

Discussion

G. Quazza (*ENEL, Milan, Italy*)

I refer to the authors' statement that in order to design the proposed load-frequency control system only information on its own area is required. However, the non-intervention (or "autonomy") criterion is not followed and a secondary control support is given to neighbours in emergency. I wonder how the choice of the best support action can be made if no information is known on neighbours' responses. One of the advantages of the autonomy principle[1] is its simple form (perfect decoupling: if G_{Ri}, G^*_{Ri} are the frequency error-to-secondary control power and exported power error-to-secondary control power transfer functions for area i, complete diagonalization of the loop matrix only requires $G^*_{Ri} = G_{Ri}G_{Ni}$ where G_{Ni} is the power injection-to-frequency transfer function of area i). It is true that when "uniform system response",[2] in terms of Δf to load changes, is also required to avoid a "weaker" partner affecting all its neighbours with larger Δf, the additional condition

$$G^*_{R1} = G^*_{R2} = \cdots = G^*_{Rn}$$

is added, which implies that each partner knows the others' transfer functions.

I should also like to ask what type of power system models should be used and whether network configuration and "elasticity" of interconnecting or intra-connecting lines must be included.

1 G. Quazza, Regolazioni non interagenti di reti interconnesse, *L'Energia Elettrica*, **40** (1963) 627.

2 G. Quazza, Criteria for equitable participation of areas in tie-line power frequency control of interconnected power systems, *Automazione e Strumentazione*, **14** (1966) 98.

H. GLAVITSCH

In reply to Dr. Quazza's comments, reference is made to the reduced model of the interconnected system where a bias term $P_{m2} - \hat{P}_{L2}$ is used to characterize the power unbalance in the other area. The estimator identifies this term. Together with the deviations in frequency and tie-line power there is substantial information available about the over-all system. Frequency and tie-line power deviations are purposely not combined to result in non-intervention.

The reduced model involves uncertain terms and does not work with exact transfer functions. Emphasis is placed on predicting tie-line power and frequency behaviour.

In the design of the controls for the support of the other areas much weight is put on the frequency deviations and less weight on the tie-line power deviations in the object function (performance index). This results in a high gain for the frequency deviation. Hence there is some information available in the base area and this is used to best advantage.

As to the second question regarding elasticity, we have found it necessary to include a model equation for the tie-line power. Intraconnecting lines would have to be modelled if sub-areas were to be distinguished.

A. H. EL-ABIAD (*Purdue University, Lafayette, U.S.A.*)

The authors are to be congratulated for considerably improving upon previously published work applying modern control theory to LFC. In particular the following two contributions deserve repeating.

(1) Recognition of separate classes of system disturbances and designing the control signals with corresponding optimal gains.

(2) Representation of prime movers and energy sources in modelling the controlled power system.

However, these same two points raise some important questions concerning the practical real-time applications of the proposed LFC scheme. Since the power system is both non-linear and time-varying it is not enough to classify the different possible disturbances; it must also be important to recognize the many sets of model parameters corresponding to the different system operating conditions resulting from changes in load, network configuration and unit commitments. Also, present thermal plant dynamics are very sluggish and rate limited, to the extent that their responses practically filter out any expected benefit from any LFC signals, especially in the case of large disturbances such as the loss of a unit.

In view of these remarks, I would like to know the authors' opinion on:

(1) the role of LFC as a means of improving system dynamic security; and

(2) if LFC has little or no value in this respect, the practical advantages and worth of the proposed method over the conventional one.

H. GLAVITSCH

It is true, as Professor El-Abiad has pointed out, that the system has many non-linearities. However, they are eliminated to a large degree by the feedback loops of the unit controllers and the linearized model is valid over a wide range. Otherwise several classes have to be introduced. A basic requirement of LFC as conceived here is that dead-bands are very narrow, as is the case for the modern electrical governors. Rate limitation, on the other hand, may be taken into account in the extension of the model.

As to the specific questions of Professor El-Abiad, it may be stated that LFC in class II definitely improves system security as deviations are smaller in magnitude and shorter in duration. The responses are well damped and the capabilities of the units are utilized in the best possible way.

J. P. BOWLES (*Hydro Quebec Institute of Research, Canada*)
Does the change of the gain (depending on the disturbances) require operator intervention or is it done automatically?

H. GLAVITSCH
The setting of gains, as queried by Mr. Bowles, is conceived to be automatic.

S. K. MITTER (*Massachusetts Institute of Technology, Cambridge, Mass., U.S.A.*)
If I want to design a Kalman feedback controller in a completely integrated way, then the theory tells me that under controllability and observability I have a stable system. However, in your two-area case you want to design local feedback controllers. Is stability then necessarily guaranteed?

Could you do a root locus study to represent the interaction between the areas?

H. GLAVITSCH
Professor Mitter raised the interesting question of whether stability is guaranteed if we apply our controllers just to our area. According to the theory there is no guarantee for stability since, seen from our area, we are dealing with a non-controllable and non-observable system. However, the assumption is justified that the other areas have some reasonable LFC concept, *e.g.* minimization of a similar objective function, and therefore the system will be stable if we avoid extreme values for the gains.

A root locus investigation could be done if the models of the areas were known and their control systems were linear.

H. BLOCH (*Brown, Boveri & Co. Ltd., Baden, Switzerland*)
From a practical standpoint the presented diagrams for simulation with LFC and OED respectively show no impressive differences.

For disturbances of less than 2% in the base area only a difference in the control variable U_1 was observable. The governor–turbine system acts as a low pass filter for the frequencies of the electromechanical oscillations. Therefore these oscillations present in the control variable of LFC are considerably damped in the mechanical power output.

For disturbances of less than 5% in the base area the tie-line power resumes its original value faster with OED than with LFC. However, considering the normal time ranges of secondary control (Europe about 100 s, U.S.A. about 600 s) this is of little importance.

For disturbances of less than 5% in the other area the frequency resumes its original value faster with OED because more power is delivered at the beginning than corresponding to the normal droop. However, it can be doubted whether this is desirable; too much decrease of stability of the whole system may result from it. In my opinion the benefits of OED compared with LFC are not obvious enough to justify the implementation of a more sophisticated and more expensive system.

H. Glavitsch

Replying to the remarks of Mr. Bloch it should be pointed out that by observation of the scales of the diagrams for class I and class II quite a difference will be noticeable. It reflects itself not only in the control variables but also in the frequency and tie-line power. Numerical values are given in the paper. The time behaviour of the response may not be of importance at present, but it will be if the security requirements are more stringent.

The problem of supporting the other areas as brought up by other discussers is a matter of control strategy and system security. Here, it was shown that the proposed scheme offers the possibility of giving support to the neighbour with no loss of stability, since the optimal control scheme yields stable solutions. If LFC is implemented on a process computer, as is the modern trend, the system should not be more expensive.

F. C. Schweppe
(*Massachusetts Institute of Technology, Cambridge, Massachusetts, U.S.A.*)

In the U.S.A. there is a time delay (about 2 sec) between the measurement of the tie-line power and the moment when this signal reaches the power plant to make a change. Do you have a similar effect in Europe and how would such a time delay affect your results?

H. Glavitsch

Professor Schweppe touches upon some hardware differences of LFC systems in the U.S.A. and Europe. At present in Europe continuous power and frequency measurements are taken. With the introduction of process computers a sampling effect will be encountered and the time delay will be noticeable at the beginning of the response. Hence, some deterioration in the results is to be expected, unless it is compensated by stronger control action. However, the principal concepts are not affected by a sampled system.

H. A. Nour Eldin (*Eidg. Technische Hochschule, Zürich, Switzerland*)

I would like to raise the following points.

(1) The optimal frequency time and power time curves show a higher frequency oscillation (about 8 times the basic frequency of the system). Why were these frequencies not compensated through the Riccati feedback?

(2) A continuous observer and a continuous feedback were used. It may be reasonable to use a discrete state estimator and discrete feedback to facilitate local digital control.

F. D. Galiana

The points raised by Dr. Nour Eldin may be answered in the following way.

(1) There are two basic reasons for this.

(a) The weighting matrices could be chosen such that the large frequency ripples are attenuated but this requires very large control gains, an impossibility in a thermal unit with limiters.

(b) One is mainly interested in alternating the larger and longer transients. The smaller ripples are not critical.

(2) This will have some advantages, but it remains to be investigated.

J. L. DINELEY (*University of Newcastle upon Tyne, England*)
For a disturbance on the system which affects many machines, what would be the effect of OED on the capability of each machine to contribute to dealing with the disturbance?

H. GLAVITSCH
The question of Professor Dineley can be answered by referring to the model in use. Better use of each machine can be made if its properties and limits are reflected in the model and if the participation coefficient is adjusted accordingly. In making use of energy storage and the knowledge of limits of the machines their capabilities are utilized in an improved manner.

J. EHLERT-KNUDSEN (*NESA, Hellerup, Denmark*)
First I want to compliment you for an excellent paper presented in a very elegant form. You stated at the beginning that you have limited your investigations to a purely thermal system. In the Scandinavian system we are gradually moving from a predominantly hydro to a mixed hydro–thermal system with a large amount of nuclear units being installed in the seventies. It would be of great interest to know if you have extended, or intend to extend, your investigations to a mixed system, where you have fast response from the hydro units and relatively slow response from the thermal units.

Do you think that your OED system might bring advantages in a mixed hydro–thermal system, where we have experienced line drop-outs due to a fast response from the hydro parts when disturbances occur in the thermal system, with a resulting increase in the drop of frequency and thus a more severe disturbance than expected?

H. GLAVITSCH
The problem raised by Mr. Ehlert-Knudsen may be treated by the concept explained in our paper. The main point is that the model has to take areas of predominantly hydro and thermal generation into account. The theory of estimation and optimal control may then be easily extended to several areas where response characteristics are different. Critical tie-lines between these areas can be controlled individually.

W. B. JERVIS (*Merz and McLellan, Newcastle upon Tyne, England*)
Would the authors agree that the scheme proposed in their paper is primarily an error adaptive type of control system?

Regarding the momentary load-frequency control aspect of using the stored steam energy, I consider this is a useful hidden bonus of which few, if any, authorities have yet taken full advantage. Possibly the reason for this is the difficulties experienced with coal-fired boilers where "follow-up" response of their controls may be relatively slow in action. Nevertheless, if a sufficient number of machines are programmed to utilize the stored energy, this can play an important part in controlling the short-time energy balance of an electrical network.

Would the authors comment on whether or not they consider it necessary for any special arrangement to be made to obtain faster opening times of the steam valves than at present, bearing in mind that at no time must water be pulled over from the boiler?

Regarding the practical aspects of implementing the proposed adaptive controls to existing machines employing mechanical flyball governors, personally I see no serious objection to retaining the existing governor and introducing finite adaptive controls to the hydraulic amplifier with flyball action override should speed rise, say, on load rejection become a major control action.

I would be grateful for the authors' comments on the observations here.

H. GLAVITSCH

The scheme is essentially an error adaptive scheme as pointed out by Mr. Jervis.

The most stringent constraints in the response of the units stem from the boiler and from the reheater, so the question of decreasing the opening times of steam valves as raised by Mr. Jervis has to be checked in the light of these conditions. This applies specifically to large disturbances, *e.g.* a three-phase short circuit. There, shorter opening times would be desirable. For small disturbances there is apparently no need for a change of the valve characteristics as far as LFC is concerned. However, for the steady state stability of the unit a different reasoning applies. This brings us to Mr. Jervis's second question. Uprating a flyball governor may be perfectly satisfactory for LFC purposes, neglecting the larger dead-band etc. For other tasks, however, the flyball governor may not be practical any more.

A. H. GLATTFELDER (*Eidg. Technische Hochschule, Zürich, Switzerland*)

No transients have been discussed for "class III disturbances", where saturations on boiler and turbine dynamics are dominant. Does the OED approach also result in better performances or do you believe a different approach will be necessary?

Are the controller gains set up by using precalculations only, or are they automatically adjusted (self-adaptation)?

How much do these gains differ for class I to III disturbances? And do these differences not cause unwanted discontinuities, *i.e.* jumps on the control signal, especially for non-step disturbances?

H. GLAVITSCH

Class III disturbances as referred to by Dr. Glattfelder cannot, in our opinion, be attacked by the OED approach since the units have to be used up to their limits and the control loops become open. It is not the control gains but the unit output which has to be predetermined. With these values precalculated it is a matter of detecting the class and putting them into effect. In this way discontinuities will certainly be initiated since they are required to achieve the optimum response. Non-step disturbances will probably not fall into class III and for them drastic changes are not required.

Power System Dynamic Energy Balance

K. NEIL STANTON

Systems Control, Inc., Palo Alto, California, U.S.A.

SUMMARY

An interconnected power system is divided into areas and a center of angle is associated with each area. The dynamic energy balance technique previously described by the author will be reviewed and extended to the multi-area case. With careful choice of areas, the multi-area dynamic energy balance simulation displays the motion of centers of angle and defines the important features of system dynamic behavior without introducing synchronizing oscillations between every pair of generators. The technique has applications in both planning and operation. Its use in transient security monitoring and simulation of load-frequency control is discussed in this paper.

1. INTRODUCTION

Dynamic energy balance is developing as a flexible approach to power system dynamic simulation. The advantages of the technique for study of power frequency transients were discussed in an earlier paper[1]. Work has continued and it is now clear that dynamic energy balance supplements transient stability studies for a broader range of problems than was claimed in ref. 1. It is safe to predict that single-area dynamic energy balance will run at 30 times the speed of a transient stability study. Multi-area dynamic energy balance will vary from the speed of the single-area case to the same speed as transient stability depending on the number of areas used. Of particular significance in this presentation is the usefulness of dynamic energy balance in the area of power system control.

This paper will introduce dynamic energy balance by reviewing the single-area case previously described in ref. 1. The methodology will then be extended to the multi-area case. A comparison between dynamic energy balance simulation and transient stability studies is presented for a 23-generator example. Applications to transient security monitoring and load-frequency control are discussed.

Reference p. 163

2. REVIEW OF SINGLE-AREA DYNAMIC ENERGY BALANCE

Generator rotor motion can be expressed as the sum of movement with respect to the system's center of inertia plus motion of the center of inertia with respect to a synchronously rotating reference.

For the rotor of generator j,

$$\theta_j = \phi_j + \theta_0 \tag{1}$$

θ_0 is the position of the system center of inertia and is referred to as the system center of angle. A velocity ω_0 and acceleration $\dot{\omega}_0$ are also associated with the system center of inertia. Motion of the system center of inertia can be expressed in terms of individual rotor angles using eqns. (2):

$$\left.\begin{aligned} \theta_0 &= \frac{1}{M_T} \sum_{j=1}^{n} M_j \theta_j \\ M_T &= \sum_{j=1}^{n} M_j \end{aligned}\right\} \tag{2}$$

These equations can be regarded as the definition of the center of angle. Differentiation yields similar equations for ω_0 and $\dot{\omega}_0$.

The objective of dynamic energy balance is to simulate the behavior of the center of inertia without performing a transient stability study to determine the behavior of individual rotors. The equations of motion for the rotor of generator j are

$$\left.\begin{aligned} M_j \dot{\omega}_j &= P_{mj} - P_{ej} \\ \dot{\theta}_j &= \omega_j - \omega_s \end{aligned}\right\} \tag{3}$$

A transient stability study integrates these equations forward in time to determine rotor motion. The equations of motion of the system center of inertia are obtained by adding the equations of motion for the n generators in the system:

$$\sum_{j=1}^{n} M_j \dot{\omega}_j = \sum_{j=1}^{n} P_{mj} - \sum_{j=1}^{n} P_{ej}$$

Defining

$$P_{mT} = \sum_{j=1}^{n} P_{mj}$$

$$P_{eT} = \sum_{j=1}^{n} P_{ej}$$

and using eqns. (2)

$$\left.\begin{aligned} M_T \dot{\omega}_0 &= P_{mT} - P_{eT} \\ \dot{\theta}_0 &= \omega_0 - \omega_s \end{aligned}\right\} \tag{4}$$

The single-area dynamic energy balance simulation integrates these equations forward in time to determine the trajectory of the system center of inertia.

Turbine inputs P_{mj}, $j = 1, 2, \ldots, n$, are defined symbolically by eqn. (5):

$$\dot{P}_{mj} = G_j(P_{mj}, \omega_j) \tag{5}$$

Models of any degree of complexity can be handled by adding extra equations and variables but eqn. (5) will serve the purpose of explanation.

Equation (5) combines more conveniently with eqns. (4) if ω_j is replaced by ω_0. The approximation is not serious and can be corrected when particular applications justify the added complexity.

$$\dot{P}_{mj} = G_j(P_{mj}, \omega_0) \tag{6}$$

Equations (4) and (6) describe the dynamic behavior of the system center of inertia. P_{eT} is the only input to these equations and can be interpreted as either the total electrical power developed by the generators or the sum of loads and losses (including generator stator loss).

Comparison of eqns. (3) and (4) shows why it is advantageous to work with (4). Mechanical inputs change slowly because of turbine and prime mover dynamics. Since the electrical output of an individual generator can vary rapidly during a transient, integration of eqns. (3) requires a small step size. However, P_{eT}, which is the total network load plus losses, changes slowly and therefore allows eqns. (4) to be integrated using a much larger step size than is needed for (3). This integration proceeds as follows.

Given knowledge of P_{eT} at time t, eqns. (4) and (6) are integrated forward one time step by treating P_{eT} as constant or by using some extrapolation of past behavior. Values of ω_0, θ_0 and P_{mT} at $t + \Delta t$ are obtained but the value of P_{eT} at $t + \Delta t$ must be checked to be sure that its assumed behavior is consistent with the results of integration before proceeding with the next integration step. A load flow is needed to perform this check.

Since $\dot{\omega}_j = \ddot{\theta}_j = \ddot{\phi}_j + \ddot{\theta}_0$, eqns. (3) can be written as

$$M_j(\ddot{\phi}_j + \ddot{\theta}_0) = P_{mj} - P_{ej}$$

Power transfers from one rotor to another do not significantly affect P_{eT}. (There is a minor effect since the transfer may increase losses.) Thus, it is reasonable to assume such transfers are zero. This means that no generator rotor can be accelerating or decelerating relative to another. It follows therefore that no rotor can be accelerating or decelerating relative to the center of inertia, *i.e.*

$$\ddot{\phi}_j = 0$$

Hence, generator j must be producing electrical power as indicated by eqn. (7):

$$P_{ej} = P_{mj} - M_j\ddot{\theta}_0 \tag{7}$$

The problem remains that $\ddot{\theta}_0$ is not known at $t + \Delta t$ and cannot be determined until $P_{eT}(t + \Delta t)$ is known. However, $P_{eT}(t + \Delta t)$ depends on the $P_{ej}(t + \Delta t)$ which can be computed from eqn. (7) only if $\ddot{\theta}_0(t + \Delta t)$ is specified. An iterative approach is obviously necessary at this point.

A value is assumed for $\ddot{\theta}_0(t + \Delta t)$ and eqn. (7) is used to compute the electrical power produced by each generator. The excitation system and generator electrical representation produces an equation constraining the generator internal voltage (see Appendix). Load buses are represented as they would be in a normal load flow solution (any combination of constant P, Q, constant current and power factor, or constant impedence loads). The angle of the slack bus is specified and the slack generator power will be determined from the load flow solution. Solution of the load flow is by a Newton–Raphson procedure using triangular factorization and programmed to take advantage of sparsity. The load flow solution is used as follows to obtain a better estimate of $\ddot{\theta}_0(t + \Delta t)$.

Before performing the load flow, the electrical power produced by the slack generator is computed from eqn. (7) as

$$P_{es} = P_{ms} - M_s\ddot{\theta}_0$$

However, P_{es} is left undefined in the load flow input so that the slack generator can provide the power needed to balance total network input against loads plus losses. Let P_s be the slack generator electrical power as computed by the load flow. If P_s differs from P_{es}, the slack generator will not have the same acceleration as other generators and this is contrary to the energy balance procedure. However, the load flow does provide a new value of P_{eT} that can be used to improve the accuracy of $\ddot{\theta}_0(t + \Delta t)$.

$$P_{eT} = \sum_{j \neq s} P_{ej} + P_s \tag{8}$$

$$P_{eT}^{\text{new}} = P_{eT}^{\text{old}} + P_s - P_{es}$$

$$\ddot{\theta}_0^{\text{new}}(t + \Delta t) = \ddot{\theta}_0^{\text{old}}(t + \Delta t) + \frac{1}{M_T}(P_s - P_{es})$$

Substituting in eqn. (7):

$$P_{ej}^{\text{new}}(t + \Delta t) = P_{mj}(t + \Delta t) - M_j\ddot{\theta}_0^{\text{old}}(t + \Delta t) + \frac{M_j}{M_T}(P_s - P_{es})$$

$$P_{ej}^{\text{new}}(t + \Delta t) = P_{ej}^{\text{old}}(t + \Delta t) + \frac{M_j}{M_T}(P_s - P_{es}) \tag{9}$$

Equation (9) gives new generator electrical powers which can be used as inputs to a new load flow solution. This iterative process is repeated until $P_s - P_{es}$ is less than a specified tolerance. At this stage of the computation, $P_{eT}(t + \Delta t)$, $\ddot{\theta}_0(t + \Delta t)$ as well as bus angles and voltage magnitudes have been correctedly determined. The position and velocity of the center of inertia are known at $t + \Delta t$ and the necessary information is available to integrate forward one more time step.

3. MULTI-AREA DYNAMIC ENERGY BALANCE

The single-area dynamic energy balance is well suited to study of power-frequency transients where the integrity of the system is not a problem. However, interconnected power systems frequently contain areas which are tightly coupled internally but are less strongly coupled to neighboring areas. In such circumstances it is convenient to look at the relative motion of areas rather than deal with synchronizing oscillations for a large number of generating units. Many of these oscillations will be of small amplitude and will pose no threat to the system, yet they may increase computational cost by limiting integration step size. A single-area study would eliminate valuable information on the behavior between areas. Multi-area dynamic energy balance selectively reduces the system's degrees of freedom to remove unimportant synchronizing oscillations from the simulation. Ideally, results are simpler to interpret. Computing will frequently be more economical than for a transient stability study but this is problem dependent and comparisons will be clearer after explanation of the method.

Each bus is assigned to an area although this is only important for buses where generators or inertia loads are connected. Synchronizing oscillations are avoided between generators in the same area by allowing rotors to move instantaneously to an equilibrium position relative to other generators in the area.

The center of angle for area K is defined by eqns. (10):

$$\left.\begin{aligned} \theta_{AK} &= \frac{1}{M_{AK}} \sum_{j \in A_K} M_j \theta_j \\ M_{AK} &= \sum_{j \in A_K} M_j \end{aligned}\right\} \tag{10}$$

Generator rotor position can be expressed in terms of an angle measured with respect to the center of angle for the area in which the generator is located,

$$\theta_j = \phi_{jK} + \theta_{AK} \quad \text{for} \quad j \in A_K \tag{11}$$

Area K internal equilibrium obtains when $\ddot{\phi}_{jK}$ is zero for all generator rotors in area K.

The objective of multi-area dynamic energy balance is to simulate the behavior of area centers of inertia without performing a transient stability to determine the position of individual rotors. The equation of motion of area K is obtained by adding the equations of motion for the generators located in area K:

$$\sum_{j \in A_K} M_j \dot{\omega}_j = \sum_{j \in A_K} P_{mj} - \sum_{j \in A_K} P_{ej}$$

Putting

$$P_{mAK} = \sum_{j \in A_K} P_{mj}$$

$$P_{eAK} = \sum_{j \in A_K} P_{ej}$$

and using eqns. (10)

$$\left.\begin{aligned} M_{AK}\dot{\omega}_{AK} &= P_{mAK} - P_{eAK} \\ \dot{\theta}_{AK} &= \omega_{AK} - \omega_s \end{aligned}\right\} \tag{12}$$

Turbine mechanical inputs are defined by eqn. (5) where ω_j is replaced by ω_{AK}:

$$\dot{P}_{mj} = G_j(P_{mj}, \omega_{AK}) \qquad j \in A_K \tag{13}$$

Equations (12) and (13) describe the dynamic behavior of the center of inertia for area K. Similar equations apply for other areas.

P_{eAK} is the only input to these equations and can be interpreted as either the total electrical power developed by generators in area K or the sum of loads and losses in area K plus net interchange to other areas. Unlike single-area energy balance, P_{eAK} depends on the angles between areas and can change significantly during an integration step. The variation in interchange between areas is likely to be the limiting factor in choosing the integration time step.

Given knowledge of P_{eAK} at time t, the area equations of motion can be integrated forward one time step Δt by treating P_{eAK} as constant or by using an extrapolation of past behavior. Integration produces estimates of $\omega_{AK}(t + \Delta t)$, $\theta_{AK}(t + \Delta t)$ and $P_{mj}(t + \Delta t)$. $P_{eAK}(t + \Delta t)$ must be computed to prepare for the next integration step and to check the accuracy of the assumed behavior of P_{eAK} over the interval t to $t + \Delta t$.

3.1 *Area electrical power*

The angles $\phi_{jK}(t + \Delta t)$ are initially set equal to $\phi_{jK}(t)$.

$$\begin{aligned} \theta_j^{(1)}(t + \Delta t) &= \phi_{jK}(t + \Delta t) + \theta_{AK}(t + \Delta t) \\ &= \phi_{jK}(t) + \theta_{AK}(t + \Delta t) \\ &= \theta_j(t) - \theta_{AK}(t) + \theta_{AK}(t + \Delta t) \end{aligned}$$

$$\theta_j^{(1)}(t + \Delta t) = \theta_j(t) + \Delta\theta_{AK}(t) \tag{14}$$

Thus, all buses in area K are advanced $\Delta\theta_{AK}(t)$, all buses in area m are advanced $\Delta\theta_{Am}(t)$, and so on.

A load flow is performed using the following inputs. The rotor angle θ and voltage e_q' are specified at all generator internal buses. (There is no slack bus.) The loads are specified at all load buses. This is the same type of load flow that is used in a transient stability program. It will be called LF 1. Generator electrical powers determined from the solution of LF 1 are used to compute $P_{eAK}^{(1)}(t + \Delta t)$. The first estimate of area acceleration at $t + \Delta t$ is

$$\dot{\omega}_{AK}^{(1)}(t + \Delta t) = \frac{1}{M_{AK}}[P_{mAK}(t + \Delta t) - P_{eAK}^{(1)}(t + \Delta t)] \tag{15}$$

3.2 Area internal equilibrium

Individual generator accelerations will differ from the area acceleration. However, internal equilibrium for an area is obtained when each rotor has the same acceleration as the center of inertia for its area. The objective is to position rotors to achieve internal equilibrium without moving the area centers of angle. The procedure is similar to the one used for single-area dynamic energy balance.

A "smoothed" power output is computed for each generator in area K using eqn. (16):

$$P_{sj}(t + \Delta t) = P_{mj}(t + \Delta t) - M_j\dot{\omega}_{AK}(t + \Delta t) \tag{16}$$

This is the electrical power developed by generator j if its acceleration is equal to the area acceleration. Smoothed powers are similarly computed for generators in other areas. For generator i in area m,

$$P_{si}(t + \Delta t) = P_{mi}(t + \Delta t) - M_i\dot{\omega}_{Am}(t + \Delta t)$$

A second load flow is performed using smoothed electrical powers P_{sj} and voltage e'_q at all generator internal buses except the slack generator where θ_s and e'_{qs} are specified. The loads are specified at all load buses. This is the same type of load flow that was used in the single-area dynamic energy balance. It will be called LF 2. The slack generator output initially computed by LF 2 will not agree with the value predicted by eqn. (16). The difference is distributed to all generators in the same area as the slack generator using the procedure shown in eqn. (9) and LF 2 is repeated. (Distributing the error to all generators in the system leads to the same end result.) This procedure converges as was explained in the single-area case. At this point, area internal equilibrium has been achieved but the power interchange between areas has probably been altered in the process. The output of the load flow gives rotor angles measured with respect to the slack generator. These angles must be related in some way to angles measured with respect to area centers of angle.

3.3 Transformation of angles

Let ρ_j be the angle of bus j measured with respect to the slack generator. By definition,

$$\theta_{AK} = \frac{1}{M_{AK}} \sum_{j \in A_K} M_j\theta_j$$

$$\rho_j = \theta_j - \theta_s$$

$$\theta_{AK} = \frac{1}{M_{AK}} \sum_{j \in A_K} M_j(\rho_j + \theta_s)$$

$$\theta_{AK} = \frac{1}{M_{AK}} \sum_{j \in A_K} M_j\rho_j + \theta_s$$

$$\phi_{iK} = \theta_i - \theta_{AK}$$

$$\phi_{iK} = \rho_i + \theta_s - \frac{1}{M_{AK}} \sum_{j \in A_K} M_j \rho_j - \theta_s$$

$$\phi_{iK} = \rho_i - \frac{1}{M_{AK}} \sum_{j \in A_K} M_j \rho_j \tag{17}$$

This is the necessary relationship between ρ and ϕ.

3.4 Corrected value of area electrical power

Recall that the angles $\phi_{jK}(t + \Delta t)$ were initially set equal to $\phi_{jK}(t)$. The results of LF 2 can now be used to obtain a better estimate of area electrical powers. Angles from LF 2 are transformed by eqn. (17) to give $\phi_{jK}(t + \Delta t)$. Then,

$$\theta_j^{(2)}(t + \Delta t) = \phi_{jK}(t + \Delta t) + \theta_{AK}(t + \Delta t)$$

It will be shown that this update procedure does not change the area centers of angle.

$$\sum_{j \in A_K} M_j \phi_{jK} = \sum_{j \in A_K} M_j \theta_j - \sum_{j \in A_K} M_j \theta_{AK} = 0$$

$$\begin{aligned}
\theta_{AK}^{\text{new}}(t + \Delta t) &= \frac{1}{M_{AK}} \sum_{j \in A_K} M_j \theta_j^{(2)}(t + \Delta t) \\
&= \frac{1}{M_{AK}} \sum_{j \in A_K} M_j \phi_{jK}(t + \Delta t) + \frac{1}{M_{AK}} \sum_{j \in A_K} M_j \theta_{AK}^{\text{old}}(t + \Delta t) \\
&= \theta_{AK}^{\text{old}}(t + \Delta t)
\end{aligned}$$

LF 1 is repeated using the updated rotor angles. New generator electrical powers are determined and used to compute a second estimate of $P_{eAK}(t + \Delta t)$. The second estimate of area acceleration is

$$\dot{\omega}_{AK}^{(2)}(t + \Delta t) = \frac{1}{M_{AK}} [P_{mAK}(t + \Delta t) - P_{eAK}^{(2)}(t + \Delta t)] \tag{15}$$

The accelerations of individual generator rotors in area K are now closer to $\dot{\omega}_{AK}$ than was true for the first estimate. Should the differences in accelerations still be unacceptable, new smoothed electrical powers are computed for each generator, LF 2 is repeated, and angles are updated once more. This process converges when the accelerations of all generators within an area differ from the area acceleration by no more than a specified tolerance. At this point, relative rotor positions within each area are such as to produce internal equilibrium but the area centers of angle are still fixed at the values determined by integrating the equations of motion. The acceleration of each area has been determined along with additional information needed to integrate the area equations of motion forward over the next time step.

4. COMPARISON OF DYNAMIC ENERGY BALANCE AND TRANSIENT STABILITY STUDIES

Testing of dynamic energy balance techniques has moved through several levels of development. Reference 1 described single-area results obtained with a demonstration program having 50-generator, 100-bus capability. This program has been extended to handle multi-area dynamic energy balance using the technique described in §3 of this paper. The result is a dynamic simulation capability with considerable flexibility. If all generators are assigned to one area, a single-area dynamic energy balance simulation is the result. If every generator is assigned to its own area, a transient stability study is the result. If the number of areas exceeds one but is less than the number of generators, a multi-area dynamic energy balance is the result. A comparison has been carried out between:

(1) a single-area dynamic energy balance simulation,
(2) a three-area dynamic energy balance simulation, and
(3) a transient stability study,

all applied to the same 23-generator, 99-bus example. The disturbance is loss of a large load at the high voltage bus of generator B. The system was heavily loaded and lightly damped and the network did not contain obvious weak links between strong subsystems. This made it difficult to define areas for the multi-area case. The example may not be the most suitable to demonstrate multi-area dynamic energy balance but it does represent a good practical test of the technique.

4.1 Single-area simulation

Some results from the single-area dynamic energy balance simulation are shown in Figs. 1, 2 and 3. There are no synchronizing oscillations, yet the average behavior of both generator angles and electrical powers is shown clearly and accurately. Since all generators have the same frequency in the single-area study, only one frequency can be plotted in Fig. 3. These results were obtained using a 12-cycle integration step size and required approximately 1 min of computation on the UNIVAC 1108.

4.2 Three-area simulation

After some experimentation, thirteen generators were assigned to area 1, nine to area 2 and one to area 3. Figures 4, 5 and 6 show results for direct comparison with the single-area case.

Figure 4 shows relative angles with generator B as reference and Fig. 5 shows electrical powers. Generators A and B are in area 2, generators C, E and F are in area 1 and D is the one generator in area 3. Areas 1 and 2 oscillate against each other at 1 cycle per second. Area 3 oscillates at about 1.1 cycles per second. Figure 6 shows frequency for each of the three areas. The correspondence between one- and three-area cases is excellent.

The three-area results were obtained using a three-cycle step size and required almost 9 min of computation on the UNIVAC 1108.

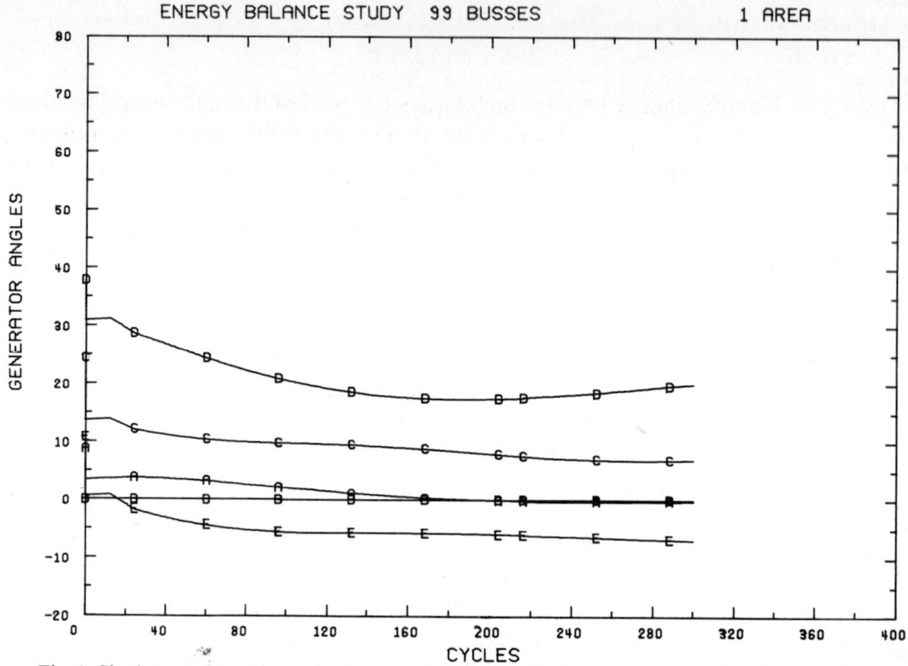

Fig. 1. Single-area dynamic energy balance. Relative angles in degrees *versus* time in cycles for generators A, B, C, D and E. B is reference.

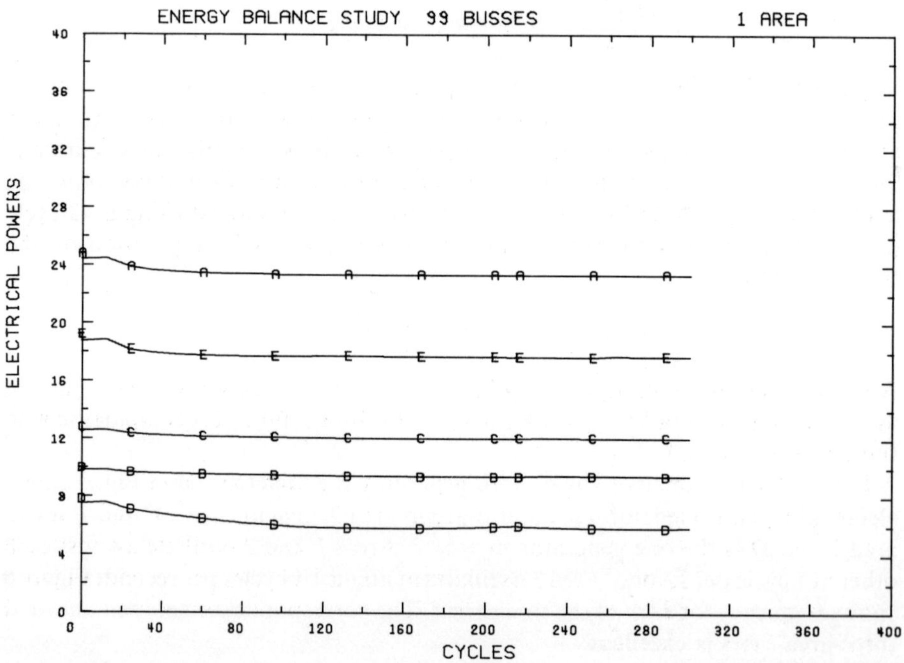

Fig. 2. Single-area dynamic energy balance. Electrical power in per unit (100 MVA base) *versus* time in cycles.

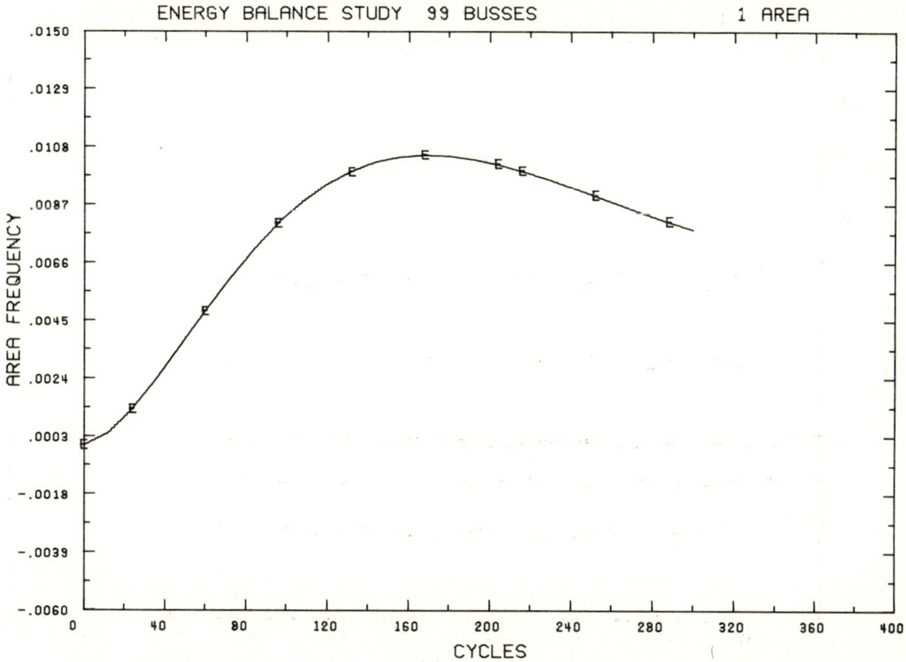

Fig. 3. Single-area dynamic energy balance. Frequency in per unit *versus* time in cycles.

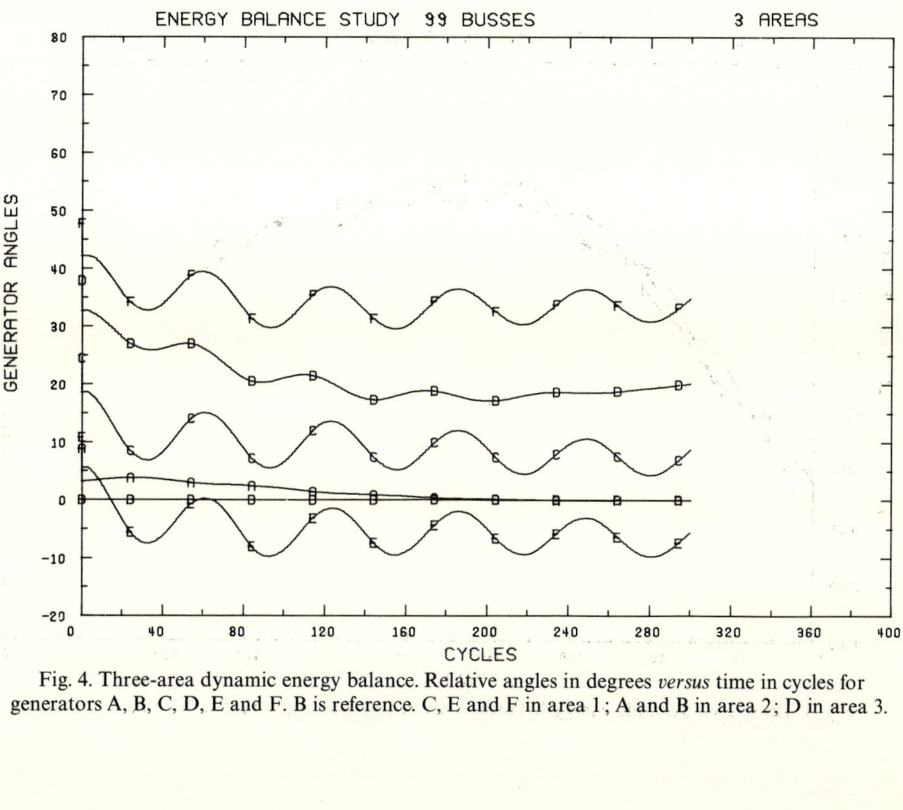

Fig. 4. Three-area dynamic energy balance. Relative angles in degrees *versus* time in cycles for generators A, B, C, D, E and F. B is reference. C, E and F in area 1; A and B in area 2; D in area 3.

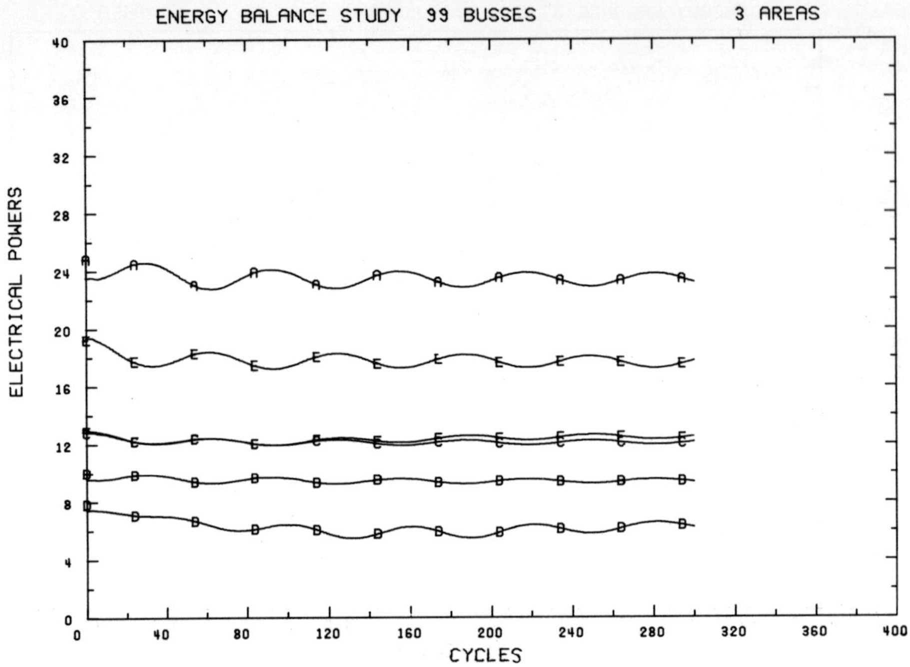

Fig. 5. Three-area dynamic energy balance. Electrical powers in per unit *versus* time in cycles.

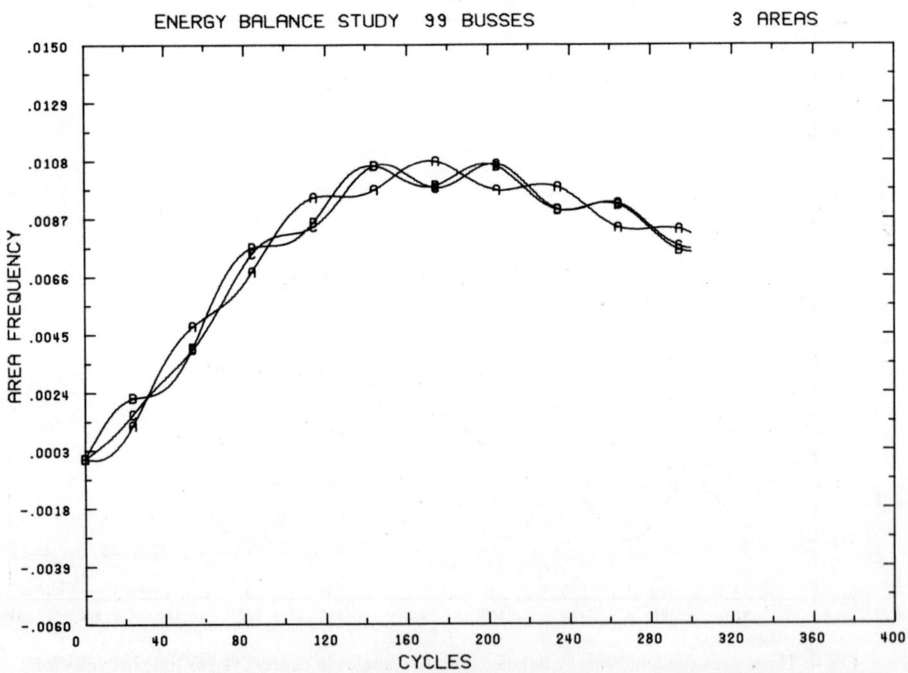

Fig. 6. Three-area dynamic energy balance. Area frequencies in per unit *versus* time in cycles.

4.3 Transient stability study

Figure 7 shows electrical powers for generators A, B, D and E as obtained by a production transient stability program. This provides an independent result against which performance of the multi-area dynamic energy balance can be measured. The broken curves shown in Fig. 7 are results from the three-area dynamic energy balance replotted for convenient comparison.

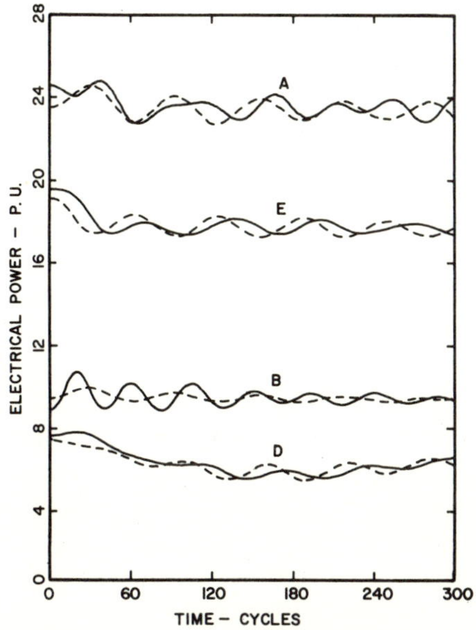

Fig. 7. Comparison of three-area dynamic energy balance with transient stability results.
——————— transient stability
- - - - - - - - - - dynamic energy balance

Including generators A and B in the same area eliminates the synchronizing oscillations between these units from the three-area study. The transient stability study shows that the disturbance, which is located at the high voltage bus of B, causes B to oscillate against A at a frequency of 1.3 cycles per second. Dynamic energy balance allows areas to reach internal equilibrium immediately after the disturbance. As a result all generators in area 2 share in dropping load and the effect on B is less pronounced.

Generator D is treated as a separate area in dynamic energy balance. Comparison with transient stability shows that it oscillates at the same frequency in both studies but there is a phase difference between the two results. This is caused by the instantaneous repositioning of rotors in dynamic energy balance which starts the transient of generator D with different initial conditions from those used in transient stability.

Generator E is the largest plant in area 1. The three-area dynamic energy balance gives the correct frequency and correspondence with transient stability is excellent except for a phase shift discussed in the previous paragraph.

Transient stability shows more complex behavior for generator A since it is oscillating against generator B at 1.3 cycles per second and against the generators in area 1 at 1 cycle per second. Since dynamic energy balance has eliminated the 1.3 cycle per second oscillation, it gives simpler behavior for generator A. The comparison is informative and shows the potential of dynamic energy balance in understanding the complex dynamic behavior of interconnected power systems.

The transient stability study used a two-cycle step size and required approximately 6 min of central processor time on a CDC 6400 computer. This same study required $4\frac{1}{2}$ min on the UNIVAC 1108 using a two-cycle step size with the transient stability option provided in the multi-area dynamic energy balance program.

4.4 Comments

The system used for the foregoing demonstration of multi-area dynamic energy balance is part of an actual power system. It does not break conveniently into subsystems, is heavily loaded and lightly damped and was chosen to demonstrate the method in a difficult situation. Comparison of the three-area dynamic energy balance simulation with the 23-generator transient stability study shows the potential of the method.

However, multi-area dynamic energy balance is still in the early stage of development and some important problems remain. Firstly, it is not easy to choose areas. The results shown here were obtained only after trying four different methods of assigning generators to areas. The transient stability study was quite helpful in this process. Multi-area dynamic energy balance results are more lightly damped than is indicated by transient stability. (This may be a function of the system used in this study.) An unsatisfactory choice of areas can produce unstable results, which illustrates why a systematic procedure for choosing areas is so important. Ideally, the procedure should avoid use of transient stability studies in defining areas.

Secondly, the multi-area dynamic energy balance reported in this paper runs more slowly than the transient stability study. Iteration around a loop involving two load flows at the end of each integration step is not an efficient way to write the program. However, the method is straightforward and can be tested by convenient modification of the single-area dynamic energy balance program. A much more efficient technique involving only one load flow has been developed but involves programming new load flow routines. When implemented, this program will have roughly the same amount of computation per time step for both multi-area dynamic energy balance and transient stability studies. The computational advantage of the dynamic energy balance will depend on the increase in step size that it makes possible.

5. APPLICATIONS IN OPERATION AND CONTROL

5.1 Transient security monitoring

Economical computation is essential for any dynamic simulation procedure used in system operation. The best course in an application such as transient security monitoring would avoid simulation altogether. Should this be impossible it can be assumed that adequate off-line studies will be performed to reduce the dimensionality of models and streamline simulation procedures to the maximum degree possible. Under these circumstances, areas can be defined very carefully by off-line studies and excitation system and prime mover models simplified. Multi-area dynamic energy balance has greater potential for development in this application than the "all-digital" transient stability approach.

5.2 Simulation of load-frequency control

The dynamic energy balance technique is ideally suited to simulation of load-frequency controls. The system is normally controlled on an area basis and the perturbations of interest are not large enough to cause loss of synchronism. Thus, the areas for the dynamic energy balance study are chosen to correspond with the areas on which load-frequency control is based. The additional equations corresponding to the controls can be added to the study. The system representation retains non-linearities and preserves the identity of the major system components, including individual transmission lines. Relaying, nonconforming loads, supplementary controls, security constraints and measurement errors can be handled in the modelling without incurring the excessive computing costs that limit the use of transient stability studies in this application.

It is expected that load-frequency control will eventually be extended to perform usefully during abnormal operating conditions. The dynamic energy balance study is expected to aid this development.

5.3 Dynamic equivalents

The need for "dynamic equivalents" has been a recurring problem for a number of years. The multi-area dynamic energy balance aggregates the equations of motion for a number of generators into an area equation of motion. At present, the network and prime mover representations are left in unreduced form. However, it appears that some aggregation of prime mover models can be performed within areas. It is possible to delete network representation entirely in areas which are strong internally and sufficiently far removed from a disturbance to have very little affect on results.

Multi-area dynamic energy balance is a flexible approach to the need for dynamic equivalents. Generators close to a disturbance are represented on a one generator per area basis; beyond this, generators are grouped into areas with no aggregation of prime movers and excitation systems; further yet from the disturbance, areas have aggregated prime mover and excitation system models; finally, network representation is deleted and remote areas are treated as single bus.

6. CONCLUSIONS

The multi-area dynamic energy balance has been described and its performance has been demonstrated for a 23-generator, 99-bus example using a program that allows any number of areas to be selected. When one area is used, results match those previously reported in ref. 1. By using one generator per area, the program produces a transient stability study. A three-area dynamic energy balance study compared well with an independently prepared transient stability study.

The program used in this study was not efficient but it has demonstrated the potential of multi-area dynamic energy balance. An efficient procedure has been developed and will be described as soon as practical results and definite computing effort has been determined.

NOTATION

θ rotor angle measured with respect to a synchronously rotating reference
M angular momentum at synchronous speed
P_m prime mover power in per unit
P_e generator electrical power in per unit
ω rotor angular velocity expressed as a fraction of synchronous speed
ω_s synchronous speed ($\omega_s = 1.0$)
E generator internal voltage phasor
V bus voltage phasor
I generator current phasor
e'_q voltage proportional to field flux linkages
e_q voltage proportional to field current
V_{FD} generator field voltage
i_d generator direct axis current
T_{do} generator field open circuit time constant
L_d generator direct axis synchronous inductance
L'_d generator direct axis transient inductance
L_q generator quadrature axis synchronous inductance
r generator stator resistance
ρ angles measured with respect to the slack generator rotor

Single-area study

θ_0 angle of the system center of inertia measured with respect to a synchronously rotating reference
ϕ angles measured with respect to the center of inertia
ω_0 angular velocity of the system center of inertia
M_T total system angular momentum

Multi-area study

θ_{AK} angle of area K center of inertia measured with respect to a synchronously rotating reference

ϕ_{jK} angle of bus j with respect to the center of inertia of area K, $j \in A_K$
ω_{AK} angular velocity of the center of inertia for area K
M_{AK} total angular momentum for area K

APPENDIX

GENERATOR REPRESENTATION

Generator representation has been simplified by eliminating damper windings and other subtransient effects. Using the conventional d–q representation with only one rotor winding results in the following equations:

$$\dot{e}'_q = \frac{1}{T_{do}}(V_{FD} - e_q) \tag{A-1}$$

$$e_q = e'_q + \omega(L_d - L'_d)i_d \tag{A-2}$$

$$|E| = e'_q + \omega(L_q - L'_d)i_d \tag{A-3}$$

$$E = V + (r + j\omega L_q)I \tag{A-4}$$

THE EXCITATION SYSTEM

The field voltage V_{FD} is specified by excitation system equations. The usual representations can be used in energy balance. A simplified representation retaining only the exciter time constant, derivative feedback and exciter saturation can be written symbolically as

$$\dot{V}_{FD} = F(V_{FD}, |V|) \tag{A-5}$$

Given knowledge of the initial conditions plus projections of $|V|$ and e_q, eqns. (A-1) and (A-5) can be integrated forward one time step. This integration must be carried out for each generator. The new values of e'_q set up the necessary constraint on the magnitude of E via eqn. (A-3).

REFERENCE

1 K. NEIL STANTON, Dynamic energy balance studies for simulation of power-frequency transients, *Proc. Power Industry Computer Applications Conference, Boston, Mass., May, 1971.*

Discussion

G. QUAZZA (*ENEL, Milan, Italy*)
I wish to congratulate the author for his interesting contributions to the analysis of electromechanical transients of multi-area systems, and previously, of multi-machine areas. The progress from the classical simple approximation of describing

the dynamics of each area by a single equivalent machine—and consequently being unable to distinguish among tie-line powers within the single "exported power"—and the simplification of the cumbersome and too detailed system descriptions in standard transient stability programs are very considerable.

However, the validity of the approach proposed by the author is perhaps strongly conditioned by a suitable division of the system areas. If I have correctly interpreted one of the author's basic statements in his mathematical formulation, the assumption is made that all machines in one area can be regarded as instantaneously following the motion of their centre of inertia, as a short cut to describe the over-all transient. Does this imply that the frequencies of their relative oscillations are supposed to be considerably higher than the frequency of the oscillations between areas? If so, did not the author find rather serious difficulties in establishing when these frequencies were separated sufficiently, and consequently in selecting the best system division, as has happened in similar studies at ENEL? Could the author kindly explain how the choice of areas is made?

K. N. STANTON

Dr. Quazza has identified one of the more difficult aspects of this project, namely, that of selecting areas in a satisfactory way. The dynamic energy balance study is applicable for any selection of areas but results may not be realistic if areas are poorly chosen. All generators can be assigned to one area if desired. At the other extreme, each generator can be assigned to its own area, and the multi-area dynamic energy balance study then corresponds to a transient stability study.

The selection of areas for the example presented in the paper was done experimentally using transient stability studies as a guide. Several selections were tried before converging on the areas used in the paper. I do not think this is very sophisticated and a more systematic procedure for selecting areas is a worthwhile subject for further investigation. I would like to see a procedure that avoids the need for transient stability studies.

The selection of areas also depends on the location of the disturbance. I can foresee that in transient security monitoring, different disturbances will be analyzed ahead of time to determine the best choice of areas for each transient contingency on the contingency list. In the interest of reducing on-line or real-time computation to an absolute minimum, it would be feasible to spend considerable time off-line performing transient stability studies or other needed analysis to determine the most effective choice of areas for each contingency.

H. GLAVITSCH (*Brown, Boveri & Co. Ltd., Baden, Switzerland*)

Could you help me to clarify your concept of energy balance? In your single-area energy balance, you made the assumption that the acceleration is zero during all the steps. This means that the frequency is constant but not necessarily equal in the various steps. In the multi-area energy balance, you do not assume that the acceleration of the individual generators is zero but that it is made equal to the area acceleration. In this way you get the relative motion of the areas as shown in your comparison. Is this correct?

K. N. STANTON

In single-area dynamic energy balance, there is one center of inertia and each generator is forced to have the same angular acceleration as the system center of inertia. It is the *relative* angular accelerations that are assumed to be zero. This means that generator rotors may move with respect to each other with constant relative angular velocity during an integration step. Angles do change but this is significant only from the standpoint of monitoring the integrity of the network.

In multi-area dynamic energy balance, each generator is forced to have the same acceleration as the center of inertia for the area in which the generator is located. Relative acceleration of generators within the same area is zero. The complicating factor is the need to compute area interchange, since this couples the behavior of angles directly into the simulation procedure. Thus, the multi-area dynamic energy balance contains an iterative procedure which forces generators to have the same acceleration, but keeps the area centers of angle in the position determined by integration of the area equations of motion. Generator rotors are allowed to move to new positions relative to the area center of angle during the load flow that follows each integration step.

H. GLAVITSCH (*Brown, Boveri & Co. Ltd., Baden, Switzerland*)

In comparing results of the power energy balance with transient stability results the question arises: How does your method work if one generator pulls out of step?

K. N. STANTON

If dealing with a case where one generator pulls out of step, that generator should be assigned to an area by itself.

H. GLAVITSCH (*Brown, Boveri & Co. Ltd., Baden, Switzerland*)

What is the computational gain of your energy balance method?

K. N. STANTON

The single-area dynamic energy balance was running ten times as fast as a production transient stability program. Further improvements are now apparent, and I feel confident in predicting a 30:1 computational advantage over transient stability studies. The computational gain of the multi-area dynamic energy balance depends on the method of defining areas. The use of one area per generator leads to the same computational requirements as transient stability. With skillful elimination of fast synchronizing oscillations through choice of areas, definite gains can be achieved through the use of larger integration step size.

G. HOSEMANN (*Technische Hochschule, Darmstadt, Germany*)

Do you usually assume constant active and reactive powers for the load representation?

K. N. STANTON

Load representation in dynamic energy balance is similar to that used in transient stability studies and suffers from the same uncertainties. The present dynamic energy balance program includes any combination of the following:

constant power and reactive loads,
constant current and power factor loads,
constant impedance loads and
frequency-sensitive loads.

A. H. El-Abiad (*Purdue University, Lafayette, U.S.A.*)
I think it is unfair to compare the results of a transient stability study with the dynamic energy balance. The purpose of a transient stability study is to see what happens to the angles within 1–3 sec after the disturbance has occurred. The main purpose of dynamic energy balance would be to look for the average frequency of each area as the time extends beyond that period, perhaps up to several minutes. In transient stability studies, one is interested in the synchronizing powers between the machines which are limited by transmission capabilities. In the dynamic energy balance, the main interest is in the frequency response and the effect of the change in frequency upon plant and load performance, which depend on adequate representation of the dynamics of energy sources. Synchronizing power oscillations are eliminated in the dynamic energy balance model in order to be able to use larger integration steps. However, ignoring the synchronous oscillations necessitates that one assumes that there is no transient instability, which can only be assured by a transient stability study following each large disturbance. In other words, transient stability and dynamic energy balance models are both needed and one does not eliminate the need of the other.

K. N. Stanton
The multi-area dynamic energy balance study can be used as a transient stability study; it can also be used to study power-frequency transients. It all depends on how the areas are defined in the input to the study. I agree that there are two classes of study but the one program now includes both within the scope of its capabilities.

U. G. Knight (*CEGB, London, England*)
The suggestion has been made that suitable groups for dynamic energy balance assessments can be determined from off-line studies. At the present state of the art, a similar off-line requirement is inherent in the exact economic dispatch models described in the paper by Brameller and Lo (p. 81): outage conditions cannot be included in the exact solution and must be translated into limiting flows for the normal state analysed.

There are potential risks and difficulties, however, both here and in the energy balance assessments described by Dr. Stanton. The greatest need for computer aids in the security field arises when the system is least like its normal state, but it is at such times that guidance from predictive studies is least likely to be appropriate. Whilst accepting the value of off-line studies, I would also urge research into analytical methods of group selection.

K. N. Stanton
I agree with Dr. Knight concerning the limitations of off-line studies and the need for analytical methods for area selection. We have been investigating ways to determine the right areas quickly so that the monitoring might proceed in two

steps: a fast algorithm to select areas, followed by a very simplified analysis on an area-type model. The feasibility of this approach to transient security monitoring is not yet determined. There are other approaches. Hybrid methods have their advocates. There are steady state restraints which are quite effective in minimizing the likelihood of transient security problems. However, the use of area-type models with dynamic energy balance or with methods such as transient security indices deserves serious consideration. We should also keep in mind that selection of a suitable contingency list is a requirement in all of the so-called "preventive mode" security schemes, both steady state and transient. If contingencies have been selected, it is certainly reasonable to perform preparatory analysis.

CLOSURE BY K. N. STANTON

The preceding discussion has emphasized the transient security monitoring problem. I would like to point out that the multi-area dynamic energy balance is also useful for simulation of load-frequency control systems. It is a non-linear simulation, it allows individual line flows to be monitored, relaying can be simulated, and effects of measurement errors and incorrect operator inputs can be studied. This is a potentially important application of both single- and multi-area dynamic energy balance.

Stability Analysis and Control of Power Systems

M. MANSOUR

Lehrstuhl für Automatik, Eidg. Technische Hochschule, Zürich, Switzerland

SUMMARY

In this paper, the important results achieved so far in the stability analysis of power systems, using the second method of Liapunov, and in the stabilization of power systems, using optimal control theory, are summarized. The energy metric algorithm[1] for the generation of Liapunov functions is applied to the two-machine and the multi-machine systems. The Liapunov function thus derived is general, in the sense that almost all previous results can be obtained as special cases. A method for stabilizing the two-machine system using digital control and capacitor switching is presented, and compared with other methods. Suggestions for further investigations are given.

1. INTRODUCTION

During the last 6 years many papers[2–16] have been published which apply the second method of Liapunov to the transient stability analysis of power systems. Various methods which have been used for the generation of the Liapunov functions are:

Trial and error method—energy considerations[2–4,6,8,15]
Variable gradient method[5,9]
Cartwright method (generalization from the linear to the non-linear case)[11]
Aizerman method[11]
Popov method (Liapunov function corresponding to Popov criterion)[10,13,14]
Zubov method[7,16]
Method of integration by parts[9]
Method of Infante and Clark[9]

Other methods for the generation of Liapunov functions which may also be applied to the same problem are:

Auxiliary exact differential equations method
Ingwerson method
Szego method
Format method
Energy metric algorithm[1]

It will be shown in §2 that the last method, with some generalization, will yield

References pp. 187–188

Liapunov functions which are general in the sense that they generate almost all Liapunov functions that have been derived before.

Another problem dealt with in this paper is the stabilization of power systems using optimal control theory. Improving the transient stability using capacitor switching was considered in refs. 17 and 18. In refs. 19 and 20 the maximum principle was used to get time optimal control using capacitor switching. The same maximum principle was used in ref. 21 to get time optimal control using the excitation voltage as the control variable. In §3 a suboptimal solution using digital control with state variable feedback and capacitor switching will be considered.

In §4 the method of stabilization of a power system using state variable feedback and quadratic performance index is considered[22,23]. Suggestions for using digital control to solve this problem are proposed.

2. STABILITY STUDY USING THE SECOND METHOD OF LIAPUNOV

To apply the second method of Liapunov, the system equations must first be put in the state variable form. The Liapunov function $V(\mathbf{x})$ obtained by any of the above methods must satisfy the following conditions:

(1) $V(\mathbf{x})$ and its first partial derivatives are continuous;

(2) $V(\mathbf{x})$ is positive definite in a region around the equilibrium point;

(3) $\dot{V}$ is negative definite in a region around the equilibrium point, or negative semidefinite in a region around the equilibrium point and not identically equal to zero on a solution of the system other than the equilibrium point.

The region of asymptotic stability is defined by the largest closed curve $V = V_c$ in the domain where $\dot{V} \leq 0$. It can be shown that the curve $V = V_c$ need not be closed if the phase variable representation is used.

2.1 Two-machine system

The system under consideration consists of a synchronous generator connected to an infinite busbar through a transmission line. The voltage of the infinite busbar is constant while the machine voltage, being proportional to the field flux linkage E_q', is variable during the transient period due to the field decay. All resistances are neglected. Their inclusion will not impose any major difficulty.

The system can be described by the equation[24]

$$M\ddot{\delta} + D(\delta)\dot{\delta} = P_m(\dot{\delta}) - P_e(E_q', \delta) \tag{1}$$

The variable damping $D(\delta)$ is given by

$$D(\delta) = E^2\left(\frac{x_d' - x_d''}{(x_e + x_d')^2} T_{do}'' \sin^2 \delta + \frac{x_q' - x_q''}{(x_e + x_q')^2} T_{qo}'' \cos^2 \delta\right)$$

$$= a_1 \sin^2 \delta + a_2 \cos^2 \delta \tag{2}$$

The electrical power output P_e considering transient saliency is given by

$$P_e(E'_q, \delta) = \frac{E'_q E}{(x_e + x'_d)} \sin\delta - \frac{E^2(x_q - x'_d)}{2(x_e + x_q)(x_e + x'_d)} \sin\delta$$

$$= b \sin\delta - c \sin 2\delta \tag{3}$$

The equation for the flux decay is

$$\frac{\mathrm{d}E'_q}{\mathrm{d}t} = \frac{E_{ex}}{T'_{do}} - \frac{E'_q}{T'_d} + \frac{E(x_d - x'_d)}{(x_e + x'_d)T'_{do}} \cos\delta = \frac{E_{ex}}{T'_{do}} - \eta_1 E'_q + \eta_2 \cos\delta \tag{4}$$

The governor action can be approximated by an equivalent first-order system as follows:

$$T_e \frac{\mathrm{d}P_m}{\mathrm{d}t} + P_m = P_{m0} - k\dot{\delta}$$

$$\frac{\mathrm{d}P_m}{\mathrm{d}t} = -\gamma_1(P_m - P_{m0}) - \gamma_2\dot{\delta} \tag{5}$$

2.1.1 System equations

Considering the origin as the point of stable equilibrium the following state variables can be used:

$$\begin{aligned} x_1 &= \delta - \delta_0 \\ x_2 &= \dot{\delta} \\ x_3 &= E'_q - E'_{q0} \\ x_4 &= P_m - P_{m0} \end{aligned} \tag{6}$$

Equations (1)–(6) can be reduced to the following system of equations:

$$\begin{aligned} \dot{x}_1 &= x_2 \\ M\dot{x}_2 &= -[a_1 \sin^2(x_1 + \delta_0) + a_2 \cos^2(x_1 + \delta_0)]x_2 - b[\sin(x_1 + \delta_0) - \sin\delta_0] \\ &\quad + c[\sin(2x_1 + 2\delta_0) - \sin 2\delta_0] - g x_3 \sin(x_1 + \delta_0) + x_4 \\ \dot{x}_3 &= -\eta_1 x_3 - \eta_2[\cos\delta_0 - \cos(x_1 + \delta_0)] \\ \dot{x}_4 &= -\gamma_1 x_2 - \gamma_2 x_4 \end{aligned} \tag{7}$$

2.1.2 The energy metric algorithm for generating Liapunov functions[1]

The method is based upon a generalization of the total energy of the system and is obtained from the line integral of a composite differential one-form derived directly from the system equations. The procedure is as follows.

(a) Describe the system as a set of first-order differential equations.

(b) Form a set of differential one-forms by eliminating dt from every two equations.

(c) By addition and substitution these differential one-forms are reduced to a single one-form.

(d) The Liapunov function is obtained by line integration.

To apply this algorithm it is necessary to modify eqns. (7). The third equation can be replaced by

$$\dot{x}'_3 = -\eta_1 x'_3 - \eta_2 x_2 \sin(x_1 + \delta_0) \tag{8}$$

where $x'_3 = \dot{x}_3$.

The differential one-forms are

$$\begin{aligned}
\alpha_1 = {} & Mx_2\,dx_2 + \{[a_1 \sin^2(x_1 + \delta_0) + a_2 \cos^2(x_1 + \delta_0)]x_2 + b[\sin(x_1 + \delta_0) \\
& - \sin\delta_0] - c[\sin(2x_1 + 2\delta_0) - \sin 2\delta_0] + gx_3 \sin(x_1 + \delta_0) \\
& + x_4\}\,dx_1 = 0
\end{aligned}$$

$$\begin{aligned}
\alpha_2 = {} & \{-[a_1 \sin^2(x_1 + \delta_0) + a_2 \cos^2(x_1 + \delta_0)]x_2 - b[\sin(x_1 + \delta_0) - \sin\delta_0] \\
& + c[\sin(2x_1 + 2\delta_0) - \sin 2\delta_0] - gx_3 \sin(x_1 + \delta_0) + x_4\}dx'_3 \\
& + \{\eta_1 x'_3 + \eta_2 x_2 \sin(x_1 + \delta_0)\}M dx_2 = 0
\end{aligned}$$

$$\alpha_3 = x_2\,dx'_3 + \{\eta_1 x'_3 + \eta_2 x_2 \sin(x_1 + \delta_0)\}dx_1 = 0$$

$$\begin{aligned}
\alpha_4 = {} & \{-[a_1 \sin^2(x_1 + \delta_0) + a_2 \cos^2(x_1 + \delta_0)]x_2 - b[\sin(x_1 + \delta_0) - \sin\delta_0] \\
& + c[\sin(2x_1 + 2\delta_0) - \sin 2\delta_0] - gx_3 \sin(x_1 + \delta_0) + x_4\}dx_4 \\
& + \{\gamma_1 x_2 + \gamma_2 x_4\}M dx_2 = 0
\end{aligned}$$

$$\alpha_5 = x_2\,dx_4 + \{\gamma_1 x_2 + \gamma_2 x_4\}dx_1 = 0$$

$$\alpha_6 = \{-\eta_1 x'_3 - \eta_2 x_2 \sin(x_1 + \delta_0)\}dx_4 + \{\gamma_1 x_2 + \gamma_2 x_4\}dx'_3 = 0 \tag{9}$$

The Liapunov function can be obtained by the line integral

$$V = \int \left(\alpha_1 + \frac{1}{\eta_2 \sin(x_1 + \delta_0)}\alpha_2 + \frac{1}{\gamma_1}\alpha_4 + \alpha_6 \right)$$

with the substitutions

$$dx_3 = -\left(\frac{\eta_1 x'_3}{x_2} + \eta_2 \sin(x_1 + \delta_0) \right) dx_1 \qquad \text{from } \alpha_3$$

$$dx_4 = -\left(\gamma_1 + \gamma_2 \frac{x_4}{x_2} \right) dx_1 \qquad \text{from } \alpha_5$$

$$dx'_3 = -\eta_1\,dx_3 - \eta_2 \sin(x_1 + \delta_0)x_2\,dx_1 \qquad \text{from eqns. (7) and (8)}$$

The redundant and indeterminate terms are neglected. The Liapunov function thus obtained will be

$$\begin{aligned}
V = {} & \tfrac{1}{2}Mx_2^2 + b[\cos\delta_0 - \cos(x_1 + \delta_0) - x_1 \sin\delta_0] \\
& - \frac{c}{2}[\cos 2\delta_0 - \cos(2x_1 - 2\delta_0) - 2x_1 \sin 2\delta_0] \\
& + gx_3[\cos\delta_0 - \cos(x_1 + \delta_0)] + \tfrac{1}{2}g\frac{\eta_1}{\eta_2}x_3^2 + \frac{1}{\gamma_1}x_4^2
\end{aligned} \tag{10}$$

$$\dot{V} = -[a_1 \sin^2(x_1 + \delta_0) + a_2 \cos^2(x_1 + \delta_0)]x_2^2 - \frac{g}{\eta_2}\dot{x}_3^2 - \frac{\gamma_2}{\gamma_1}x_4^2 \tag{11}$$

with $V > 0$ in a region near the origin and $\dot{V} \leq 0$ in the whole state space.

2.1.3 *Generalization*

To get a more general form for V complete the quadratic terms in eqn. (10):

$$V = \alpha x_1^2 + \beta x_1 x_2 + \tfrac{1}{2}Mx_2^2 + b[\cos\delta_0 - \cos(x_1 + \delta_0) - x_1 \sin\delta_0]$$
$$- \frac{c}{2}[\cos 2\delta_0 - \cos(2x_1 - 2\delta_0) - 2x_1 \sin 2\delta_0]$$
$$+ gx_3[\cos\delta_0 - \cos(x_1 + \delta_0)] + \tfrac{1}{2}g\frac{\eta_1}{\eta_2}x_3^2 + \frac{1}{2\gamma_1}x_4^2 \qquad (12)$$

The quadratic form in V is non-negative if $\alpha \geq \beta^2/2M$. In order to eliminate the terms in $x_1 x_2$ in $\dot{V}$ let

$$\alpha = \frac{\beta}{2M}[a_1 \sin^2(x_1 + \delta_0) + a_2 \cos^2(x_1 + \delta_0)] \qquad (13)$$

where $0 \leq \beta \leq \min(a_1, a_2)$.

$$\dot{V} = \frac{\beta x_1^2 x_2}{2M}(a_1 - a_2)\sin(2x_1 + 2\delta_0) - \frac{\beta x_1}{M}\{b[\sin(x_1 + \delta_0) - \sin\delta_0]$$
$$- c[\sin(2x_1 + 2\delta_0) - \sin 2\delta_0] - gx_3 \sin(x_1 + \delta_0) - x_4\}$$
$$- [a_1 \sin^2(x_1 + \delta_0) + a_2 \cos^2(x_1 + \delta_0) - \beta]x_2^2 - \frac{g}{\eta_2}\dot{x}_3^2 - \frac{\gamma_2}{\gamma_1}x_4^2 \qquad (14)$$

In order that V be positive definite and $\dot{V}$ negative semidefinite in a region around the origin, β should satisfy

$$0 \leq \beta \leq \min\left[a_1, a_2, 4M\eta_1\left(\frac{\eta_1(b\cos\delta_0 - 2c\cos\delta_0) - \eta_2 g \sin^2\delta_0}{\eta_2 g \sin^2\delta_0 + (\gamma_1/\gamma_2)\eta_1^2}\right)\right] \qquad (15)$$

2.1.4 *Determination of stability region*

To determine the stability region it is necessary to determine first the unstable equilibrium state closest to the stable equilibrium state. The surface $V = V_c$ passing through the unstable equilibrium state, together with the part of the plane perpendicular to the x_1-axis with negative x_2, will give a stability region. The union of all such regions for the admissible values of β is the required stability region, which is naturally smaller than the actual domain of asymptotic stability.

2.1.5 *Special cases*

2.1.5.1 Neglecting transient saliency. Assuming constant damping, constant field flux linkage and constant input power:

$$\begin{aligned} \dot{x}_1 &= x_2 \\ M\dot{x}_2 &= -ax_2 - b[\sin(x_1 + \delta_0) + \sin\delta_0] \end{aligned} \qquad (16)$$

The Liapunov function is

$$V = \frac{a\beta}{2M}x_1^2 + \beta x_1 x_2 + \tfrac{1}{2}Mx_2^2 + b[\cos\delta_0 - \cos(x_1 + \delta_0) - x_1 \sin\delta_0] \quad (17)$$

$$\dot{V} = -(a - \beta)x_2^2 - \frac{b\beta}{M}x_1[\sin(x_1 + \delta_0) - \sin\delta_0] \quad (18)$$

with $0 \leq \beta \leq a$. For $\beta = 0$ the results of refs. 2 and 6 are obtained. For $\beta \leq a$ one gets the results of refs. 10, 13 and 14.

Figure 1 shows the stability region which is the union of the regions obtained for the admissible values of β. The region is bounded by the vertical line through the unstable equilibrium point.

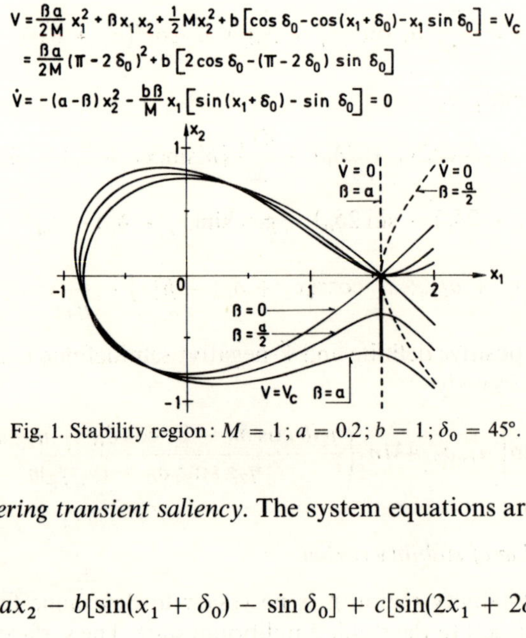

Fig. 1. Stability region: $M = 1$; $a = 0.2$; $b = 1$; $\delta_0 = 45°$.

2.1.5.2 Considering transient saliency. The system equations are

$$\begin{aligned} \dot{x}_1 &= x_2 \\ M\dot{x}_2 &= -ax_2 - b[\sin(x_1 + \delta_0) - \sin\delta_0] + c[\sin(2x_1 + 2\delta_0) - \sin 2\delta_0] \end{aligned} \quad (19)$$

The Liapunov function is

$$\begin{aligned} V = {} & \frac{a\beta}{2M}x_1^2 + \beta x_1 x_2 + \tfrac{1}{2}Mx_2^2 + b[\cos\delta_0 - \cos(x_1 + \delta_0) - x_1 \sin\delta_0] \\ & - \frac{c}{2}[\cos 2\delta_0 - \cos(2x_1 + 2\delta_0) - 2x_1 \sin 2\delta_0] \end{aligned} \quad (20)$$

$$\begin{aligned} \dot{V} = {} & -(a - \beta)x_2^2 - \frac{\beta x_1}{M}\{b[\sin(x_1 + \delta_0) - \sin\delta_0] \\ & - c[\sin(2x_1 + 2\delta_0) - \sin 2\delta_0]\} \end{aligned} \quad (21)$$

with $0 \leq \beta \leq a$. For $\beta = a/2$ and $\beta = a$ the results of ref. 11 are obtained. Figure 2 shows the stability region.

2.1.5.3 Considering transient saliency and variable damping. The system equations are

$$\dot{x}_1 = x_2$$
$$M\dot{x}_2 = -[a_1 \sin^2(x_1 + \delta_0) + a_2 \cos^2(x_1 + \delta_0)]x_2 - b[\sin(x_1 + \delta_0) - \sin \delta_0] + c[\sin(2x_1 + 2\delta_0) - \sin 2\delta_0] \quad (22)$$

The Liapunov function is

$$V = \frac{1}{2M}[a_1 \sin^2(x_1 + \delta_0) + a_2 \cos^2(x_1 + \delta_0)]\beta x_1^2 + \beta x_1 x_2 + \tfrac{1}{2}Mx_2^2 + b[\cos \delta_0 - \cos(x_1 + \delta_0) - x_1 \sin \delta_0] - \frac{c}{2}[\cos 2\delta_0 - \cos(2x_1 + 2\delta_0) - 2x_1 \sin 2\delta_0] \quad (23)$$

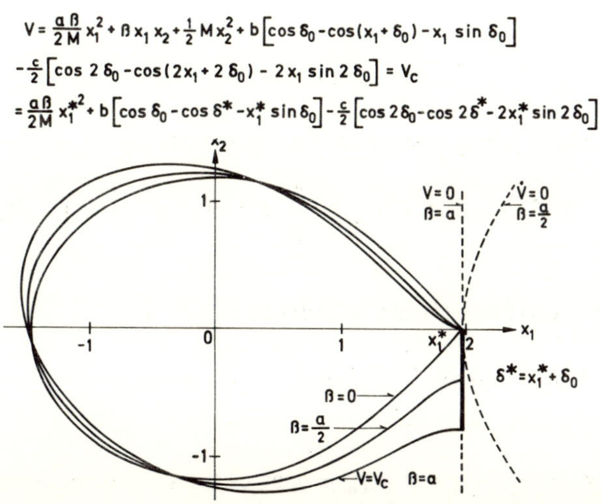

Fig. 2. Stability region considering transient saliency: $M = 1$; $a = 0.2$; $b = 1$; $c = 0.2$; $\delta_0 = 45°$.

$$\dot{V} = -\{[a_1 \sin^2(x_1 + \delta_0) + a_2 \cos^2(x_1 + \delta_0)] - \beta\}x_2^2 - \frac{\beta x_1}{M}\{b[\sin(x_1 + \delta_0) - \sin \delta_0] - c[\sin(2x_1 + 2\delta_0) - \sin 2\delta_0]\} + \frac{\beta x_1^2 x_2}{2M}(a_1 - a_2)\sin(2x_1 + 2\delta_0) \quad (24)$$

with $0 \leq \beta \leq \min[a_1, a_2]$. Figure 3 shows the stability region.

References pp. 187–188

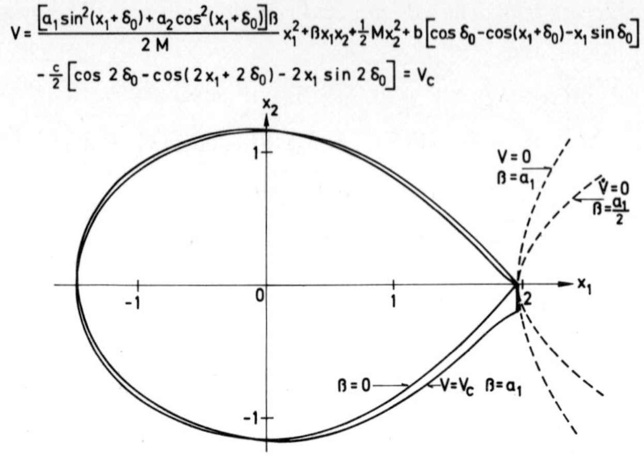

Fig. 3. Stability region considering transient saliency and variable damping: $M = 1$; $a_1 = 0.05$; $a_2 = 0.1$; $b = 1$; $c = 0.2$; $\delta_0 = 45°$.

2.2 *Multi-machine system*

The multi-machine power system can be represented by

$$M_i\delta_i + a_i\delta_i + E_i^2 G_{ii} + \sum_{\substack{j=1\\ j\neq i}}^{n} E_iE_jB_{ij}\sin(\delta_i - \delta_j) = P_{mi} \tag{25}$$

for $i = 1 \ldots n$. Equation (25) can be reduced to

$$M_i\ddot{\delta}_i + a_i\dot{\delta}_i + \sum_{\substack{j=1\\ j\neq i}}^{n} b_{ij}[\sin(\delta_i - \delta_j) - \sin(\delta_{i0} - \delta_{j0})] = 0 \tag{26}$$

Substituting $x_i = \delta_i - \delta_{i0}$ and $x_{i+n} = \dot{\delta}_i$ gives the system equations:

$$\dot{x}_i = x_{i+n}$$

$$M\dot{x}_{i+n} = -a_ix_{i+n} - \sum_{\substack{j=1\\ j\neq i}}^{n} b_{ij}[\sin(x_i - x_j + \delta_{i0} - \delta_{j0}) - \sin(\delta_{i0} - \delta_{j0})] \tag{27}$$

for $i = 1 \ldots n$. The differential one-forms are

$$M_ix_{i+n}\,\mathrm{d}x_{i+n} + \{a_ix_{i+n} + \sum_{\substack{j=1\\ j\neq i}}^{n} b_{ij}[\sin(x_i - x_j + \delta_{i0} - \delta_{j0})$$

$$- \sin(\delta_{i0} - \delta_{j0})]\}\mathrm{d}x_i = 0 \tag{28}$$

The Liapunov function is

$$V = \sum_{i=1}^{n} \tfrac{1}{2}M_i x_{i+n}^2 + \sum_{i=1}^{n-1} \sum_{j=i+1}^{n} b_{ij}[\cos(\delta_{i0} - \delta_{j0}) - \cos(x_i - x_j + \delta_{i0} - \delta_{j0})$$

$$- (x_i - x_j)\sin(\delta_{i0} - \delta_{j0})] \tag{29}$$

$$\dot{V} = - \sum_{i=1}^{n} a_i x_{i+n}^2 \tag{30}$$

2.2.1 *Generalization*

$$V = \beta[\sum_{i=1}^{n} (a_i x_i + M_i x_{i+n})]^2 + \sum_{i=1}^{n} \tfrac{1}{2}M_i x_{i+n}^2$$

$$+ \sum_{i=1}^{n-1} \sum_{j=i+1}^{n} b_{ij}[\cos(\delta_{i0} - \delta_{j0}) - \cos(x_i - x_j + \delta_{i0} - \delta_{j0})$$

$$- (x_i - x_j)\sin(\delta_{i0} - \delta_{j0})] \tag{31}$$

$$\dot{V} = - \sum_{i=1}^{n} a_i x_{i+n}^2 \tag{32}$$

for $\beta \geq 0$. β has a negative lower limit so that the quadratic form

$$\beta[\sum_{i=1}^{n} (a_i x_i + M_i x_{i+n})]^2 + \sum_{i=1}^{n} \tfrac{1}{2}M_i x_{i+n}^2 \geq 0 \tag{33}$$

It should be noted that the generalization in eqn. (31) corresponds to the generalization in eqn. (17) provided $\beta = a$ is used in eqn. (17).

2.2.2 *Determination of the stability region*

The procedure to determine the stability region is principally the same as for the two-machine system except that the determination of the unstable equilibrium state is much more complicated. Methods for its determination are given in refs. 3 and 15 but a satisfactory algorithm has not yet been found.

2.2.3 *Special cases*

$a_i = 0$:

$$\beta[\sum_{i=1}^{n} M_i x_{i+n}]^2 + \sum_{i=1}^{n} \tfrac{1}{2}M_i x_{i+n}^2 \geq 0$$

gives

$$\beta \geq - 1/(2 \sum_{i=1}^{n} M_i)$$

For $\beta = - 1/(2 \sum_{i=1}^{n} M_i)$ the result given in ref. 15 is obtained. For $\beta = 0$ the results of refs. 2 and 3 are obtained.

References pp. 187–188

3. TIME OPTIMAL CONTROL

3.1 Time optimal control using capacitor switching

The method of using series capacitors to stabilize power systems is very old. Recent investigations were published in refs. 17 and 18. Applying optimal control theory using series capacitors was recently reported in ref. 19. The maximum principle was applied to a simple power system model to bring the system to its final steady state in minimum time after a transient disturbance had occurred. The optimal control thus obtained is a bang-bang control of line reactance. Figure 4 shows the simple power system model.

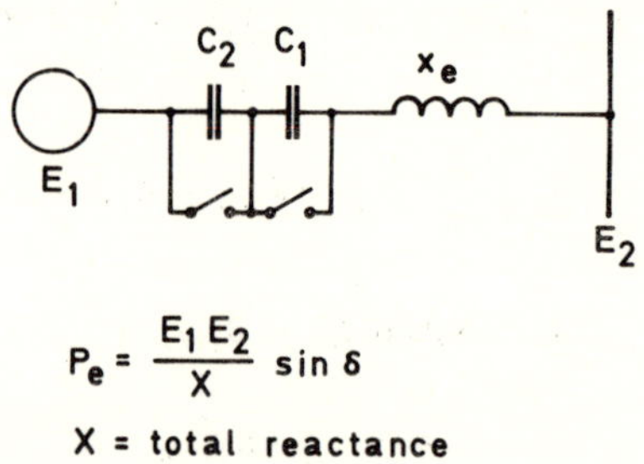

Fig. 4. Bang-bang control of reactance[19].

System equations:

$$\ddot{\delta} = -\frac{E_1E_2}{MX}\sin\delta + \frac{P_m}{M}$$

$$= -\frac{b_1}{X}\sin\delta + \frac{b_1}{X_0}\sin\delta_0$$

$$= -\frac{b_1}{X}\sin\delta + b_0\sin\delta_0$$

Substituting $x_1 = \delta - \delta_0$ and $x_2 = \dot{\delta}$ gives

$$\dot{x}_1 = x_2 \qquad (34)$$

$$\dot{x}_2 = -\frac{b_1}{X}\sin(x_1 + \delta_0) + b_0\sin\delta_0$$

Control variable:

$$u = 1/X \text{ for } u \leq u_m$$

Numerical values:

$$b_1 = b_0 = 41.888 \qquad (x_0 = 1)$$

$$\delta_0 = 0.5$$

Figures 5 and 6 show the response of the system without control of reactance, while Figs. 7 and 8 show the system response with bang-bang control of reactance. The reclosing time *RC* is taken as a parameter in each case. Figure 9 gives the critical reclosing time as a function of the final steady state load angle with the minimum reactance as parameter.

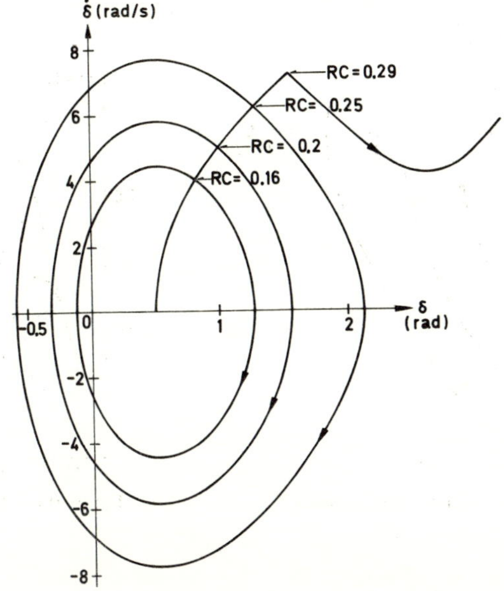

Fig. 5. Reclosing without condensers: phase plane.

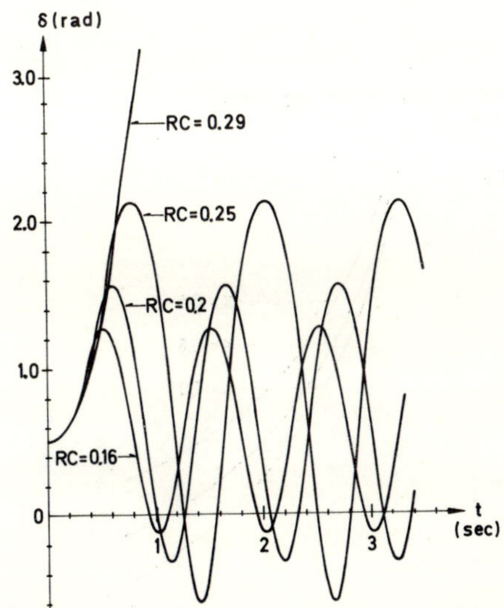

Fig. 6. Reclosing without condensers: time response.

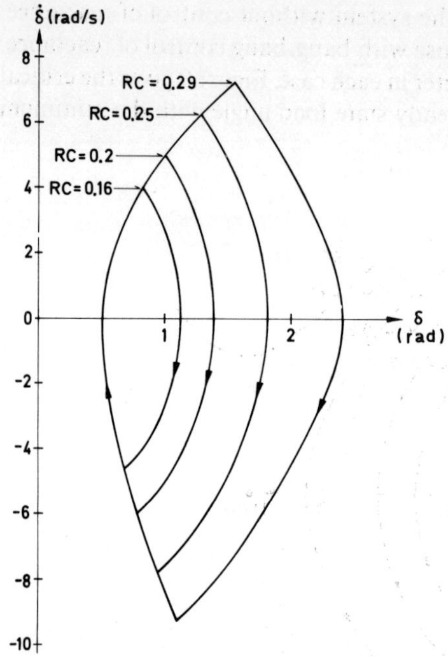

Fig. 7. Bang-bang control: phase plane.

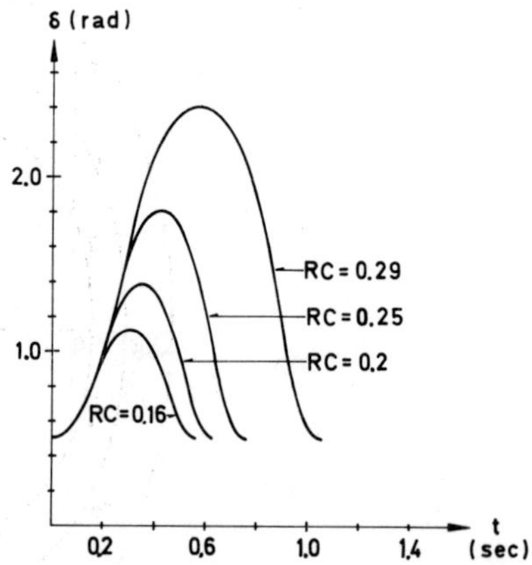

Fig. 8. Bang-bang control: time response.

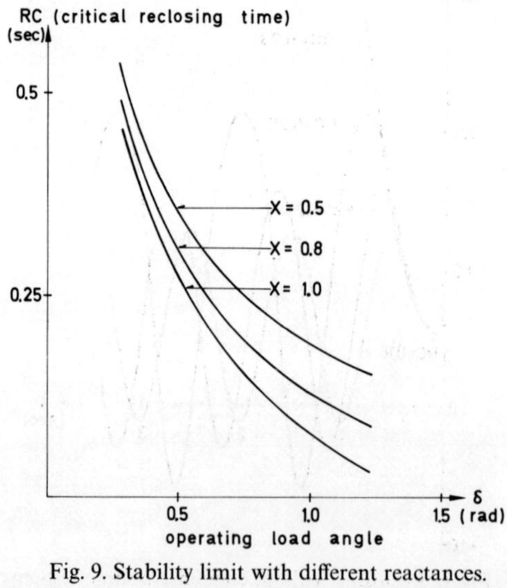

Fig. 9. Stability limit with different reactances.

3.2 *Time suboptimal digital control using capacitor switching*

Drawbacks of the method in §3.1 are the determination of the switching curve for different final steady state conditions as well as the open-loop control after the last capacitor switching. In this section it is hoped to avoid these drawbacks by using state variable feedback.

The system equations with damping are

$$\begin{aligned} \dot{x}_1 &= x_2 \\ \dot{x}_2 &= -ax_2 - \frac{b_1}{X}\sin(x_1 + \delta_0) + b_0 \sin\delta_0 \end{aligned} \tag{35}$$

The system (35) can be simplified by considering the last two terms in the second equation as the control variable u.

$$\begin{aligned} \dot{x}_1 &= x_2 \\ \dot{x}_2 &= -ax_2 + u \end{aligned} \tag{36}$$

The determination of the feedback coefficients for the time optimal digital control of the system (35) is well explained in ref. 25.

The values computed for the control variable u, which is a linear combination of the state variables, are then used to get the total reactance X'. As we can only use discrete values of reactances, the actual reactance is obtained according to a non-linear relation as shown in Fig. 10.

Figure 11 shows the hybrid computer program for the system and its digital control and Fig. 12 shows the implementation of the algorithm. The computer

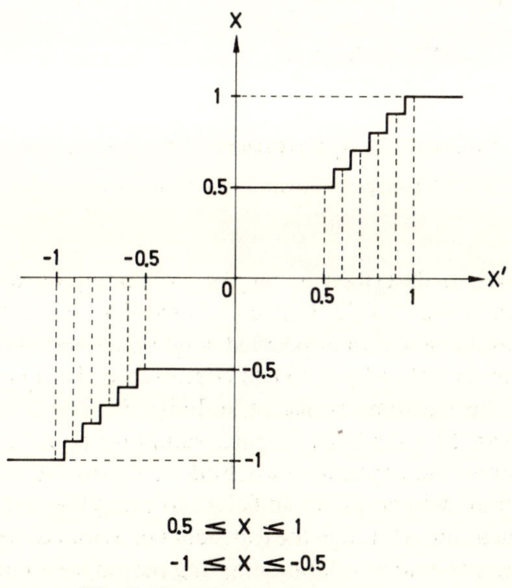

Fig. 10. Quantization of reactance.

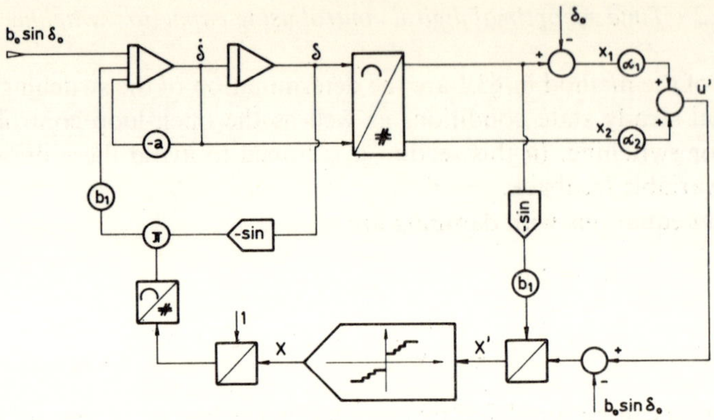

$$\dot{x}_1 = x_2$$
$$\dot{x}_2 = ax_2 - \frac{b_1}{X}\sin(x_1 + \delta_0) + b_0 \sin \delta_0$$

Fig. 11. Hybrid program.

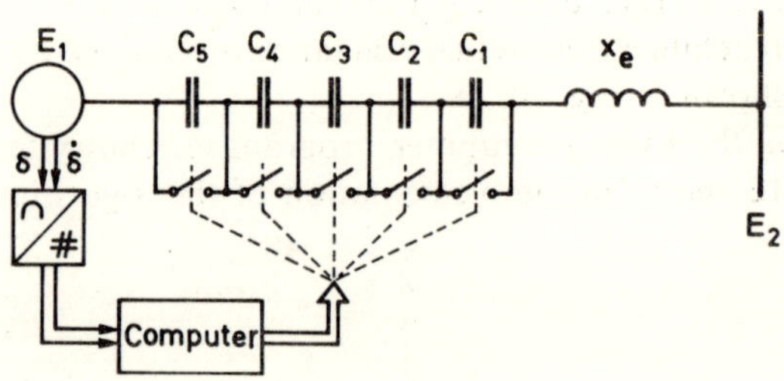

Values of $C_1 \ldots C_4$ correspond to the binary code

Fig. 12. Suboptimal control with capacitor switching.

results for the system described in ref. 19 with damping coefficient $a = 1$ and $\delta_0 = 0.5$ are given in Figs. 13–15. It is to be noted that the control should be ended when the state has come within a specified neighbourhood of the final steady state. This can be achieved either by a metric condition or by a condition making use of a measure of the real-time transient stability, *i.e.* $\xi = 1 - V/V_c$, as proposed in ref. 12. This can be calculated at each sampling point and a decision made whether to continue the capacitor switching or to stop it. In the opinion of the author three or four switchings are sufficient to bring the system sufficiently close to the equilibrium point. Although the transient behaviour depends on the sampling period, it is possible to find a suitable sampling period for all initial conditions and reclosing times.

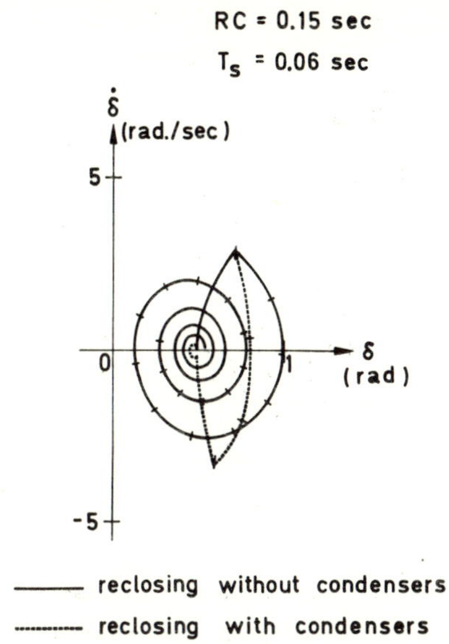

Fig. 13(a). Phase plane response: $RC = 0.15$ sec; $T_s = 0.06$ sec.

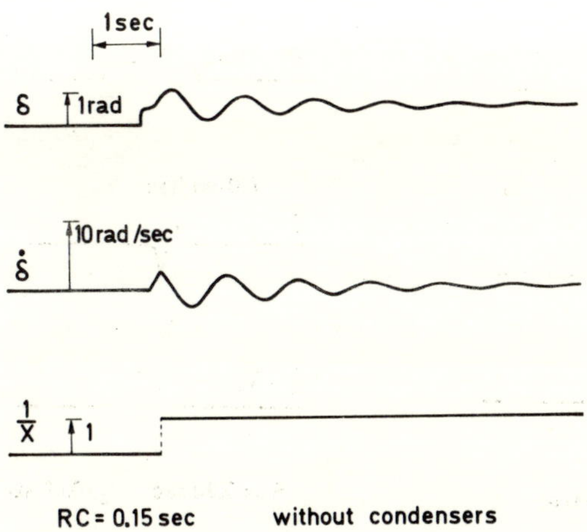

Fig. 13(b). Time response without condensers.

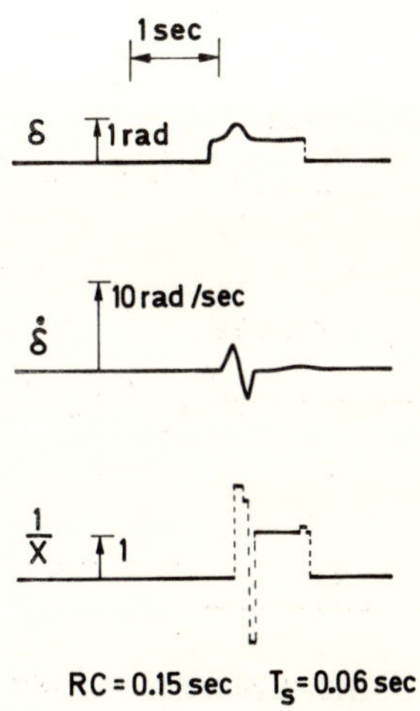

Fig. 13(c). Time response with condensers.

References pp. 187–188

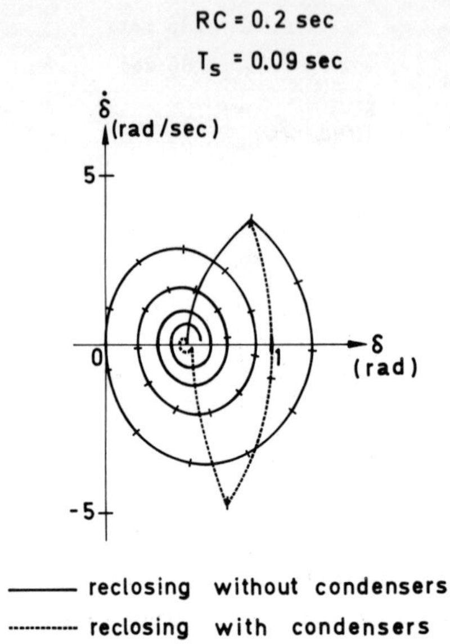

Fig. 14(a). Phase plane response: $RC = 0.2$ sec; $T_s = 0.09$ sec.

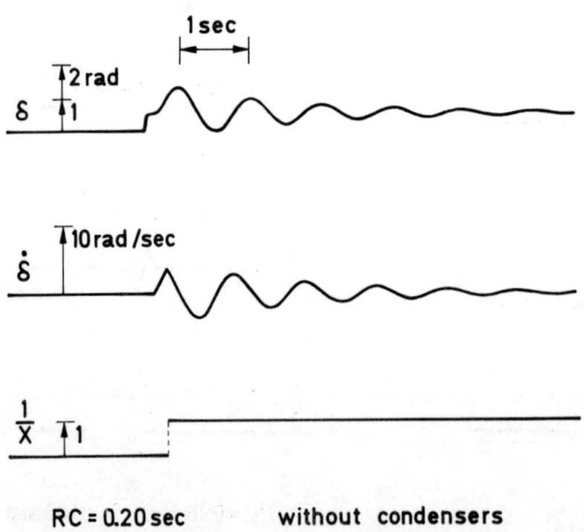

Fig. 14(b). Time response without condensers.

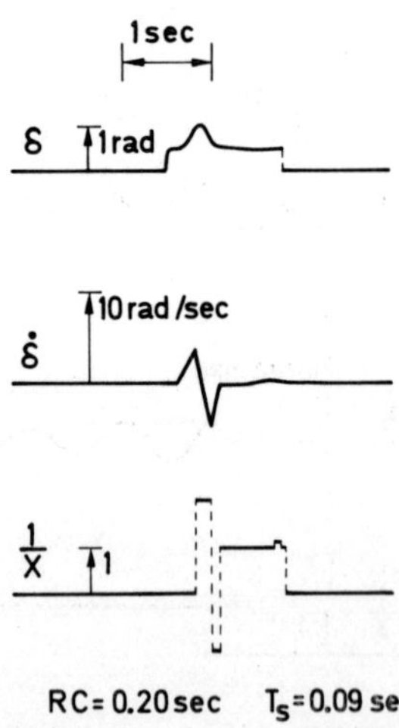

Fig. 14(c). Time response with condensers.

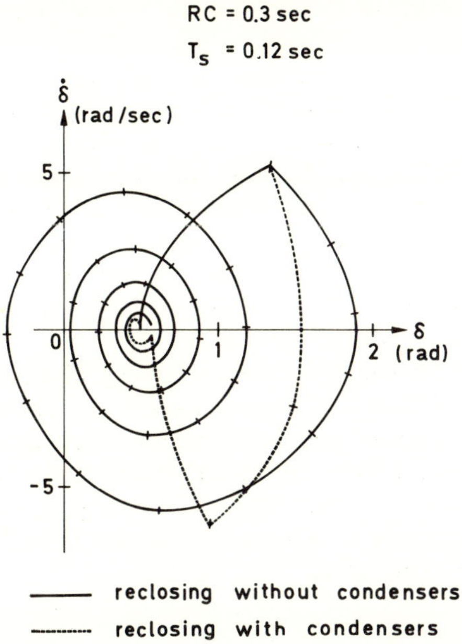

Fig. 15(a). Phase plane response: $RC = 0.3$ sec; $T_s = 0.12$ sec.

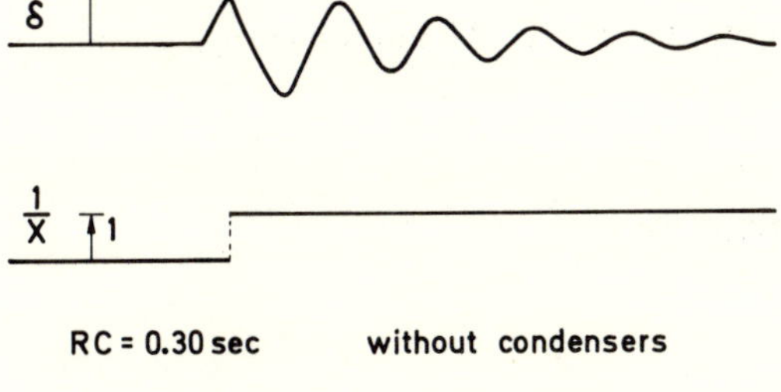

Fig. 15(b). Time response without condensers.

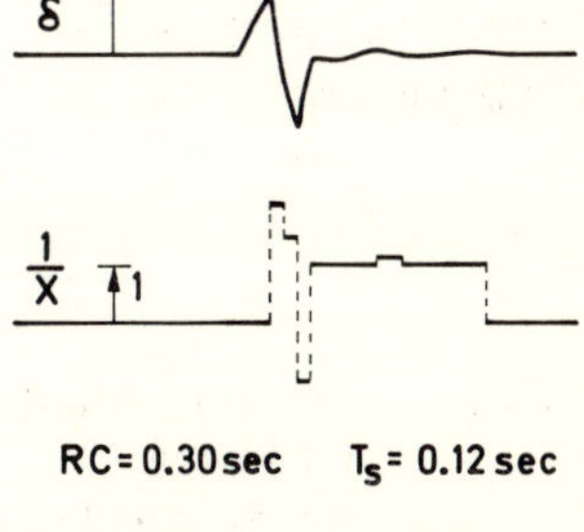

Fig. 15(c). Time response with condensers.

References pp. 187–188

4. OPTIMAL CONTROL WITH QUADRATIC PERFORMANCE INDEX

In recent years some work has been done on applying a certain class of optimal control theory to improve the dynamic response of a power system. The matrix Riccati solution of the state regulator problem was used. It requires the linearization of the system equations, and this was done at the operating point[22,23].

Yu *et al.*[22] deal with a synchronous generator connected through a transmission line to an infinite busbar. The resulting system, considering exciter–voltage regulator and governor–hydraulic systems, is of the eighth order. Two control signals applied to the exciter–voltage regulator and governor–hydraulic systems are obtained through state variable feedback.

The method used by Anderson[23] is almost similar. However, a better solution to the problem can be obtained by using digital control. The linearization should be done at each sampling instant and the feedback coefficients should be calculated accordingly. Tou[25] gives the theory for the calculation of the feedback coefficients. The linearization around the operating point can give unacceptable results, especially when the amplitude of swing is high enough to bring the system on the other side of the power angle curve.

5. CONCLUSION

This paper shows the application of modern control theory to the power system stability problem. Liapunov functions for the two-machine and the multi-machine problems are obtained using the energy metric algorithm. The Liapunov functions obtained are generalized by completing a quadratic form, so that almost all previously obtained results can be considered as special cases. Some methods of optimal control theory which were recently used for stabilization are discussed, together with some propositions to use digital control instead of analogue control. This requires the use of on-line computers. These propositions should be further studied for multi-machine systems to discover their practical applications.

ACKNOWLEDGEMENT

I would like to thank Mr. P. Wegmann and Mr. C. Albani for carrying out the computations.

NOTATION

M, M_i	inertia constants
δ	load angle
δ_0	steady state load angle
δ^*	load angle at the closest unstable equilibrium

D, a, a_i	damping coefficients
P_m	mechanical power input
P_{m0}	steady state mechanical power input
P_e	electrical power output
E'_q	voltage proportional to field flux linkage
E'_{q0}	steady state voltage proportional to field flux linkage
E_{ex}	exciter voltage referred to the armature circuit
E	infinite busbar voltage
E_i	internal voltage of the ith machine
x_d, x_q	direct-axis and quadrature-axis synchronous reactances
x'_d, x'_q	direct-axis and quadrature-axis transient reactances
x''_d, x''_q	direct-axis and quadrature-axis subtransient reactances
x_e	external reactance
X	total reactance
G_{ii}	short-circuit conductance of ith machine
B_{ij}	short-circuit transfer susceptance between the ith and the jth machine
t	time
T'_{d0}	open-circuit transient time constant
T'_d	short-circuit transient time constant
T_e	equivalent time constant of the governor system
RC	reclosing time
T_s	sampling period
b	amplitude of the fundamental of power angle curve
c	amplitude of the second harmonic of power angle curve
g	coefficient of power output due to flux decay
η_1	$1/T'_d$
η_2	$\dfrac{(x_d - x'_d)E}{(x_e + x'_d)T'_{d0}}$
γ_1	$1/T_e$
γ_2	k/T_e
k	governor amplification factor
V	Liapunov function

Dots over a symbol denote differentiation with respect to time. All values are in per unit.

REFERENCES

1 E. T. Wall, The generation of Liapunov functions in control theory by an energy metric algorithm, *Preprints of JACC*, 1968, pp. 172–179.

2 G. E. Gless, Direct method of Liapunov applied to transient power system stability, *IEEE Trans.*, **PAS-85** (1966) 159–168.

3 A. H. El-Abiad and K. Nagappan, Transient stability region of multimachine power systems, *IEEE Trans.*, **PAS-85** (1966) 169–179.

4 A. G. J. McFarlan, M. A. Laughton and G. F. Shannon, Generalized non-linear network analysis applied to the dynamic behaviour of synchronous machine systems, *Intern. J. Control*, **3** (1966) 97–111.

5 F. FALLSIDE AND M. R. PATEL, On the application of Liapunov method to synchronous machine stability, *Intern. J. Control,* **4** (1966) 501–513.
6 O. SCHULER, Die Anwendung der zweiten Methode von Liapunov auf die Stabilität des Parallelbetriebes, *Scientia Electrica,* **13** (1967) 1–10.
7 Y. N. YU AND K. VONGSURIYA, Nonlinear power system stability study by Liapunov function and Zubov's method, *IEEE Trans.,* **PAS-86** (1967) 1480–1485.
8 M. W. SIDDIGEE, Transient stability of an a.c. generator by Liapunov's direct method, *Intern. J. Control,* **8** (1968) 131–144.
9 N. D. RAO, Generation of Liapunov functions for the transient stability problem, *Trans. Eng. Inst. Canada,* **11** (1968) I–XI.
10 J. L. WILLEMS, Improved Liapunov function for transient power system stability, *Proc. IEEE,* **115** (1968) 1315–1317.
11 N. D. RAO, Routh–Hurwitz conditions and Liapunov methods for the transient-stability problem, *Proc. IEE,* **116** (1969) 539–547.
12 R. D. TEICHGRAEBER, F. W. HARRIS AND G. L. JOHNSON, New stability measure for multimachine power systems, *IEEE Trans.,* **PAS-89** (1970) 233–239.
13 M. A. PAI, M. A. MOHAN AND J. G. RAO, Power system transient stability regions using Popov's method, *IEEE Trans.,* **PAS-89** (1970) 788–794.
14 J. L. WILLEMS AND J. C. WILLEMS, The application of Liapunov methods to the computation of transient stability regions for multimachine power systems, *IEEE Trans.,* **PAS-89** (1970) 795–801.
15 G. A. LÜDERS, Transient stability of multimachine power systems via the direct method of Liapunov, *IEEE Trans.,* **PAS-90** (1971) 23–36.
16 A. K. DE SARKER AND N. D. RAO, Zubov's method and transient stability problems of power systems, *Proc. IEE,* **118** (1971) 1035–1040.
17 E. W. KIMBARK, Improvement of system stability by switched series capacitors, *IEEE Trans.,* **PAS-85** (1966) 180–188.
18 O. J. M. SMITH, Power system transient control by capacitor switching, *IEEE Trans.,* **PAS-88** (1969) 28–35.
19 N. RAMARAO AND D. K. REITAN, Improvement of power system transient stability using optimal control: Bang-bang control of reactance, *IEEE Trans.,* **PAS-89** (1970) 975–984.
20 W. AEBLI AND H. BOSSART, Bang-Bang Reaktanzsteuerung, *Diplomarbeit Lehrstuhl für Automatik der ETH,* 1971.
21 H. GLAVITSCH, Möglichkeiten der Verbesserung der Stabilität und der Spannungsregelung von Synchronmaschinen mit Hilfe der Gleichrichtererregung, *E und M,* **85** (1968) 58–65.
22 Y. N. YU, K. VONGSURIYA AND L. N. WEDMAN, Application of an optimal control theory to a power system, *IEEE Trans.,* **PAS-89** (1970) 55–62.
23 J. H. ANDERSON, The control of a synchronous machine using optimal control theory, *Proc. IEEE,* **59** (1971) 25–35.
24 E. W. KIMBARK, *Power System Stability,* Vol. III, Wiley, New York, 1956.
25 J. T. TOU, *Modern Control Theory,* McGraw-Hill, New York, 1964.

Discussion

H. GLAVITSCH (*Brown, Boveri & Co. Ltd., Baden, Switzerland*)
Did you find criteria which enable us to know whether the Liapunov functions used represent the stability region sufficiently accurately?

M. MANSOUR
It is in general impossible, when applying Liapunov theory, to get the total region of asymptotic stability analytically. This is because Liapunov stability conditions are only sufficient. In this paper the aim was to try using some sort of generalization to enlarge the asymptotic stability region. How far the result is from the actual stability region can be obtained by simulation.

J. L. DINELEY (*University of Newcastle upon Tyne, England*)
Does Professor Mansour agree that time optimal control is at present probably the most useful of the methods he has described? Does he also agree that, at present, excitation control is perhaps the most economic and interesting application of this method?

M. MANSOUR
As far as economy is concerned, it is clear that excitation control is the most economical method to achieve time optimal control. However, its improvement of the transient stability is limited relative to that of capacitor switching. The latter method should be further studied concerning its economic feasibility.

H. GLAVITSCH (*Brown, Boveri & Co. Ltd., Baden, Switzerland*)
The last discussion has brought up the more general area of time optimal control, which is a theoretical concept applicable to three different things:

(1) excitation control by switching the field voltage,
(2) switching capacitors and
(3) switching resistors.

In the first case, we can only affect the transient stability limit to a certain extent but it has a nice effect on smoothing rotor oscillations. In the other two cases, we are extending the transient stability limit quite extensively. They can be compared, in view of economic aspects, with building another line or changing the reactance of a machine. We have to see time optimal control in this wider sense as far as applications and economics are concerned.

M. MANSOUR
I completely agree with the statements of Mr. Glavitsch and I state again that the economics of capacitor switching should be investigated to find out how practical its use may be.

H. A. NOUR ELDIN (*Eidg. Technische Hochschule, Zürich, Switzerland*)
I would like to make a remark which may be of interest. We heard that about a thousand Liapunov functions may be generated. This number can be infinite, because if V_1 and V_2 are Liapunov functions, then any convex combination is also a Liapunov function. In addition, any product of V_1 and V_2 is also a Liapunov function.

M. MANSOUR
The aim of the first part of this work is to enlarge the asymptotic stability region obtained using the Liapunov method. Any Liapunov function which achieves this is welcomed; however, I do not know whether the combinations suggested by Dr. Nour Eldin will lead to such an enlarged region.

J. G. BALCHEN (*Technical University of Norway, Trondheim, Norway*)
In your objective functional for deterministic optimal control you have the term $\mathbf{u}^T\mathbf{R}\mathbf{u}$. This term does not make much sense if $\mathbf{u}$ does not swing around $\mathbf{0}$, or

$$\lim_{t\to\infty} \mathbf{u}(t) \to \mathbf{0}.$$

In such a case it is more reasonable to apply a term $\dot{\mathbf{u}}^T\mathbf{R}\dot{\mathbf{u}}$, which weights the speed of the control action rather than the magnitude. The theory becomes the same as before except that the state of the system will now be increased by a number of variables equal to the number of control variables.

M. MANSOUR
The control vector $\mathbf{u}$ used in deterministic optimal control with quadratic performance index is actually the deviation from the equilibrium values corresponding to the desired machine terminal voltage and bus power at any specified bus voltage. The same is true for the state vector $\mathbf{x}$. In this case, it is not necessary to consider the speed of control action, which leads to complication of the problem without any assurance of improving the performance.

F. C. SCHWEPPE (*Massachusetts Institute of Technology, Cambridge, Mass., U.S.A.*)
From the point of view of security, I wonder if you could give us any indication of whether you think the Liapunov technique for transient stability security monitoring is practical today, or when do you predict it might actually be used on-line?

M. MANSOUR
I have indicated in the paper a possibility of applying the Liapunov technique on-line for stopping capacitor switching when the state of the system reaches an acceptable region around the equilibrium point. At the moment I have no idea about further on-line application.

K. J. ÅSTRÖM (*Lund Institute of Technology, Lund, Sweden*)
(1) In your paper you derived a set of Liapunov functions depending upon a parameter β, and you proposed to obtain a lower bound for the maximum region of asymptotic stability by taking the union of the stability regions obtained for each Liapunov function corresponding to a particular value of β. How do you actually compute this estimate of the stability region for a system of high order?
(2) Can the minimum time strategy you described, based on capacitor switching, be implemented realistically with the technology existing today? Are there any side effects of the switching that are not apparent in the simplified model?

M. MANSOUR
It was found for the second-order system that it is sufficient to consider only the lower and upper limit of the parameter β to get a larger stability region. In my opinion the same can be done for higher-order systems to get two surfaces, the union of which is a candidate for the asymptotic stability region. In this case, as you have only two surfaces, it is not necessary to get the union explicitly.

As to the second question, there is no problem in implementing the minimum time strategy based on capacitor switching other than the economic justification and the voltage transients appearing during the negative power mode. These are harmonic overvoltages which do not affect the fundamental frequency voltage. Before any practical application of the method, this problem should be investigated through system simulation.

Steady State Security Monitoring

WILLIAM F. TINNEY

Bonneville Power Administration, Portland, Oregon, U.S.A.

AND

NORRIS M. PETERSON

Systems Control, Inc., Palo Alto, California, U.S.A.

SUMMARY

A fast, approximate a.c. power flow solution can be used as the basis for a security monitoring scheme. Each cycle of security monitoring determines the effects of an appropriately selected series of line and generator outage contingencies on the steady state operation of the system. The injections for each outage simulation correspond to the current operating state of the system. The results are checked against predetermined constraints on line power flows, reactive sources and voltage magnitudes, and when an outage causes constraint violations this information is transferred to other functions in the dispatch center for appropriate action.

The power flow algorithm is a compromise in speed, storage and accuracy between the familiar linear (or "d.c.") power flow and the Newton–Raphson a.c. power flow. It requires about half the storage of the Newton–Raphson method and for line outages obtains a satisfactory a.c. solution in about 20% of the time required by the Newton–Raphson method.

The algorithm is based on the triangular factorization of the two major submatrices of the system Jacobian matrix evaluated at the no-load point. Each submatrix, which is real and symmetric, is triangularized only once at the beginning of each contingency evaluation cycle. The effect of line outages is obtained without modifying the triangularization. Non-linearity and the coupling between real and reactive power are introduced iteratively. One of the main obstacles to this and related schemes is the representation of neighboring systems because their injections are unknown and their passive equivalents tend to degrade the sparsity of the monitored system. The paper describes how to mitigate these difficulties.

1. INTRODUCTION

A major justification for on-line computer control of power systems is improvement of security. The simplest security monitoring concept is to determine the effects of a list of specific transmission and generator outage contingencies on the steady state performance of the system at its current operating condition. Although this concept ignores the dynamic transition of the system from its operating state to its post-outage steady state, it is generally accepted as a useful assessment of security.

Two basic approaches have been proposed for this application. One involves the use of pre-computed distribution factors[1]; the other is based on an approximate a.c. power flow solution for each contingency. The distribution factor approach has poor accuracy and gives only a real power solution. The a.c. power flow solution approach fulfills all requirements but it is usually regarded as being too slow. This paper describes a new power flow solution algorithm for security monitoring that is fast, sufficiently accurate and does not require excessive computer memory. The algorithm is a composite of methods used in other applications[2–4] but their use in security monitoring presents new problems. The work is the result of a cooperative research contract between the Bonneville Power Administration and Systems Control, Inc.

2. NOTATION

Symbols are defined as they are introduced. Matrices and vectors are printed in bold type, *e.g.* **V**. Complex quantities are indicated by a bar, *e.g.* $\bar{V}$. Complex conjugate is indicated by an asterisk, *e.g.* $\bar{V}^*$. The summation notation $m \in \mu_k$ means m runs over the set of nodes of class μ adjacent to node k.

3. FUNCTION AND ORGANIZATION OF STEADY STATE SECURITY MONITOR

In this paper the Security Monitor is defined as an on-line process using real-time data for analyzing the effects of a series of outage contingencies on the steady state performance of the system. Each contingency is solved approximately as a steady state a.c. power flow problem. Except for simulated outages, the network is the same as the actual operating system and the injection schedule corresponds to the currently estimated state of the system. The results of each solution are checked against predetermined constraints. If a contingency causes a constraint violation or if a solution for a contingency is impossible, this information is transferred from the Security Monitor to another function in the control center. The action from this point on is not properly a part of the Security Monitor and is not considered in this paper. However, portions of the Security Monitor program could also be utilized to determine corrective action and other dispatch functions.

The Security Monitor will be one of several interrelated programs in an automated dispatch center. The ways in which it can be integrated with other functions are not considered in this paper. However, a simplified schematic diagram for the flow of information is shown in Fig. 1. The most critical input is the state vector from the State Estimator, a program which processes raw measurements to produce an estimate of the actual state of the system[5]. This estimate is transmitted in the

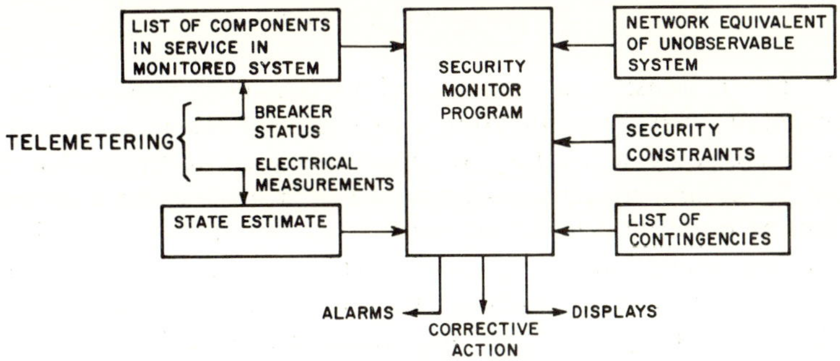

Fig. 1. Flow of information.

form of the state vector $\overline{\mathbf{V}}_s$ consisting of complex voltages at each node of the monitored system. Other essential inputs are briefly described.

3.1 *Network data*

The passive network is modelled by the nodal admittance matrix which is developed from a detailed list of basic network components including transmission lines, transformers, capacitors and reactors. It is essential to have real-time information on the state of these components at the beginning of each solution cycle. A solution cycle is the solution and checking of results for all contingent outages in a specified contingency list. The change of state of every network component is transmitted to the control computer and whenever a component is switched in or out, its effect is reflected by a change in the admittance matrix.

3.2 *Nodal injections*

For a line outage the injections should correspond to the actual state of the system. The injection schedule is computed once at the beginning of each solution cycle based on the network admittances and the state vector. The injection at every node k is computed as

$$P_k + jQ_k = \overline{V}_k \sum_{m\in\alpha_k} \overline{Y}^*_{km}\overline{V}^*_m \tag{1}$$

where P_k and Q_k are real and reactive power, $\overline{V}_k$ and $\overline{V}_m$ are elements of the state vector $\overline{\mathbf{V}}_s$, $\overline{Y}_{km}$ is an element of the nodal admittance matrix and α_k is the set of all nodes adjacent to node k.

For physical as well as mathematical reasons it is necessary to fix the voltage angle at the slack node and allow the variation in losses to be supplied by the injection at this node. For a generator outage it may also be necessary to revise the injection schedule for some generators to correspond to the normal reallocation. The allocation formula itself can be included in the power flow solution but this increases the solution time.

References pp. 210–211

3.3 Security constraints

The constraints are transmission line power flows, bus voltages and reactive limits. These constraints may originate from customer requirements, relay settings, insulation levels, equipment ratings or other sources. Constraints can also be established by off-line simulation studies. An approximate check of transient stability can be made by determining appropriate power flow constraints from off-line transient stability studies. In some systems it may be desirable to alter the constraints according to the state of the system, with different constraints applying under different operating conditions or contingencies. Establishing appropriate constraints for the Security Monitor is another important subproblem that requires more investigation.

3.4 Contingency list

Each contingency consists of the outage of a line or generator. The development of a suitable contingency list is a major subproblem that can only be described briefly here. Contingencies should be processed in an order that takes into account their probability and severity. Since the time for a complete solution cycle is proportional to the length of the contingency list, it should be kept as short as possible. The number of contingencies to be tested can be reduced by conditional testing wherein the outcome of one contingency affects the decision of what contingency to check next. The list should be conditional on the state of the system. For each outage on the actual system the list should be modified to include in priority order those second contingencies which off-line studies or prior experience have indicated to be most severe for this state of the system.

4. SOLUTION APPROACH

It is desirable to check system voltages and reactive sources as well as the flow of real power. Therefore the algorithm should produce a real and reactive power flow solution. Since accuracy criteria for security monitoring have not been established, the decision concerning accuracy depends to some extent on judgment. The two extremes of accuracy are the "d.c." or linear power flow[6] and the exact a.c. power flow as solved by the Newton–Raphson method[7]. The linear power flow is about 50 times faster but its accuracy is poor and it does not give a complex voltage solution. On the other hand the time and memory requirements of the Newton–Raphson method are excessive for this application.

4.1 Accuracy criteria

Accuracy criteria can be established by off-line studies in which the exact solution is compared with approximate solutions obtained from different algorithms. The absolute error in power flows on heavily loaded lines is the most critical index of accuracy. Large percentage errors on lightly loaded lines do not degrade the security assessment. It is advantageous to use a method with controllable accuracy so that, if necessary, greater accuracy can be obtained at the expense of more running time. The objective is to find a method that is the best compromise between speed and accuracy.

4.2 Choice of solution method

Methods that could be used for this application include: (1) Gauss–Seidel or successive displacement methods; (2) explicit matrix inversion methods; and (3) direct solution methods that exploit matrix sparsity by ordered triangular factorization.

Gauss–Seidel methods have minimum memory requirements but they are too slow and problem dependent. Matrix inversion methods provide direct solution possibilities, but their time and memory requirements increase as the square of system size, making them unsuitable for this application. Sparse matrix methods are rapidly displacing the other methods not only in power network applications but in all fields where sparse matrices arise. The over-all superiority of the factored direct solution with sparsity exploitation has been clearly established. Therefore the problem is to find the most suitable approximate power flow solution that can be implemented by triangular factorization.

5. DIRECT SOLUTIONS BY ORDERED TRIANGULAR FACTORIZATION

The algorithm to be described is based on the direct solution of a sparse system of linear algebraic network equations by ordered triangular factorization and sparse matrix programming techniques. This solution method has been explained in detail elsewhere and is only briefly reviewed here[2, 8–13].

5.1 Triangular factorization

The following matrix equation is used to describe the method:

$$\mathbf{A}\mathbf{x} = \mathbf{y} \tag{2}$$

$\mathbf{A}$ is a non-singular matrix, $\mathbf{y}$ is the independent vector and $\mathbf{x}$ is the dependent vector. $\mathbf{A}$ can be factored into the product of a lower triangular matrix $\mathbf{L}$ and an upper unit triangular matrix $\mathbf{U}$ by Gaussian elimination:

$$\mathbf{L}\mathbf{U}\mathbf{x} = \mathbf{y} \tag{3}$$

Equation (2) can be solved directly for $\mathbf{x}$ in two steps. First solve for the intermediate vector $\mathbf{x}'$ by forward substitution on $\mathbf{L}$

$$\mathbf{L}\mathbf{x}' = \mathbf{y} \tag{4a}$$

Then solve for $\mathbf{x}$ by back substitution on $\mathbf{U}$

$$\mathbf{U}\mathbf{x} = \mathbf{x}' \tag{4b}$$

References pp. 210–211

If **A** is symmetric only **L** is needed. In factorizing a symmetric matrix only terms on and below (or above) the diagonal need to be computed, thereby saving approximately half the computer time and storage of the non-symmetric case. For the direct solution (indicated by eqns. (4a) and (4b)), however, the number of operations is unaffected by symmetry.

If **A** is sparse, the Gaussian elimination can be ordered in such a way as to approximately minimize the fill-in of non-zero terms in **L** and **U** and thereby conserve much of the sparsity of the original system. If sparsity is exploited in programming, the number of operations for obtaining **x** is proportional to the number of non-zero terms in **L** and **U**. For power network equations this number is approximately proportional to system size.

5.2 Sparsity-directed ordering

An algorithm is used to determine an order of elimination for the factorization that tends to minimize the fill-in of non-zero terms in **L** and **U**. Since the security monitor is to be used for many direct solutions with essentially the same network, an extensive search for the optimal order is justified. The ordering algorithm simulates the fill-in effects of triangularization by maintaining tables that describe the changing connection topology. A typical algorithm operates on the following simple principle.

> At each step of the triangularization eliminate the node with the fewest connected equivalent branches. If more than one node has the same number of connected branches, select any one.

5.3 Sparsity programming

The sparsity of **L** and **U** can be exploited by special programming techniques that process and store only the non-zero matrix elements. The object is to eliminate all non-trivial operations. The development of sparse matrix programming for the direct solution of systems of linear equations is now receiving much attention. Approaches vary depending on the application and computer system and whether storage or speed is the more important. In order to realize the full potential of the algorithm described in this paper, it is essential that the most effective sparse matrix techniques are used. The statements in this paper concerning the speed of the method are given in comparison with the Newton–Raphson method, assuming that comparable sparse matrix programming techniques are used for both methods. The statements concerning storage are based on programming schemes described in refs. 12 and 13.

In addition to storing the non-zero matrix elements, most sparse matrix techniques also require the storage of indexing information to identify the elements and facilitate their retrieval. In general, there is a trade-off between speed and storage. It has not yet been determined whether special sparse matrix methods are needed for control computers.

6. DEVELOPMENT OF SOLUTION ALGORITHM

In this section an algorithm for solving the power flow for the Security Monitor is presented. The algorithm is based on two symmetric real matrices that remain constant for a complete solution cycle. Therefore each matrix needs to be triangularized only once. A similar formulation has been used for the exact solution[3]. The effect of a line outage is simulated without changing the matrices.

6.1 Transformer representation

The representation of transformers with off-nominal tap ratios requires special consideration. The definition of terms is shown in Fig. 2.

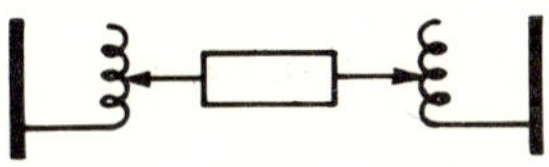

Fig. 2. Transformer model.

The transfer admittance for a transformer between nodes k and m is given by

$$\bar{Y}_{km} = \bar{Y}_{mk} = -\frac{1}{t_k t_m}\bar{Y}^0_{km} \tag{5}$$

The contributions of the transformer to the driving point admittances at nodes k and m are as follows:

$$\bar{Y}_{kk}(\bar{Y}^0_{km}) = (1/t_k^2)\bar{Y}^0_{km} \tag{6a}$$

$$\bar{Y}_{mm}(\bar{Y}^0_{mk}) = (1/t_m^2)\bar{Y}^0_{mk} \tag{6b}$$

By defining $t_{km} = t_m/t_k$ the contributions to the driving point admittances become

$$\bar{Y}_{kk}(\bar{Y}_{km}) = -t_{km}\bar{Y}_{km} \tag{7a}$$

$$\bar{Y}_{mm}(\bar{Y}_{mk}) = -\frac{1}{t_{km}}\bar{Y}_{mk} \tag{7b}$$

6.2 Basic expressions

Expressions for the real and reactive power injections at node k can be written as

$$P_k = V_k^2 G_{kk} + V_k \sum_{m\in\alpha_k} V_m(G_{km}\cos\psi_{km} + B_{km}\sin\psi_{km}) \tag{8a}$$

$$Q_k = -V_k^2 B_{kk} + V_k \sum_{m\in\alpha_k} V_m(G_{km}\sin\psi_{km} - B_{km}\cos\psi_{km}) \tag{8b}$$

where

$P_k + jQ_k$ is the power injection at node k
V_k, V_m are voltage magnitudes at node k and node m
ψ_{km} is the voltage phase angle across branch km

References pp. 210–211

$G_{km} + jB_{km}$ is the transfer admittance of branch km
$G_{kk} + jB_{kk}$ is the driving point admittance of node k
α_k is the set of branches connected to node k

The B_{kk} term can be separated into its components

$$B_{kk} = \sum_{m \in \alpha_k} (-t_{km}B_{km} + By_{km}) + Bc_k \tag{8c}$$

where

B_{km} is the transfer susceptance of branch km
t_{km} is the tap ratio when branch km is a transformer (when branch km is a transmission line, $t_{km} = 1.0$)
By_{km} is the charging susceptance of the k leg of the equivalent pi of the transmission line km
Bc_k is the susceptance of the shunt capacitor or reactor at node k

6.3 *Derivation of real power model*

The system of equations for solving for real power flows will be developed first.

The following substitutions are made in eqn. (8a):

$$\sin\psi = \psi + \sin\psi - \psi \tag{9a}$$

$$\cos\psi = 1 + \cos\psi - 1 \tag{9b}$$

This gives

$$P_k = V_k^2 G_{kk} + V_k \sum_{m \in \alpha_k} V_m[G_{km}(1 + \cos\psi_{km} - 1) + B_{km}(\sin\psi_{km} - \psi_{km})]$$
$$+ V_k \sum_{m \in \alpha_k} V_m B_{km} \psi_{km} \tag{10}$$

Equation (10) can be rearranged as

$$V_k \sum_{m \in \alpha_k} V_m B_{km} \psi_{km} = P_k - V_k^2 G_{kk} - V_k \sum_{m \in \alpha_k} V_m G_{km}$$
$$- V_k \sum_{m \in \alpha_k} V_m[G_{km}(\cos\psi_{km} - 1) - B_{km}(\sin\psi_{km} - \psi_{km})] \tag{11}$$

The left-hand side (LHS) of eqn. (11) can be arranged in the form of a system of N linear equations with N voltage phase angles as variables by making the substitution

$$\psi_{km} = \theta_k - \theta_m \tag{12}$$

where θ_k and θ_m are the voltage phase angles at nodes k and m. The left side can then be written in matrix notation as $\mathbf{A\theta}$ where $\mathbf{\theta}$ is an N vector of node voltage phase angles and $\mathbf{A}$ is an $N \times N$ matrix with elements a_{km} defined as follows:

$$a_{kk} = V_k \sum_{m \in \alpha_k} V_m B_{km} \tag{13a}$$

$$a_{km} = -V_k V_m B_{km} \tag{13b}$$

On the right-hand side (RHS) of eqn. (11) the first and second terms of the Taylor's series expansions of the sine and cosine functions are substituted:

$$\sin \psi_{km} = \psi_{km} - \frac{\psi_{km}^3}{6} \tag{14a}$$

$$\cos \psi_{km} = 1 - \frac{\psi_{km}^2}{2} \tag{14b}$$

The RHS of eqn. (11) can then be written in terms of two new vectors $\mathbf{P}'$ and $\mathbf{P}''$ whose components are defined as follows:

$$P_k' = P_k - V_k^2 G_{kk} - V_k \sum_{m \in \alpha_k} V_m G_{km} \tag{15a}$$

$$P_k'' = V_k \sum_{m \in \alpha_k} V_m \left(G_{km} \frac{\psi_{km}^2}{2} + B_{km} \frac{\psi_{km}^3}{6} \right) \tag{15b}$$

The complete system of equations for real power can then be written as

$$\mathbf{A\theta} = \mathbf{P}' + \mathbf{P}'' \tag{16}$$

6.4 Derivation of reactive power model

Making the substitution indicated by eqns. (9) in eqn. (8b) gives the following expression for reactive power:

$$\begin{aligned} Q_k = {} & V_k^2 \sum_{m \in \alpha_k} (t_{km} B_{km} - By_{km}) - V_k^2 Bc_k - V_k \sum_{m \in \alpha_k} V_m B_{km} \\ & + V_k \sum_{m \in \alpha_k} V_m [G_{km} \sin \psi_{km} - B_{km}(\cos \psi_{km} - 1)] \end{aligned} \tag{17}$$

Equation (17) can be rearranged as follows:

$$\begin{aligned} \sum_{m \in \alpha_k} (V_k t_{km} B_{km} - V_m B_{km}) = {} & \frac{Q_k}{V_k} + V_k \sum_{m \in \alpha_k} (By_{km} + Bc_k) \\ & - \sum_{m \in \alpha_k} V_m [G_{km} \sin \psi_{km} - B_{km}(\cos \psi_{km} - 1)] \end{aligned} \tag{18}$$

The next step is to reduce the number of equations represented by eqn. (18). By defining a PQ node as a node where P and Q are scheduled and a PV node as a node where P and V are scheduled, eqn. (18) is reduced to include only PQ nodes. The number of PQ nodes is N'.

Two new sets of nodes are introduced:

η_k is the set of all PQ nodes connected to node k

μ_k is the set of all PV nodes connected to node k

Equation (18) can be rewritten for the N' nodes of PQ type by transferring the terms involving PV nodes to the RHS:

$$V_k \sum_{m\in\alpha_k} t_{km}B_{km} - \sum_{m\in\eta_k} V_m B_{km} = \frac{Q_k}{V_k} + V_k \sum_{m\in\alpha_k} (By_{km} + Bc_k)$$
$$- \sum_{m\in\alpha_k} V_m[G_{km} \sin \psi_{km} - B_{km}(\cos \psi_{km} - 1)]$$
$$+ \sum_{m\in\mu_k} V_m B_{km} \tag{19}$$

The LHS of eqn. (19) can be written in matrix form as **CV** where **V** is an N' vector of node voltage magnitudes and **C** is an $N' \times N'$ matrix with elements c_{km} defined as follows:

$$c_{kk} = \sum_{m\in\alpha_k} t_{km}B_{km} \tag{20a}$$

$$c_{km} = -B_{km} \quad \text{for} \quad m \in \eta_k \tag{20b}$$

The substitutions for the sine and cosine terms indicated by eqns. (14) are made on the RHS of eqn. (19), and it is written in terms of two new vectors $\mathbf{Q}'$ and $\mathbf{Q}''$ whose components are defined as follows:

$$Q'_k = \frac{Q_k}{V_k} + V_k \sum_{m\in\alpha_k} (By_{km} + Bc_k) + \sum_{m\in\mu_k} V_m B_{km} \tag{21a}$$

$$Q''_k = -\sum_{m\in\alpha_k} V_m \left[G_{km}\left(\psi_{km} - \frac{\psi^3_{km}}{6}\right) + B_{km}\frac{\psi^2_{km}}{2} \right] \tag{21b}$$

The complete system of equations for reactive power can then be written as

$$\mathbf{CV} = \mathbf{Q}' + \mathbf{Q}'' \tag{22}$$

6.5 Branch outage simulation

For maximum efficiency it is desirable to triangularize matrices **A** and **C** only once for each solution cycle. It is inefficient and unnecessary to represent a branch outage by changing **A** and **C** as this would require a new triangularization for each outage. The scheme described in this section simulates the effect of a branch outage on the solution without changing the triangularized matrix to reflect the branch change. It is a special application of a general method for modifying inverted matrices[4, 14].

An outage of a line in a branch km causes four changes in matrix **A** of eqn. (16) as follows:

$$\Delta a_{kk} = \Delta a_{mm} = -\Delta a_{km} = -\Delta a_{mk} \tag{23}$$

The change in **A** can be expressed in matrix notation as

$$\mathbf{A}' = \mathbf{A} + \Delta a_{km}\mathbf{M}_A\mathbf{M}_A^T \tag{24}$$

where $\mathbf{M}_A$ is an N vector that is all zeros except element k which is $+1$ and element m which is -1.

The inverse of $\mathbf{A}'$ is given by

$$[\mathbf{A} + \Delta a_{km}\mathbf{M}_A\mathbf{M}_A^T]^{-1} = \mathbf{A}^{-1} - \left(\frac{1}{\Delta a_{km}} + \mathbf{M}_A^T\mathbf{A}^{-1}\mathbf{M}_A\right)^{-1}\mathbf{A}^{-1}\mathbf{M}_A\mathbf{M}_A^T\mathbf{A}^{-1} \tag{25}$$

For a branch outage eqn. (16) can be rewritten as

$$[\mathbf{A} + \Delta a_{km}\mathbf{M}_A\mathbf{M}_A^T][\boldsymbol{\theta} + \boldsymbol{\Delta\theta}] = \mathbf{P}' + \mathbf{P}'' \tag{26}$$

where $\boldsymbol{\Delta\theta}$ is a vector of phase angle corrections that account for the line outage. A similar expression can be written for eqn. (22) for the voltage correction:

$$[\mathbf{C} + \Delta c_{km}\mathbf{M}_C\mathbf{M}_C^T][\mathbf{V} + \boldsymbol{\Delta}\mathbf{V}] = \mathbf{Q}' + \mathbf{Q}'' \tag{27}$$

Here Δc_{km} is the change in element km of $\mathbf{C}$ resulting from the line outage and $\mathbf{M}_C$ is an N' vector that is all zeros except for element k which is $+1$ and element m which is -1.

Substituting the result of eqn. (25) into eqn. (26) gives

$$\boldsymbol{\theta} + \boldsymbol{\Delta\theta} = \left[\mathbf{A}^{-1} - \left(\frac{1}{\Delta a_{km}} + \mathbf{M}_A^T\mathbf{A}^{-1}\mathbf{M}_A\right)^{-1}\mathbf{A}^{-1}\mathbf{M}_A\mathbf{M}_A^T\mathbf{A}^{-1}\right][\mathbf{P}' + \mathbf{P}''] \tag{28}$$

Making a similar substitution in eqn. (27) gives

$$\mathbf{V} + \boldsymbol{\Delta}\mathbf{V} = \left[\mathbf{C}^{-1} - \left(\frac{1}{\Delta c_{km}} + \mathbf{M}_C^T\mathbf{C}^{-1}\mathbf{M}_C\right)^{-1}\mathbf{C}^{-1}\mathbf{M}_C\mathbf{M}_C^T\mathbf{C}^{-1}\right][\mathbf{Q}' + \mathbf{Q}''] \tag{29}$$

Using the results of eqns. (16) and (22), eqns. (28) and (29) can be simplified as

$$\boldsymbol{\Delta\theta} = -\left(\frac{1}{\Delta a_{km}} + \mathbf{M}_A^T\mathbf{A}^{-1}\mathbf{M}_A\right)^{-1}\mathbf{A}^{-1}\mathbf{M}_A\mathbf{M}_A^T\boldsymbol{\theta} \tag{30a}$$

$$\boldsymbol{\Delta}\mathbf{V} = -\left(\frac{1}{\Delta c_{km}} + \mathbf{M}_C^T\mathbf{C}^{-1}\mathbf{M}_C\right)^{-1}\mathbf{C}^{-1}\mathbf{M}_C\mathbf{M}_C^T\mathbf{V} \tag{30b}$$

For notational convenience two new vectors are defined as follows:

$$\mathbf{Z}_A = \mathbf{A}^{-1}\mathbf{M}_A \tag{31a}$$

$$\mathbf{Z}_C = \mathbf{C}^{-1}\mathbf{M}_C \tag{31b}$$

Although indicated as computed by an explicit inverse, $\mathbf{Z}_A$ and $\mathbf{Z}_C$ should be obtained from the triangular factorizations of $\mathbf{A}$ and $\mathbf{C}$. Equations (30a) and (30b) can be written as

$$\boldsymbol{\Delta\theta} = -\left(\frac{1}{\Delta a_{km}} + z_{Ak} - z_{Am}\right)^{-1}(\theta_k - \theta_m)\mathbf{Z}_A \tag{32a}$$

$$\boldsymbol{\Delta}\mathbf{V} = -\left(\frac{1}{\Delta c_{km}} + z_{Ck} - z_{Cm}\right)^{-1}(V_k - V_m)\mathbf{Z}_B \tag{32b}$$

The scalars z_{Ak} and z_{Am} are elements of $\mathbf{Z}_A$; the scalars z_{Ck} and z_{Cm} are elements of $\mathbf{Z}_C$; θ_k and θ_m are elements of $\boldsymbol{\theta}$, the solution vector of eqn. (16) without the line outage; V_k and V_m are elements of $\mathbf{V}$, the solution vector of eqn. (22) without the line outage. Thus by solving eqn. (16) as a linear system to obtain the solution $\boldsymbol{\theta}$, the solution for $\mathbf{\Delta\theta}$, the correction to account for a line outage, can be obtained by first solving for $\mathbf{Z}_A$ by a repeat solution and then solving eqn. (32a). A parallel statement applies for the voltage solution in eqn. (32b).

The following additional considerations apply to the reactive equation solution.

(1) If the outage is a transformer, the non-zero entries in the vector $\mathbf{M}_C$ are $\sqrt{t_{km}}$ for element k and $-1/\sqrt{t_{km}}$ for element m, for $k < m$. This effects the removal of the equivalent pi for the off-nominal tap ratio transformer.

(2) Only a ± 1 in element k of $\mathbf{M}_C$ is used if the line outage connects a PQ node to a PV node or to the slack node.

(3) If the line outage connects two PV nodes, it is not necessary to compute $\mathbf{Z}_A$ or solve eqn. (32b).

6.6 *Iterative solution for real and reactive power*

The coefficient matrices $\mathbf{A}$ and $\mathbf{C}$ of eqns. (16) and (22) are symmetric and have strong diagonals. The sparsity structure of $\mathbf{A}$ is identical to the nodal admittance matrix, and the sparsity structure of $\mathbf{C}$ corresponds to the rows and columns of PQ nodes in the nodal admittance matrix. The matrices can be triangularized in an order that conserves sparsity and used to obtain efficient direct solutions to eqns. (16) and (22) for given RHS. Several iterative schemes for using these equations to solve the power flow are possible. The one described here was found to be a good compromise between speed and accuracy for the Security Monitor. The simulation of the line outage is described in §8.

Equation (16) can be solved iteratively for $\mathbf{\theta}$ for a given schedule of power injections assuming $\mathbf{V}$ is constant. The coefficient matrix $\mathbf{A}$ is a function of $\mathbf{V}$ but not $\mathbf{\theta}$. The vector $\mathbf{P}'$ is not a function of $\mathbf{\theta}$ but $\mathbf{P}''$ is a function of $\mathbf{\theta}$ due to terms $\psi_{km} = \theta_k - \theta_m$. Therefore it is necessary to solve for $\mathbf{\theta}$ iteratively. The RHS is first evaluated using an estimated value of $\mathbf{\theta}$. Then eqn. (16) is solved directly for an improved value of $\mathbf{\theta}$ by the triangularization scheme. Using the new value of $\mathbf{\theta}$, $\mathbf{P}''$ is updated and eqn. (16) is solved again for a new value of $\mathbf{\theta}$. The cycle could be repeated until the change in $\mathbf{\theta}$ between successive iterations is negligible, but experiment indicates that for a single-line outage two iterations produce a solution that is accurate enough for security monitoring. Convergence in a line outage simulation is checked by observing the magnitude of the change in angle, θ_k or θ_m, at the power receiving end of the line prior to the outage. Convergence for a generator outage simulation is checked by observing the change in angle at the node where the generator was lost. If the change in angle under observation on the second iteration is larger than an empirically determined tolerance, this is assumed to indicate that no solution is possible.

After solving for $\mathbf{\theta}$, an approximation for $\mathbf{V}$ can be obtained from eqn. (22).

The coefficient matrix $\mathbf{C}$ is not a function of $\mathbf{V}$ or $\mathbf{\theta}$. However, $\mathbf{Q}'$ is a function of $\mathbf{V}$, and $\mathbf{Q}''$ is a function of both $\mathbf{V}$ and $\mathbf{\theta}$. The RHS of eqn. (22) is evaluated as a function of $\mathbf{\theta}$ obtained from the previous solution of eqn. (16) and an estimated value of $\mathbf{V}$. Then eqn. (22) is solved directly for an improved value of $\mathbf{V}$ by the triangularization scheme. The new value of $\mathbf{V}$ could then be used to compute a new

RHS and the cycle repeated, but for a single line outage one iteration produces a solution that is accurate enough for security monitoring.

Other iteration schemes have been tried but this one seems to be the best compromise between speed and accuracy for a single line or a single generator outage and a starting approximation that is the solution prior to the outage.

7. REPRESENTATION OF UNOBSERVABLE SYSTEM

The power flow solution in the Security Monitor should include the influence of the entire interconnected system even though some portions may be unobservable to the State Estimator. Representation of the unobservable system requires special considerations. The following conditions concerning the unobservable system are assumed: (1) its network configuration and parameters are known; (2) none of its nodes are included in the state vector and none of its injections are known; and (3) conditions within the unobservable system are of no concern to the Security Monitor.

The observable system as represented by eqns. (16) and (22) is a pseudo-linear model. It seems reasonable, therefore, that the unobservable system can be represented satisfactorily by a linear model. If a linear model is accepted in the unobservable system and its network configuration and parameters are known, its influence on the observable system can be represented exactly (assuming linearity). The nodal admittance matrix of the unobservable system is reduced (by Gaussian elimination in optimal order by sparse matrix programming) to an equivalent retaining only the boundary nodes of the observable network. This equivalent is then added to the observable network. Although the injections at the eliminated nodes are unknown, they can be represented by equivalent injections at the boundary nodes computed from the state vector of the observable system.

The linear relationship between nodal currents and voltages for the entire system can be represented by

$$\begin{bmatrix} \bar{\mathbf{Y}}_{AA} & \bar{\mathbf{Y}}_{AB} \\ \bar{\mathbf{Y}}_{BA} & \bar{\mathbf{Y}}_{BB} \end{bmatrix} \begin{bmatrix} \bar{\mathbf{V}}_A \\ \bar{\mathbf{V}}_B \end{bmatrix} = \begin{bmatrix} \bar{\mathbf{I}}_A \\ \bar{\mathbf{I}}_B \end{bmatrix} . \tag{33}$$

Equation (33) is partitioned into unobservable and observable parts, subscripted A and B respectively. Eliminating the A axes gives

$$[\bar{\mathbf{Y}}_{BB} - \bar{\mathbf{Y}}_{BA}\bar{\mathbf{Y}}_{AA}^{-1}\bar{\mathbf{Y}}_{AB}]\bar{\mathbf{V}}_B = [\bar{\mathbf{I}}_B - \bar{\mathbf{Y}}_{BA}\bar{\mathbf{Y}}_{AA}^{-1}\bar{\mathbf{I}}_A] \tag{34}$$

Substituting new symbols eqn. (34) can be written as

$$\bar{\mathbf{Y}}_E\bar{\mathbf{V}}_B = \bar{\mathbf{I}}_E \tag{35}$$

where $\bar{\mathbf{Y}}_E$ is the equivalent nodal admittance matrix and $\bar{\mathbf{I}}_E$ is the vector of equivalent current injections. Since $\bar{\mathbf{V}}_B$, the state vector of the observable system, is given, it is obvious that $\bar{\mathbf{I}}_E$ can be computed directly and that it accounts for the unobservable injections $\bar{\mathbf{I}}_A$.

References pp. 210–211

7.1 *Effect of equivalent on sparsity*

The use of a network equivalent for the unobservable system can greatly decrease the sparsity of the resultant network. In sparsity programming network reduction should be carried out only to the extent that it reduces the computational burden. The principle can be demonstrated by a simple example.

Figure 3 shows a network and the corresponding sparsity structure of its nodal admittance matrix. It is assumed nodes 1, 2, 3 forming subnetwork A are to be eliminated. The effect of this elimination on the sparsity of the retained subnetwork B is a fill-in of 10 equivalent branches (10 new terms above the diagonal). One more fill-in occurs in the triangularization of B which is performed in optimal order after the reduction. The triangularization of the resulting system requires 25 upper triangle locations, 81 multiply–add operations and 10 divide operations. Figure 4 shows the improvement resulting from eliminating only two of the three nodes. If only the two indicated nodes are eliminated, the triangularization of the matrix of the retained network can be performed in an order that conserves more sparsity. The elimination of two nodes causes a fill-in of only two branches. The triangularization of the remaining matrix B adds two more fill-ins for a total of 4 as compared with 11 for the previous case. The triangularization of this system requires 22 upper triangular locations, 50 multiply–add operations and 11 divide operations.

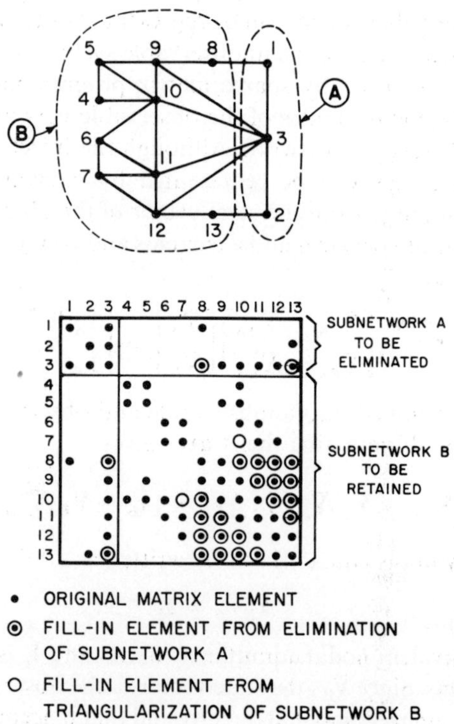

Fig. 3. Network and matrix structure with adverse reduction scheme.

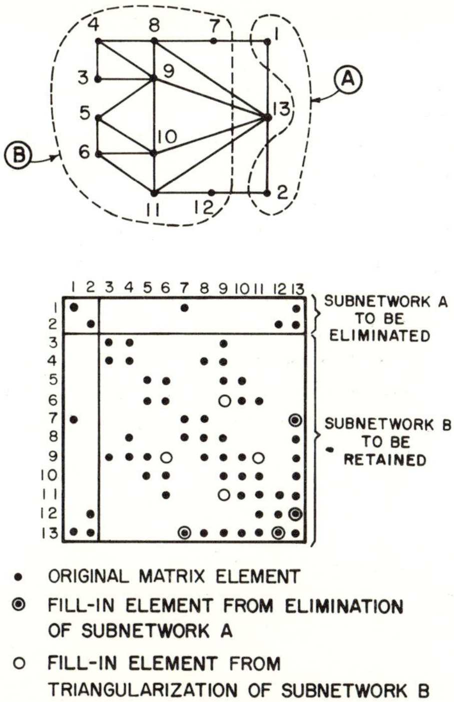

Fig. 4. Network and matrix structure with optimum reduction scheme.

7.2 Sparsity-directed network reduction

From the foregoing the objective in network reduction for sparsity programming should be clear. Given a network A that may be reduced to a specific equivalent subnetwork C, find the equivalent subnetwork B which contains C and is also the most advantageous from the standpoint of sparsity exploitation. Letting N_A, N_B, N_C be the number of nodes in network A and equivalent subnetworks B and C respectively, it is clear that

$$N_A \geq N_B \geq N_C \tag{36}$$

Finding the specific subnetwork B which would have the minimum fill-in from triangularization is an unsolved problem. A more realizable objective is to minimize the number of branches in subnetwork B.

A suboptimal scheme for finding the equivalent subnetwork B which contains C and has the minimum number of branches is as follows.

(1) Number the nodes of network A, excluding subnetwork C, by a sparsity-directed ordering algorithm.

(2) Simulate the fill-in effects of network reduction performed in the order produced in step (1) keeping track of the number of branches in the equivalent resulting from each nodal elimination step.

References pp. 210–211

(3) Continue until the network has been reduced to an equivalent on the border of subnetwork C.

The equivalent subnetwork corresponding to the point where the number of branches is minimum is an approximation of the desired subnetwork B.

The results of applying this scheme to a practical power network are shown in Fig. 5. In the original network there were 348 nodes which could be eliminated and 45 nodes on the boundary of the network to be retained. (Nodes on the interior of the retained network are irrelevant and therefore are ignored.) The 348 + 45 = 393 nodes, which were interconnected by 573 branches, were numbered by a sparsity-directed ordering algorithm and the fill-in effects of nodal elimination were simulated as described. The elimination continued to be beneficial until 329 nodes were eliminated. At this point the number of branches in the equivalent reached a minimum of 284. As the elimination of the remaining 19 nodes continued, the number of branches in the equivalent increased to 958.

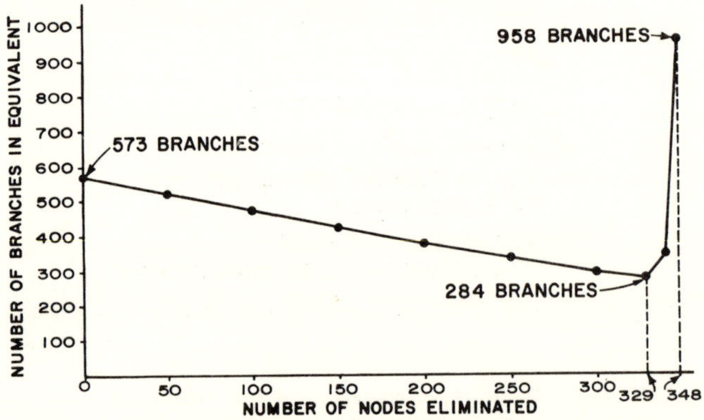

Fig. 5. Sparsity effect of network reduction.

The equivalent of 45 + 19 = 64 nodes and 284 branches is much denser than a typical power network, but it has less than 30% as many terms as the equivalent retaining only boundary nodes. The equivalent found for this example is not necessarily the optimum and a better algorithm could probably find a better equivalent. Other steps could be taken to increase the sparsity of the equivalent but they would involve approximations. Sparsity exploitation in network reduction is a new subject and results indicated here are only preliminary.

7.3 Computing injections with equivalent

The use of an equivalent that includes non-boundary nodes with unknown injections, as described in the previous section, affects the computation of the nodal injections at the boundary. The state vector $\overline{\mathbf{V}}_s$ includes only nodes of the observable system. As shown in Fig. 6 the vector includes boundary nodes but does not include any nodes in the equivalent. The state vector can be used to compute injections at all of the interior nodes of the observable system from eqn. (1) where $\overline{V}_k$ and $\overline{V}_m$ are elements of $\overline{\mathbf{V}}_s$.

The relationship between nodal currents and voltages for the subsystem consisting of the boundary nodes (subscripted A) and the passive equivalent (subscripted B), as shown in Fig. 6, can be written as in eqn. (33). All interior nodes of the equivalent are considered as passive; therefore $\overline{\mathbf{I}}_B = \mathbf{0}$. The vector $\overline{\mathbf{V}}_A$ of node voltages is a subvector of $\overline{\mathbf{V}}_s$, the state vector. Therefore the node voltages $\overline{\mathbf{V}}_B$ of the passive equivalent can be found by solving the following equation:

$$\overline{\mathbf{Y}}_{BB}\overline{\mathbf{V}}_B = -\overline{\mathbf{Y}}_{BA}\overline{\mathbf{V}}_A \tag{37}$$

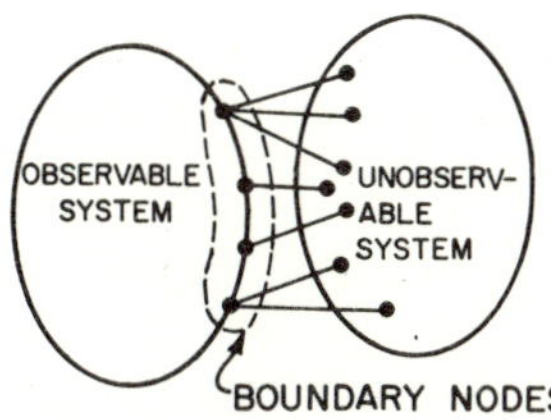

Fig. 6. Schematic of observable and unobservable systems.

This equation can be solved by optimally ordered elimination and back substitution. The injections at the boundary nodes can now be computed by eqn. (1) since the voltages of nodes within the equivalent adjacent to the boundary nodes are known from $\overline{\mathbf{V}}_B$. Since only elements of $\overline{\mathbf{V}}_B$ corresponding to nodes adjacent to boundary nodes are needed, it may not be necessary to compute all of $\overline{\mathbf{V}}_B$.

8. APPLICATION OF ALGORITHM

The main steps for a cycle of security monitoring are as follows.

(*A*) *Initialization*

(1) Read the latest state vector $\overline{\mathbf{V}}_s$ and the latest equivalent of the unobservable system.

(2) Read the list of network components of the observable system that are currently in operation.

(3) Form the equivalent nodal admittance matrix of the entire system.

(4) Compute the real and reactive injection schedules as indicated by eqns. (1) and (37).

(5) Form and triangularize matrix $\mathbf{A}$ of eqn. (16) as a function of the voltage magnitudes $\mathbf{V}_s$ of $\overline{\mathbf{V}}_s$.

(6) Form and triangularize matrix $\mathbf{C}$ of eqn. (22).

(7) Compute $\mathbf{P}'$ as function of $\mathbf{V}_s$.

(*B*) *Solve for a line outage*

(1) Read the first line outage in branch *km*.

(2) Remove the effect of line *km* from elements P'_k and P'_m of $\mathbf{P}'$.

(3) Form the vector $\mathbf{M}_A$ corresponding to the outage of the line in branch *km*. Solve for the vector $\mathbf{Z}_A$ of eqn. (31a) using the triangularization of **A**.

(4) Using elements θ_k and θ_m of $\boldsymbol{\theta}_s$, the vector of voltage angles of $\overline{\mathbf{V}}_s$, compute $\Delta\boldsymbol{\theta}^{(0)}$ from eqn. (32a). Then compute the initial voltage angle solution $\boldsymbol{\theta}^{(0)}$ as

$$\boldsymbol{\theta}^{(0)} = \boldsymbol{\theta}_s + \Delta\boldsymbol{\theta}^{(0)} \tag{38}$$

(5) Compute $\mathbf{P}''$ as a function of $\mathbf{V}_s$ and $\boldsymbol{\theta}^{(0)}$. Remove the effect of line *km* from elements P_k'' and P_m'' of $\mathbf{P}''$.

(6) Solve eqn. (16) to obtain the intermediate solution vector $\boldsymbol{\theta}^{(1)}$. Using elements of $\boldsymbol{\theta}^{(1)}$ compute $\Delta\boldsymbol{\theta}^{(1)}$ from eqn. (32a). Then the solution vector $\boldsymbol{\theta}^{(2)}$ of the first iteration is obtained as

$$\boldsymbol{\theta}^{(2)} = \boldsymbol{\theta}^{(1)} + \Delta\boldsymbol{\theta}^{(1)} \tag{39}$$

(7) Recompute $\mathbf{P}''$ as a function of $\boldsymbol{\theta}^{(2)}$ instead of $\boldsymbol{\theta}^{(0)}$.

(8) Solve eqn. (16) to obtain the intermediate solution vector $\boldsymbol{\theta}^{(3)}$. Using elements of $\boldsymbol{\theta}^{(3)}$ compute $\Delta\boldsymbol{\theta}^{(3)}$ from eqn. (32a). Then the solution vector $\boldsymbol{\theta}^{(4)}$ of the second iteration is obtained as

$$\boldsymbol{\theta}^{(4)} = \boldsymbol{\theta}^{(3)} + \Delta\boldsymbol{\theta}^{(3)} \tag{40}$$

If the convergence check is satisfactory, continue to the voltage solution, step (9). If the check indicates non-convergence, there is no solution.

(9) Form $\mathbf{M}_C$ corresponding to the outage of line in branch *km*. Solve for the vector $\mathbf{Z}_C$ of eqn. (31b) using the triangularization of **C**.

(10) Using elements V_k and V_m of $\mathbf{V}_s$ compute $\Delta\mathbf{V}^{(0)}$ from eqn. (32b). Then compute the initial voltage magnitudes solution $\mathbf{V}^{(0)}$ as

$$\mathbf{V}^{(0)} = \mathbf{V}_s + \Delta\mathbf{V}^{(0)} \tag{41}$$

(11) Compute $\mathbf{Q}'$ and $\mathbf{Q}''$ as functions of $\boldsymbol{\theta}^{(4)}$ and $\mathbf{V}^{(0)}$. Remove the effect of line *km* from elements Q_k', Q_k'', Q_m', Q_m'' of $\mathbf{Q}'$ and $\mathbf{Q}''$.

(12) Solve eqn. (22) to obtain the intermediate solution vector $\mathbf{V}^{(1)}$. Using elements of $\mathbf{V}^{(1)}$ and $\mathbf{Z}_C$ compute $\Delta\mathbf{V}^{(1)}$ from eqn. (32b). Then the voltage solution vector $\mathbf{V}^{(2)}$ is obtained as

$$\mathbf{V}^{(2)} = \mathbf{V}^{(1)} + \Delta\mathbf{V}^{(1)} \tag{42}$$

The vectors $\boldsymbol{\theta}^{(4)}$ and $\mathbf{V}^{(2)}$ give an approximate solution in polar form for the line outage. The analysis steps which would follow are not considered in detail here but their efficient execution is just as important as the solution. Line flows can be checked most efficiently by converting flow constraints to angle constraints. Since angles are solved directly, the angle check is very fast. Voltage magnitudes can also be checked directly. Reactive source limit checking requires an additional computation of **Q** at each *PV* node.

9. EVALUATION OF PERFORMANCE

9.1 Solution accuracy

The solution algorithm has been tested on a series of typical problems ranging in size from 100 to 500 nodes. Approximate solutions obtained by the algorithm were compared with the exact solutions. The mean error in real power flows was less than 1 %. (On the same problems the mean error of the conventional "d.c." power flow was more than 8 %.) A few lightly loaded lines had errors as high as 20 %, but since these lines were not critical for security monitoring this error was judged to be acceptable. The changes in line flows predicted by the algorithm were also compared with the exact changes. The mean error in predicted changes was less than 4 % and the maximum error in predicted changes in critical lines was less than 10 %. The voltage magnitude solution was very accurate with the largest error less than 0.2 %.

9.2 Solution speed

The total time for a complete solution cycle can be divided into five parts: (1) input, (2) initialization, (3) outage solutions, (4) solution analysis and (5) output. This paper has been concerned mainly with the outage solutions. The total iterative time for each outage solution is about half the time required for one iteration of the Newton–Raphson method. The input and output speeds depend on hardware and system requirements. The speed of the solution analysis depends mainly on the number and kind of constraints to be checked. The initialization time, which includes forming the branch admittances and triangularizing the matrices **A** and **C**, is about equal to the time for one iteration of the Newton–Raphson method.

9.3 Storage

The amount of central and auxiliary storage required depends on the computer configuration and programming scheme. The triangularized matrices **A** and **C** and their associated vectors can be overlayed alternately in the same storage area if necessary. If both matrices are stored simultaneously, the storage requirement is about half that of the Newton–Raphson method. The most critical decision affecting storage is the way the solution analysis and output are handled.

10. CONCLUSIONS

A solution algorithm for steady state security monitoring has been described that fulfills the requirements for this application. It solves for both real and reactive power flows, its storage requirements are not excessive and it is fast for its accuracy.

The accuracy of the solution for the recommended iteration scheme is judged to be satisfactory for security monitoring purposes. In tests the mean error in power

References pp. 210–211

flows was less than 1 % and the predicted changes in power flows due to outages was less than 4 % in critical lines. The maximum voltage error was less than 0.2 %. The accuracy of the solution is controllable, but for security monitoring any increase in accuracy does not appear to be worth the additional time required. The method is more accurate than linear approximation methods.

The complete iterative solution for a line outage takes about half the time of one iteration of the Newton–Raphson method and one-half the storage. If a voltage magnitude solution is not needed, its computation can be omitted and the running time reduced by about one-third.

For the on-line security monitor application, a network equivalent must be used to represent the unobservable system. A method is described for conserving the sparsity of this equivalent, thereby minimizing the time and storage burden of the equivalent.

The efficiency of the method is due mainly to the following: (1) use of triangular factorization and sparse matrix programming; (2) simulating line outages by the matrix inversion lemma instead of retriangularizing; (3) conserving sparsity in the network equivalent; and (4) employing an experimentally determined iteration scheme that avoids excessive computation for non-essential accuracy.

The paper has been concerned almost exclusively with the outage solution algorithm. Important related problems are the development of the list contingencies to be tested and determining the appropriate constraints to be used. The paper also omits discussion of efficient methods for analyzing the outage results to determine if violations exist.

The method is based on a linearization of the real and reactive nodal power equations at the no-load operating point. The no-load condition results in symmetric matrices. A closely related algorithm based on linearization at the operating condition prior to the outage is also being investigated. It should be faster and/or more accurate, but it will require more storage because the matrices will be non-symmetric.

REFERENCES

1 A. Limmer, Techniques and application of security calculations applied to dispatch computers, *Proc. 3rd Power Systems Computation Conference, Rome, 1969.*

2 W. F. Tinney and J. W. Walker, Direct solution of sparse network equations by optimally ordered triangular factorization, *Proc. IEEE,* **55** (1967) 1801–1809.

3 S. T. Despotovic, B. S. Babic and V. P. Mastilovic, A rapid and reliable method for solving load flow problems, *IEEE Trans. Power Apparatus Syst.,* **PAS-90** (1971) 123–130.

4 W. F. Tinney, Compensation methods for network solutions by optimally ordered triangular factorization, *PICA Conference Record, Boston, May 24–26, 1971.*

5 R. E. Larsen, W. F. Tinney and J. Peschon, State estimation in power systems, Parts I and II, *IEEE Trans. Power Apparatus Syst.,* **PAS-89** (1970) 345–362.

6 D. G. Taylor, A linear programming model suitable for powerflow problems, *Proc. 3rd Power Systems Computation Conference, Rome, 1969.*

7 W. F. Tinney and C. E. Hart, Power flow solution by Newton's methods, *IEEE Trans. Power Apparatus Syst.,* **PAS-86** (1967) 1449–1460.

8 N. Sato and W. F. Tinney, Techniques for exploiting the sparsity of the network admittance matrix, *IEEE Trans. Power Apparatus Syst.,* **82** (1963) 944–950.

9 J. Carpentier, Ordered eliminations, *Proc. Power System Computation Conf., London, 1963.*

10 H. Edelmann, Ordered triangular factorization of matrices, *Proc. Power System Computation Conf., Stockholm, 1966*.

11 D. Feingold and D. Spohn, Direct methods for linear networks of load flow computation, *Proc. Power System Computation Conf., Stockholm, 1966*, Sec. 4.4.

12 F. Gustavson, W. Liniger and R. Willoughby, Symbolic generation of an optimal Crout algorithm for sparse systems of linear equations, *Research Dept. RC 1852*, T. J. Watson Research Center, IBM Corp., Yorktown Heights, N.Y., June 17, 1967.

13 R. Willoughby (Ed.), *Sparse Matrix Proceedings, Symposium on Sparse Matrices, IBM Watson Research Center, Yorktown Heights, N.Y., Sept. 9–10, 1968*.

14 A. Ralston and H. Wilf, *Mathematical Methods for Digital Computers*, Vol. 1, Wiley, New York, 1960, 73–77.

Discussion

F. Ariatti (*ENEL, Rome, Italy*)

If I understand correctly, you say that the distribution factors calculated on the basis of a d.c. approximation are not accurate enough. In the case of a line outage you carry out a load flow calculation. I would be interested to know what assumptions you make on the voltage magnitudes for at least some nodes. Do you suppose that the voltage magnitude remains unchanged? I think there is a list of possibilities of the number of nodes on which you consider the voltage magnitude to remain unchanged and you perform a load flow computation only for one part of the system.

W. F. Tinney

In the scheme we propose, the voltages at busses with voltage control are held at their normal values during an outage simulation. After each outage solution, the reactive power required at each voltage control bus is computed and checked to see if it has exceeded its limit. If some reactive limits have been exceeded under the outage condition, this information is transferred to another program for further analysis and possible remedial action. Since all controllable voltages are held at their normal values, reactive violations in the outage simulation do not necessarily mean the case is impossible. If the reactive violations are not too large and there is sufficient reactive reserve elsewhere in the system, it may be possible to readjust the voltage control to overcome the reactive violations. This also applies to voltage constraint violations. The analysis to determine remedial action is performed after the Security Monitor has completed its complete cycle of contingency checking. The purpose of the Security Monitor is to identify the critical outages as quickly as possible.

M. A. Laughton (*Queen Mary College, London, England*)

I would like to point out that the method presented by Mr. Tinney has been previously used for low-voltage networks[1].

1 M. A. Laughton, Thevenin and Norton theorems—generalisations and dualities with particular reference to the analysis of simultaneous faults, *Proc. PSCC, Rome, June 1969*.

W. F. TINNEY

As Dr. Laughton points out, the scheme used for simulating the effect of a line outage in the iterative algorithm of this paper has been used previously by himself and others in other applications. The scheme is essentially the compensation theorem of network analysis. It can also be viewed as the link-at-time algorithm applied to the dependent voltage vector instead of the impedance matrix. From the mathematical viewpoint, with complete generality for linear equations, it is the matrix inversion lemma applied to the dependent vector.

To our knowledge this scheme has not been used before in connection with triangular factorization. It should be noted that the first approximation of the solution vector $\boldsymbol{\theta}^{(0)}$ obtained by applying the matrix inversion lemma via eqns. (31a), (32a) and (38) is the linear power flow solution. The lemma is applied again in each of the two succeeding iterations to obtain much improved accuracy over the initial linear approximation.

U. G. KNIGHT (*CEGB, London, England*)

Concerning the accuracy of approximate network calculations, some years ago I compared the results of a.c. and d.c. model solutions to a power flow problem and also actual system and d.c. model flows. I concluded that at the reasonable values of circuit flows in which one is interested for security, and perhaps transmission loss economics, the accuracy of the d.c. model in predicting active power flows was good, particularly when weighed in the context of uncertainties from metering or prediction errors. Circuit voltage drops calculated from voltage drop = $\mathbf{PR} + \mathbf{Qx}$ have been found to be in excellent agreement with those from the full equations.

I would like Mr. Tinney's views on the relative advantages and disadvantages of an "incremental approach" to security assessment, *i.e.* one in which changes in flow due to credible plant outages are calculated and added to existing telemetered flows.

Mr. Tinney touched on pattern recognition in his lecture. Although not pattern recognition in the accepted sense, I have wondered whether the inclusion of simple parameters of the credible post-fault systems in a series of state estimation functions would be advantageous as one approach to on-line stability assessments.

W. F. TINNEY

Tests showed that the d.c. or linear power flow solution has a mean error of $\pm 8\%$ for our system. We expect to obtain better accuracy than this from state estimation and we feel that the Security Monitor should have commensurate accuracy. The accuracy of the linear power flow solution may be much better for some other systems. However, even if the linear solution for real power were accurate enough, we would still want the voltage magnitude solution as it is often critical for our system. The decisions concerning how much accuracy is possible, or necessary, or economically justifiable are important because they significantly affect the software and hardware requirements for real-time control of power systems.

As I understand it, the "incremental approach" uses precalculated distribution factors to determine the change in flow on each line due to various outages. These precalculated factors are obtained from a linear approximation and would have the usual errors of this approximation. The main objection to this scheme, however, is its poor adaptability. The factors are precalculated with the network in its normal state. When there are one or more actual outages on the system, the factors cannot be used to determine the effect of an additional postulated outage until they have been recalculated taking into account the actual outages. Thus the scheme would be delayed exactly when it was most needed. We feel that for our system the accuracy and adaptability of the on-line simulation outweigh the speed of the precalculated incremental approach.

The many possibilities of pattern recognition should not be ignored. Some effort has been made to discover key parameters, either in the operating state or in credible post-fault states, that would provide clues to stability. Much more work needs to be done before the feasibility of this approach to transient security can be determined.

A. H. EL-ABIAD (*Purdue University, Lafayette, U.S.A.*)
Do you have a time estimate for applying your solution on-line, given a certain computer to do the analysis, and then, after you have found insecure situations, for finding out the required remedial measure? How often would the contingency evaluation have to be done?

W. F. TINNEY
These questions are of fundamental importance, but we cannot provide definite answers until our system is in operation. It is difficult to give an absolute time estimate for the algorithm because the time is a function of the computer, programming scheme and the sparsity of the network, including the equivalent. Our prototype program for the control computer solves a typical 100 bus problem and analyzes the results for violations in about 1 sec. The frequency of the contingency checking cycles has not yet been firmly established. It will probably be performed about once every 20 min or whenever there is a significant change in the network or loading. For a 100 bus system, 50 line outages and 10 generator outages would make a satisfactory contingency list. Thus, the Security Monitor would perform a cycle of checking in 1 min.

A. H. EL-ABIAD (*Purdue University, Lafayette, U.S.A.*)
What I had in mind was comparing simulations *versus* pattern recognition. If you have, for example, a hundred outages to check, how long will it take to do the complete study and be able to say that the over-all system is secure?

W. F. TINNEY
Pattern recognition would be practical if, among other things, the off-line computation required for learning the pattern compared favourably with the on-line computation for the simulation approach to security monitoring. If the on-line scheme requires 3 min out of every hour, this would be 438 hours a year. Present indications are that it would take much more time than this to perform the necessary simulations of different loading schedules and outages to learn the pattern of secure operating conditions. The hope for pattern recognition is that once the pattern is learned for a particular system, the learning can be updated for system

changes with only a small fraction of the initial learning effort. Since pattern recognition would be much faster on-line than a simulation scheme, it would be worth a large amount of off-line time to achieve it. At present, it is not known if pattern recognition is even remotely competitive. The simulation approach is practical now. It should be noted, however, that learning and pattern recognition always have been used in an informal way in power system operation. Also, the constraints imposed on the simulation type Security Monitor are obtained from an off-line learning process and the constraint check following a simulation solution in the Security Monitor is a form of pattern recognition.

W. B. JERVIS (*Merz and McLellan, Newcastle upon Tyne, England*)
The author mentions the difficulty of identifying the state of operation of unobservable areas having no telemetering facilities. I would like to know if the author has any views on the use of random noise technique, *i.e.* injecting known levels of disturbances of power and voltage at particular points in the inter-tie lines and monitoring the effect of these changes to assess the characteristics of the unobservable network.

W. F. TINNEY
The idea of injecting known disturbances on the inter-tie lines and monitoring their effect to obtain the characteristics of the unobservable system is intriguing. We have not studied it, but we understand that a project is being initiated by others to perform analytical studies and system tests of its feasibility. If one may assume, as we have in this paper, that the network configuration of the unobservable system is known, then the problem is somewhat simplified. In most situations the unobservable network will be known except for a few possible outages. Knowledge of the injections in the unobservable system is of lesser importance than knowledge of the network configuration. We are looking forward with interest to developments in this approach.

Role of Power Station Control in Overall System Operation

G. QUAZZA AND E. FERRARI
ENEL, Centro Ricerca di Automatica, Milan, Italy

SUMMARY

The role of generating station controls in power system operation, with reference to quality efficiency and continuity of service in normal and abnormal operating conditions, is first analyzed.

Typical responses to active power demand of hydroelectric units and steam plants in different control modes, including coordinated boiler–turbine–generator control, are illustrated and the problem of establishing performance criteria and specifications is briefly discussed.

Finally, the requirements of reactive power and voltage regulation and the influence on stability of both speed and excitation controls, with use of suitable feedback signals, are concisely examined.

1. INTRODUCTION

1.1

The statement that performance specifications for generating unit controls should be strictly related to the requirements of the electric power system may appear to be too obvious, since power generators are there just to satisfy the demands of the electric network. However, if one looks closely at how specifications are currently formulated, the general impression is that their relationship to system requirements is often far from being evident and much remains to be done in order to make them more complete, quantitative and suitable for expression in an analytical or logical form.

When proposals for alternative designs are to be evaluated, or existing control systems compared, the need is felt for compact performance criteria and cost functions. If any real significance is to be given to optimal control designs when based upon quadratic performance indices, the relative weights of deviations of controlled variables or states from desired values should be given for all chosen operating conditions. Establishing tolerance ranges may offer more freedom to the designer, but does not give enough clues to comparative evaluation of control systems coping with such tolerances, unless of course the final judgment is determined only by cost and all cost functions needed are available.

Weighting coefficients or, more generally, performance indices for generating unit controls should be established on the basis of the effects of controlled variable deviations upon power system performance.

1.2

As a preliminary contribution to prepare the ground for more complete and consistent performance criteria, let us review the role of generating unit controls in overall system operation.

Initially we recall that, following the usual well-known definition, the *task of a power system* can be synthetically expressed as one of supplying the electric power demanded by consumers with satisfactory continuity and quality of service at minimum price, with the additional obligation, for interconnected systems, to fulfill exchange power contracts with neighbours.

The *role of generating units and their controls* stems directly from the basic goals of the power system. Firstly, the requirements of *quality of service*, *i.e.* keeping short- and long-term deviations of frequency and consumer voltages (including wave shapes) within tolerances in the face of the changing demands, call for:

(1) unit flexibility in following an active and reactive power generation schedule, often expressed in terms of loading rates (MW/min);

(2) speed and accuracy of response in voltage regulation; and

(3) speed and accuracy of response to frequency changes and dispatching commands, as a contribution to system-frequency and exported power control.

Avoiding major system upsets, thereby contributing both to *continuity* and to *quality of service*, calls for:

(1) fast response to sudden positive or negative demands of power in emergencies, including changes of operating modes for the case of pumping–generating hydrostations;

(2) ample stability margins within a wide range of operating conditions, in the face of transmission line faults or severe perturbations;

(3) positive damping of electromechanical oscillations; and

(4) best reliability of operation, to keep faults causing abrupt losses of generating power, through runbacks and generator trips, to a minimum.

Ready *restoration of service*, after a generator trip or a major system disturbance has led to area or power station separation from the system, requires the unit to be capable of:

(a) fast restart and resynchronization after a trip;

(b) "saving" the power station after full load rejection due to line trip, by keeping "house-load" for a sufficiently long time for later fast restart; and

(c) wide limits of under-frequency (to some extent also over-frequency) operation in emergency conditions—isolated small or large area—and in area resynchronization.

Finally, *minimizing the price of electric energy* supply to the consumer implies minimization of weighted investment and operation costs, including fuel consumption for spinning reserve, *i.e.* (omitting the overall system planning and economic dispatching aspects of the problem) for the single power stations:

(1) direct and indirect minimization of fuel consumption, not only by closely following the mean economic schedule, but also by keeping transient heat losses (steam, nuclear or gas units) to a minimum and avoiding transient waste of water in hydrostations;

(2) full exploitation of available sources of active power (through better boiler and turbine controls) and of reactive power (by widening generator operating limits by suitable excitation control);

(3) search for best operating reliability, to reduce spinning reserve requirements and associated operation and investment costs (as a consequence of improved forced outage ratios in planning) and also to decrease replacement costs and increase average long-term achievable efficiency through better material conditions and longer unit life.

1.3

Evaluating the performance of generating unit controls with reference to their influence upon power system operation and establishing a consistent set of specifications requires, for the sake of simplicity, that a *choice* be made among the many possible system and unit *operating conditions*, in order to limit analysis and tests to the most significant situations only. Such a choice should be based on the one hand upon the actual burden posed to control systems—in the sense that more attention should be given to heaviest conditions—and on the other on how serious the consequences of control deficiencies may be on system and unit operation and integrity. For example, there should be no need to specify governor speed of response to the small and slow frequency deviations occurring during normal interconnected operation if the requirements in emergencies and especially in isolated area operation are much more demanding. On the other hand, it is imperative to have explicit guarantees for the situations when heavy material stresses may be engendered, *e.g.* large abrupt or very fast changes in unit output power, or when lack of prompt support from the unit to the power system would cause a general breakdown or at least a serious upset. Suitable specifications for the almost steady regimes that may last, for example, 95 % of the time, when disturbances are small and random, should however not be overlooked, since achievement of high efficiency and minimization of equipment wear may be very important here.

Different operating conditions often pose nearly independent requirements (*e.g.* values of ceiling voltage and of maximum rate of change in field voltage are determined with regard to large disturbances, while the small perturbation performance of excitation control is defined with reference to voltage regulation in normal system operation) but sometimes must be considered together, especially when their demands are conflicting (*e.g.* a larger frequency bandwidth in voltage regulation may have a negative effect on damping of electromechanical oscillations, see §5 and §6).

Determining these most significant operating conditions and, where possible, arriving at quantitative assessments implies a detailed analysis for each type of generating plant. Only part of it will be carried through here, since several aspects of the problem would deserve more thought and further data.

1.4

Reference will be made, in reviewing system operating conditions and the associated requirements for generating unit controls and in discussing quality criteria and performance of typical control designs, to

(1) *frequency and active power control, i.e.* turbine speed governors and boiler or nuclear reactor or hydro supply control systems, and

(2) *voltage and reactive power control, i.e.* generator excitation control.

The interaction between voltage and frequency controls is generally weak enough to justify their analysis as separate single-variable systems. This is no longer so with *stability and damping of power system oscillations*, the influence on which of both excitation and turbine mechanical power controls, including "fast valving", will be briefly examined.

2. SYSTEM AND UNIT OPERATING CONDITIONS *versus* FREQUENCY AND ACTIVE POWER CONTROL REQUIREMENTS

2.1

The demands posed to generating unit controls by the power system requirement of keeping frequency and tie-line powers close to their desired values in normal and abnormal operating conditions, including perturbations within the power station itself, can be expressed in terms of maximum active power rate for large output changes ("loading rate", especially relevant with regard to thermal stresses), speed governor small-signal transfer functions and large signal responses (especially with reference to primary frequency control loop stability and to system secondary control requirements, and also in emergencies).

Inspection of *daily system load versus time diagrams* and the results of inquiries[1] show that for rather large systems maximum loading rate is somewhat lower than 1% of peak load per min on average, with exceptional values around 2%; higher values are found for systems below 1000 MW. Although the case of small systems should not be ignored, since it may give valid indications for power stations operating on separated areas, and several recent big steam and nuclear units for large output changes do not tolerate rates in excess of 1.5%/min, utilities appear to have had no serious difficulties so far in generation scheduling from unit loading rate limitations.

It should be recognized, however, that economic dispatching may well call for non-uniform loading rates among generators in the system; consequently, some units will be subject to a rate higher than that of the overall demand and possibly close to the limit posed by thermal stress in the turbine metal. Thus, maximum "pick-up" rate under programmed loading is a significant parameter in unit performance, to be explicitly stated in specifications.

Considerable flexibility, as is well known, is added to the system capability of facing rapidly increasing demands if hydroelectric generators, with normal rates in excess of 100%/min, are available, the advantage however being greater, obviously, in emergency operation.

2.2

A second aspect of normal interconnected system operating conditions, especially in connection with speed governor desired performance, concerns *random demand deviations* with respect to predicted hourly averages. The present wide interconnections have reduced their effects on system frequency f (see Fig. 1 for a typical power density spectrum of Δf, from measurements on the Western European system[2,3]) to such an extent that no serious problem is posed to speed governors, provided dead-bands are low, stability is good and the number of units participating in primary frequency control is large enough[4,5]. Notice that the standard deviation of f from reference corresponding to the situation in Fig. 1 was 8 mHz, and the amplitude of its very low frequency components did not exceed 50 mHz, yielding only a 2% maximum variation in output power of no dead-band, 5% regulation units.

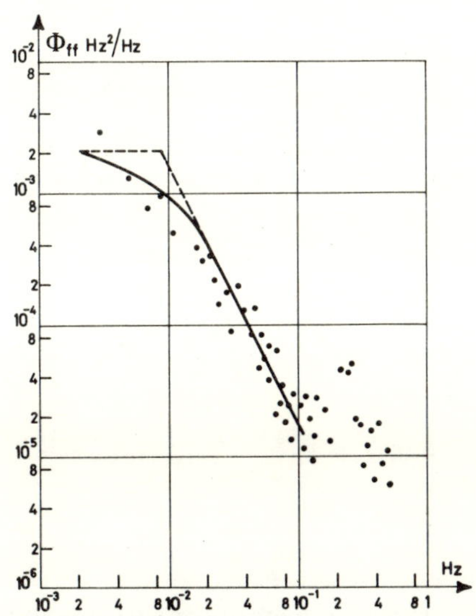

Fig. 1. Spectral density of frequency variations. Trials on Western Europe System, Nov. 23, 1969.

Knowledge of power-to-frequency governor transfer functions $G_{ri}(s)$ (see Fig. 2, together with Fig. 1) would allow determination of spectra for unit output power changes due to primary governing. This is not so significant in itself, since primary frequency control demands to speed governors in interconnected system operation appear to be so small. However, specifying governor responses to Δf (with a possible adaptive feature *vs.* network configuration) is important not only because of their

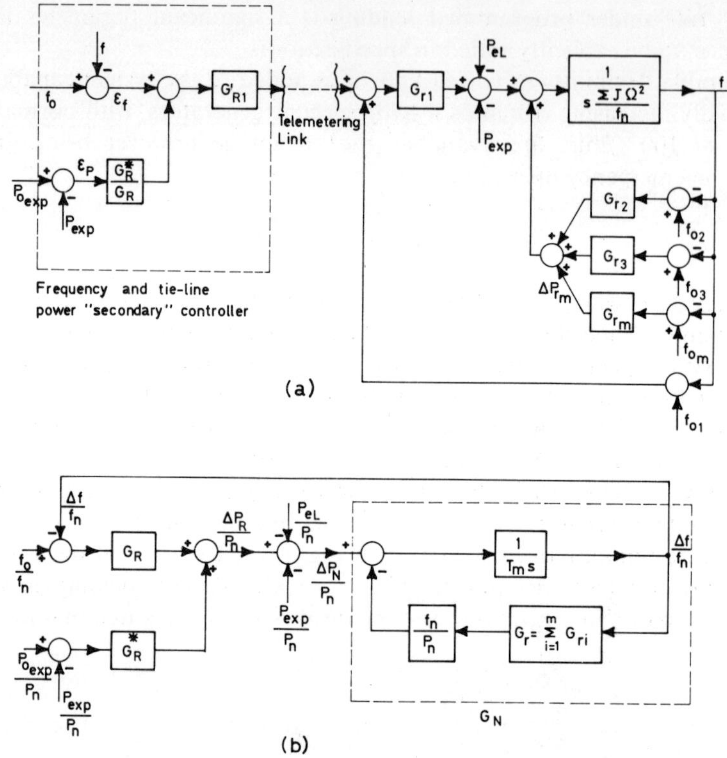

Fig. 2. Block diagrams of primary and secondary controls; no electrical power feedbacks are explicitly considered, only station 1 in secondary control.

$$G_R = \frac{f_n}{P_n} G'_{R1} G_{r1}$$

$$G_N = \frac{P_n}{f_n} \frac{\Delta f}{\Delta P_N} = \frac{1}{T_m s + (f_n/P_n) G_r}$$

$$\sigma = \frac{P_n}{f_n} \frac{1}{\Sigma E_{ri}}$$

If all $G_{ri} = E_{ri} \dfrac{1 + sT_a}{1 + sT_1} G_{HF}$ are equal,

$$G_N \doteqdot \sigma \frac{1 + sT_1}{1 + s(T_a + \sigma T_m) + s^2 \sigma T_m T_1}.$$

role in limiting turbine overspeed after a trip, or ensuring stability and good damping during the wide frequency excursions when the unit operates in a small isolated area, or making resynchronization faster and easier, but also because their behaviour may strongly affect, through $G_N(s)$, in $G_N G_R$ (see Fig. 2), the overall performance of tie-line and frequency "secondary" control in interconnected system operation.

The ideal situation is reached in secondary control when the transfer functions of all governors in the system are similar, and their dynamics is fast with respect to secondary control bandwidth. Actually, the variety in types of generating plant makes it quite difficult to obtain similar governors within each subsystem, and economic considerations would suggest keeping most efficient units at maximum load, with no contribution towards sustaining frequency. On the other hand, equity would call for a uniform policy by all partners in the system, *i.e.* keeping as many units as possible with enough output margin for primary governing, minimizing dead-bands and possibly avoiding poor performance speed changer drive servomechanisms, and compensating the responses of slower units by faster ones, in order to achieve from all partners the same dynamic contribution, relative to their spinning powers, to primary frequency control.

2.3

The main burden on governors in normal operating conditions derives from the demands by the dispatching centre in *secondary control*, since fewer units take part in it and greater power changes are needed to make up for most of the low frequency components of the unbalance between actual demand and scheduled generation, and to respond to tie-line control requirements. While the advantage of distributing control action among as many units as possible is recognized, minimizing costs of most efficient steam and nuclear units and otherwise reducing thermal stresses and control equipment wear, when sufficient flexible hydro is available, frequently leads to restricting the number of generators in secondary control to much less than 50% of spinning reserve.

If participation coefficient and range (with unlimited loading rate) are known, the control effort of each unit can be evaluated on the basis of the maximum excursion of the tie-line power and frequency controller output signal—covering, for example, 4% of spinning power—and its spectral density diagram. See, for example, Fig. 3, reported in ref. 2, which is roughly representative of a typical situation of Western Europe partners operating not too far from the conditions for autonomy ($G^*_{Rj} = G_{Rj}G_{Nj}$: each secondary controller acts only on load changes in its own area j) and for uniform system response[6] ($G^*_{R1} = G^*_{R2} = \ldots = G^*_{Rn}$: all members contribute equally to frequency control and, no matter where a given load change occurs, its effect on system frequency is the same).

Estimating the control effort of regulating units in a given situation is, however, less interesting than exploring the effect of different unit responses on the standard deviations of frequency, exported power and control power rate. It can easily be shown[3], after recognizing that random load changes corresponding to Fig. 3 are substantially Poisson-distributed, that slower responses (higher main lag time constant T_1) would yield poorer performance, higher control power rate and consequently heavier material stresses, mainly because they would somewhat reduce stability, as could be confirmed by Bode diagrams[4].

From the above, indications can be derived on the desirable, or acceptable, ranges for unit response time constants—and especially T_1—for given values of tie-line power–frequency controller and system transfer function parameters. Quite often, however, the freedom in choosing the demand-to-output power transfer

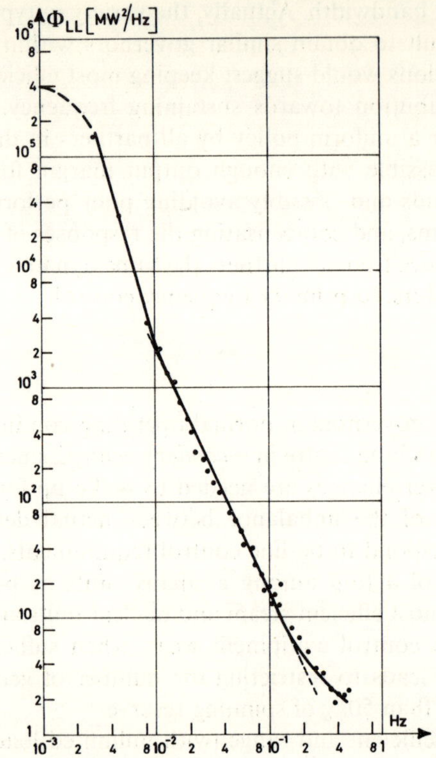

Fig. 3. Spectral density of Italian network power–frequency controller output, Nov. 3, 1969 trials.

function, which generally acts inside the speed governor loop, is limited by the need for ensuring stability of primary frequency control.

On the other hand, a good steady state linearity between dispatching command and regulating unit power output, such as can be achieved by linearizing devices on valve opening or feedbacks from impulse chamber pressure or electric power, is a prime requirement for good performance of secondary control and uniform distribution of effort among regulating units.

2.4

Figure 3 shows that the high-frequency content of the dispatching command signal is very small; abrupt and large disturbances in the system are filtered by the integral action of the secondary controller. Nevertheless, *step responses* yield essentially all the information required to identify regulating unit transfer functions within the linear range. Therefore they can be considered as *representative of random perturbations in normal operating conditions.*

Performance specifications may then be limited to cover responses to small steps, not exceeding the participation band, plus static characteristics, ramps of suitable rates being chosen to simulate programmed loading with wide output power variations.

Large steps, or ramps of faster rates, should be employed to describe the dispatcher requests for quick changes of generation, such as may be caused by unexpectedly fast load changes exceeding the capabilities of automatic secondary control, *e.g.* at midday, or by relatively large system disturbances. Units having enough flexibility and spinning reserve are here subject to step or fastest loading rate dispatcher demands.

2.5

Steps and ramps, while not always suitable for the description of unit response to *transmission line faults*, where turbine power fast control may play a relevant role but excitation control (see ref. 5) is involved to a major extent, still give enough evidence of governor responses to the "boost" or steam extraction closing commands in "alert" situations and also during the frequency changing periods of isolated area operation and system restoration. *System abnormal operating conditions* deserve in general more attention in establishing specifications, since disturbances are larger and the need for fast and efficient controls to avoid wide upsets and save and restore the system is strongly felt.

Special consideration must be given to the problem of saving power stations on auxiliary load for a long enough time after a full load rejection to have them ready for fast service restoration[7].

The limits of under-frequency operation should be carefully examined and chosen as a compromise between machine stresses and system flexibility[7a].

2.6

To come to a complete and safe definition of generating unit active power controls, in addition to direct system requirements *internal power station perturbations* should also be properly taken into account. A few faults[7–9], disturbances or programmed operations posing more or less heavy demands on the control system if it is to protect the unit from serious damage or material deterioration are listed here:

loss of one feedwater pump, circulating water pump, fan, mill or burner
boiler tube failure or nuclear cartridge burst
unit cold or not start-up, normal or emergency shut-down, hot restart
generator stator or rotor winding faults or excitation system failures
hydrogen cooling system faults
excessive turbine vibrations
steam admission valves stuck open

Some of these major emergencies call for fast runback to 50% load, others for immediate lighting off of boiler fires, and others for generator circuit-breaker trip with consequent turbine overspeed. All should be considered in unit control specifications, with regard mainly to unit protection but also to the system upsets they engender.

Likewise, minor disturbances, *e.g.* in a steam unit, feedwater heater cut-off, furnace purge, soot blowing, steam blowdown for ejector supply or other purposes, burner scavenge and burner fast oil supply, should be checked against their effects on control performance.

3. ESTABLISHING PERFORMANCE CRITERIA AND ASSOCIATED DIFFICULTIES

The most desirable way to state performance criteria, with reference to unit protection and system operational requirements, would be (as mentioned in §1.1) to use weighted deviations from desired conditions. Availability of an analytic expression for these, plus cost functions associated with (a) constraints, (b) cost of control equipment and (c) cost of control operation (*e.g.* energy required, wear etc.), would allow objective and automatic computer evaluation of alternative designs.

Unfortunately, the difficulties which must be overcome to arrive at this result have until now discouraged this approach. For example, estimating the cost of a higher loading rate involves the computation of thermal stresses[10] in several sections of boiler, turbine and pipes, with complicated geometries, and some assumptions on how the unit will be operated in its life and to what disturbances it will probably be subjected, in order to determine the integral effects of the stresses due to the higher loading rate, *i.e.* the total damage to the unit. As another example, tracing the benefits of increasing superheat and reheat steam temperature set-points in terms of improved unit efficiency, as a basis to judge the value of a higher performance but more costly temperature control system with lower transient errors and therefore higher acceptable set-points, calls for the rather complex heat-balance calculations needed to arrive at an "economic sensitivity function" of the multivariable boiler–turbine system.

Consequently, several questions concerning costs and benefits are still waiting for an answer, and performance criteria have so far been expressed only as tolerances on controlled variables, without indications on the relative weights of deviations, or on suitability for some specified operating conditions. For example, one of the most complete specifications for drum boiler controls sets tolerances on transient maximum deviations of superheat and reheat temperatures, T_s and T_R, drum level y, superheated steam pressure p_s, oxygen content in stack combustion gases O_2, furnace pressure p_c, and settling times after load changes, for the following operating conditions:

(1) approximately constant load (loading rate $< 1\%$/min)
(2) slowly varying load (loading rate $< 5\%$/min)
(3) fast varying load (loading rate $< 25\%$/min)
(4) emergency fuel flow increase rate ($<4\%$/sec)
(5) full load rejection

with different tolerance levels depending upon the operating condition and the output power range.

In other cases, precise requirements may not be posed but the expected performance must be stated in advance and is later subjected to experimental check: see, for example, transient responses of speed governors, minimum resynchronization time in a hot restart, output power range with stable operation.

4. ACTIVE POWER CONTROL PERFORMANCE OF TYPICAL GENERATING UNITS

4.1

Most conditions in §2.1–2.5 and §3 can be described as variations in active power demand (including the participation in primary governing, where the demanded power is caused by Δf). Hence the response to active power demands will be examined as being the most representative, taking into account the relevant effect of the water or steam supply system and its controls[11]. Reference will be made to the results of trials on typical plants in different operating modes. Present practices and trends will be briefly recalled, with regard to existing and recently proposed control designs.

4.2 Steam plants

4.2.1 Boiler-following mode

(The power change is basically achieved by opening turbine admission valves, with boiler controls re-establishing perturbed p_s, y, T_s, T_R and O_2 back to reference values.) Different solutions are currently employed:

(I) with valve position θ_t feedback only, without impulse chamber pressure p_{cR} or electric power P_e feedbacks;

(II) with p_{cR} feedback around θ_t loop;

(III) with P_e feedback (Fig. 4).

Figure 5 shows the step response of a 320 MW drum boiler–turbine generator, single reheat, Piacenza station, solution I. The main discrepancy with respect to the conventional approximate interpretation of the single-reheat unit response as a step change in HP turbine power plus a simple exponential change in IP–LP turbine output, with time constant τ_R due to reheater capacity, corresponding to the well-known transfer function:

$$\frac{\Delta P_e/\bar{P}}{\Delta q_s/\bar{q}_s} \doteq \left(\alpha + \frac{1 - \alpha'}{1 + s\tau_R}\right) \tag{1}$$

is caused by p_s variations.

The usual approximation of considering steam flow rates Δq_s as proportional to $\Delta\theta_t$ cannot be accepted, since in

$$\frac{\Delta q_s}{\bar{q}_s} \doteq \frac{\Delta\theta_t}{\bar{\theta}_t} + \frac{1}{r_t}\frac{\Delta p_s}{\bar{p}_s} \tag{2}$$

turbine equivalent resistance r_t is much less than 1 p.u., and $\Delta p_s/\bar{p}_s$ is relatively large. Pressure and temperature transient and steady state changes certainly to some extent also affect enthalpy drops in HP and IP–LP turbines, and hence ΔP_e, but their influence is not as relevant as the steam flow rate change due to Δp_s. Figure 5 indicates that the decrease in power output due to Δp_s is as high as 45% for the HP turbine and 26% in total output power from $t = 40$ sec to 2.5 min and that it remains considerable for a long time, since the p_s control loop is relatively slow.

References pp. 251–253

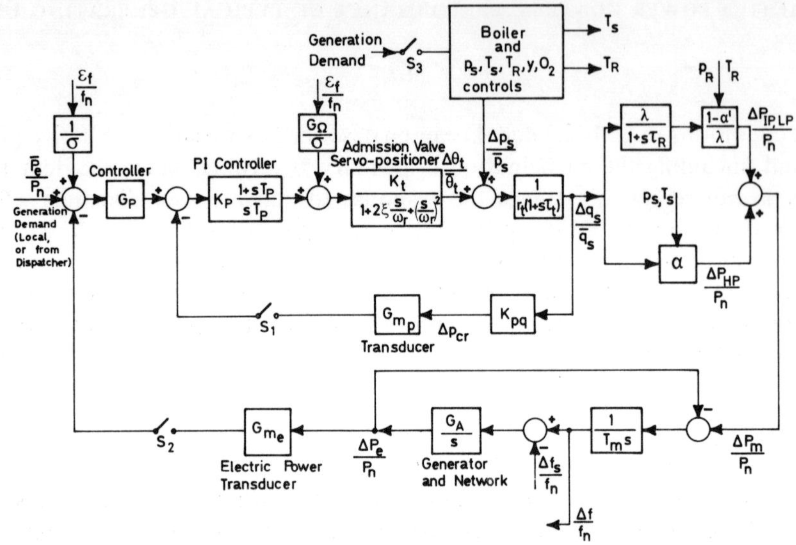

Fig. 4. Block diagram of the speed governor and active power control loop for a boiler–turbine generator:

S_3 open: boiler-following mode

S_3 closed: coordinated control mode

I	S_1, S_2 open
II	S_1 closed, S_2 open
III	S_1, S_2 closed

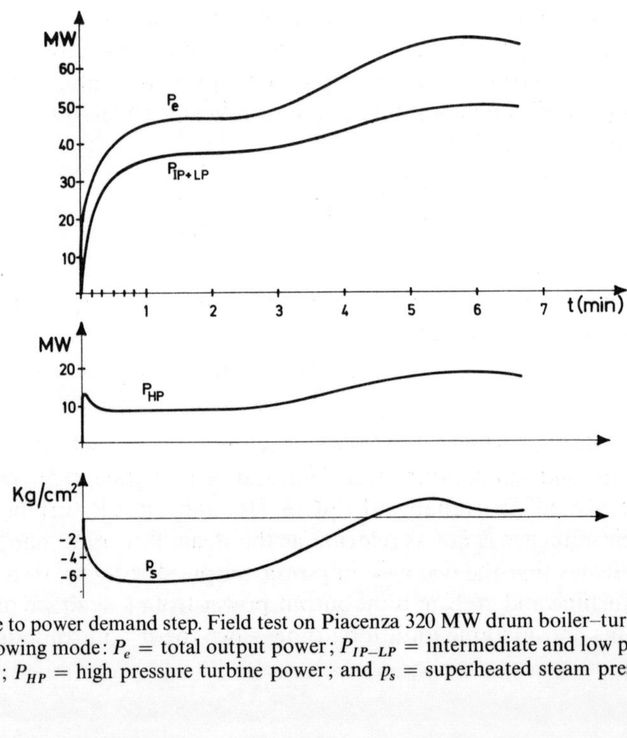

Fig. 5. Response to power demand step. Field test on Piacenza 320 MW drum boiler–turbine–generator unit, boiler-following mode: P_e = total output power; P_{IP-LP} = intermediate and low pressure turbine power; P_{HP} = high pressure turbine power; and p_s = superheated steam pressure.

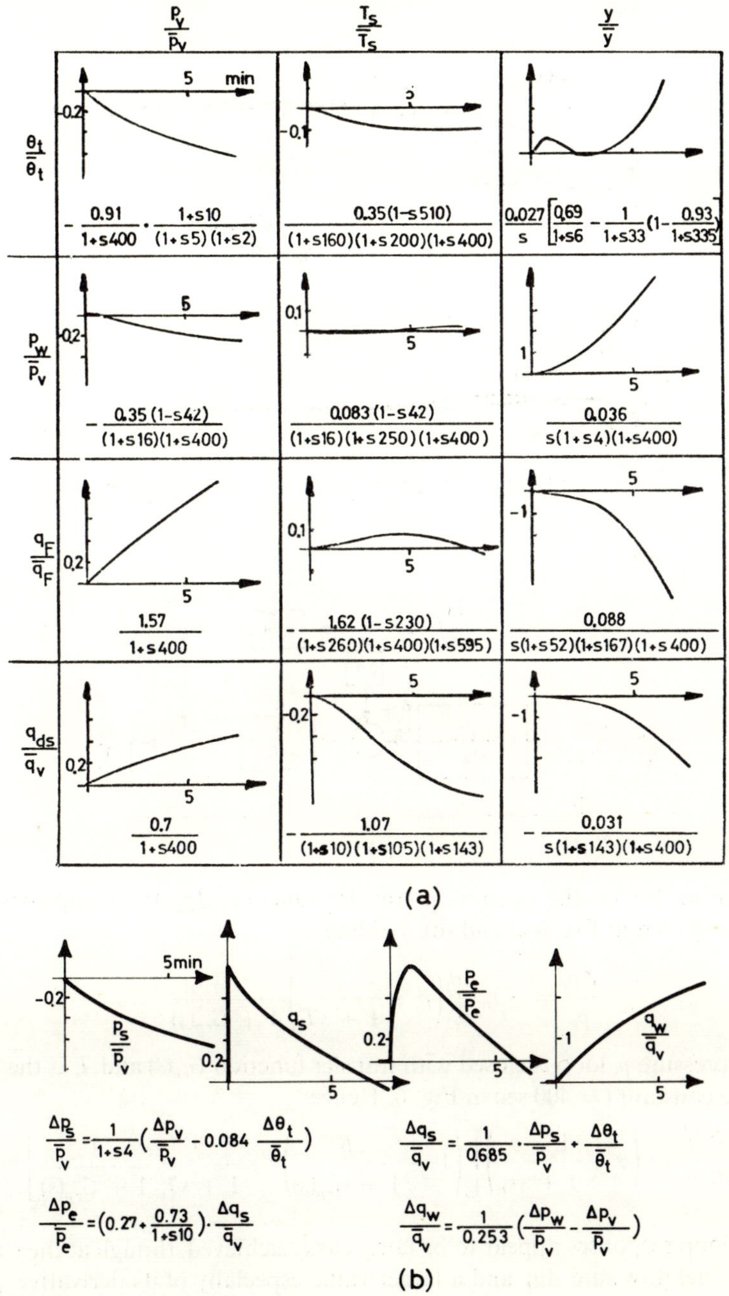

Fig. 6. Drum boiler, 160 MW, at 2/3 load.
(a) Transfer functions and unit step responses.
(b) Responses to a unit step in the turbine valve position.
(p_v drum pressure; p_w feedwater pump "internal pressure"; q_w, q_F, q_{ds} flow rates of feedwater, fuel and desuperheat of superheated steam.)

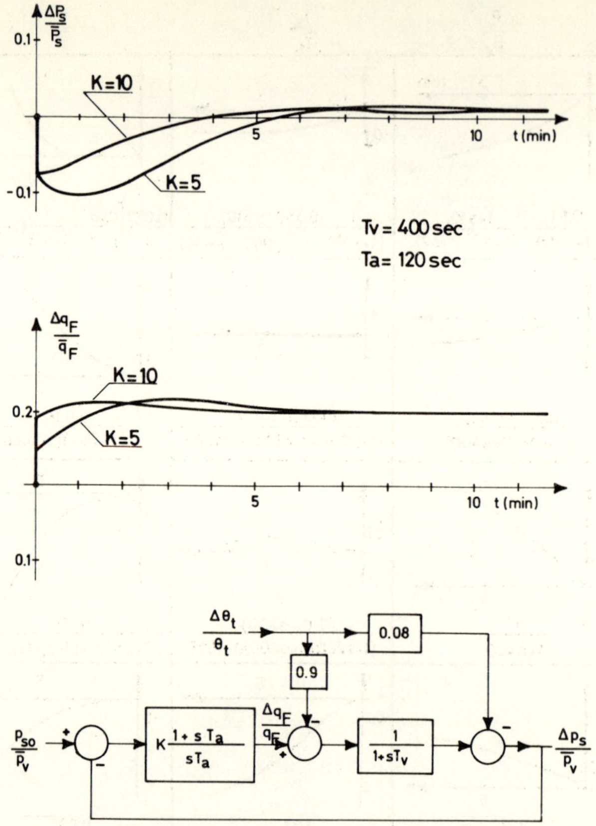

(a)

For a drum boiler, the open-loop transfer function $\Delta p_s/\Delta\theta_t$ is approximately of the type shown in Fig. 6(a) and (b), yielding

$$\frac{\Delta p_s}{\bar{p}_s} \doteqdot -r_t \frac{\Delta\theta_t}{\bar{\theta}_t}\left(\beta + \frac{\gamma}{1+sT_v}\right)\frac{1}{1+G_{rp}(s)} \tag{3}$$

when the pressure p_s loop is closed with transfer function $G_{rp}(s)$ and T_v is the boiler main time constant (= 400 sec in Fig. 6. Hence

$$\frac{\Delta P_e/\bar{P}}{\Delta\theta_t/\bar{\theta}_t} \doteqdot \left(\alpha + \frac{1-\alpha'}{1+s\tau_R}\right)\left[\left(1 - \frac{\beta}{1+G_{rp}(s)}\right) - \frac{\gamma}{1+sT_v}\frac{1}{1+G_{rp}(s)}\right] \tag{4}$$

Faster p_s loop responses appear to be fairly easily achieved, though at the expense of greater fuel flow rate Δq_F and a larger value especially of its derivative $\dot{q}_F$ (see Fig. 7, where the $\Delta p_s/\Delta q_F$ transfer function also has been approximated by a single lag of time constant T_v (see Fig. 6) the air flow to p_s transfer function being proportional to $\Delta p_s/\Delta q_F$, with p.u. gain -0.57, and very fast air-to-fuel ratio control). The situation $K = 10$, $T_a = 120$ sec, indicates a reasonable solution.

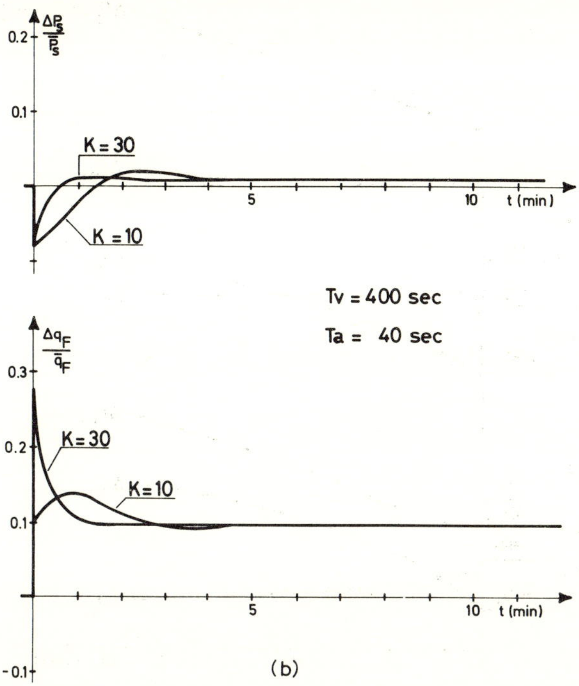

Fig. 7. Responses of the steam pressure control loop to a unit step in the turbine valve position—drum boiler, oil firing.

Figure 8 refers to a 630 MW once-through boiler–turbine generator, double reheat, solution I. Figure 9 shows the once-through unit response for case II, with a p_{cR} feedback loop. Such a loop should ensure that the steam flow rate is kept nearly constant for $t > 0$, the transfer function between power demand and power output now becoming nearly equal to the second term of eqn. (1), at the expense of a larger θ_t as long as $\Delta p_s < 0$.

Notice that the p_s response of the once-through unit is somewhat different from the drum boiler one, and slightly more oscillatory, consistent with open-loop boiler behaviour (see Fig. 10).

4.2.2 Turbine-following mode

(This is admission valve control just to move the throttle pressure back to the reference value when firing is cut down to 50% on pump loss, etc.) This mode has an essential role in cases of runback but is not used in normal operating conditions or in secondary control, due to its poorer response to power demands.

4.2.3 Coordinated boiler–turbine control mode

Typical examples of different control designs for this mode are shown in Fig. 12, La Spezia 630 MW once-through supercritical unit (see in Fig. 11(a) and (b) its responses for the cases with electric power feedback and with or without p_{cR} feedback) and in Fig. 13, Brindisi 320 MW once-through subcritical unit. The wide

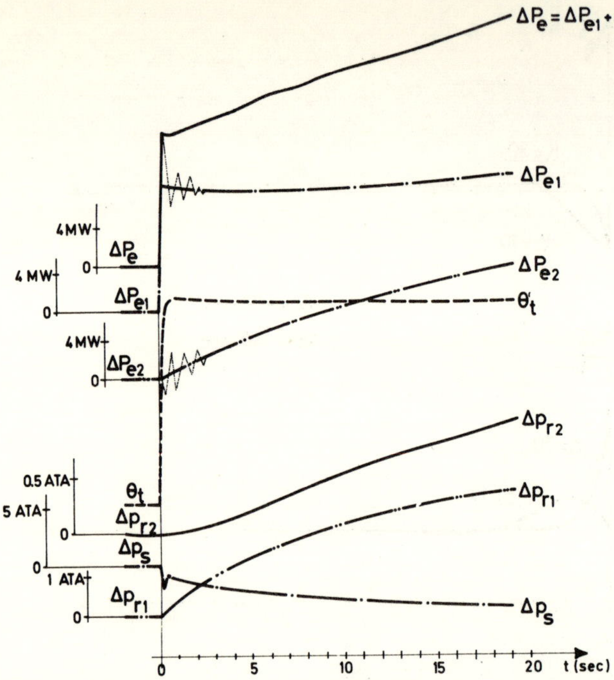

$P_o = 480$ MW

(a)

use of "load-demand" feedforward action is apparent in both [12], but a systematic resort to non-linear actions for ratio adjustments is characteristic of the second control structure.

In terms of the "non-interaction" concept, rather than of the current intuitive "energetic balance" interpretations, coordinated control can be easily defined as follows. If the only disturbance is power demand $\bar{P}_e$, then controllers G_{h1}, h $\neq$ 1, operating on error $\bar{P}_e - P_e$ can be chosen, except for possible realizability difficulties, such that always

$$\Delta p_s = 0, \qquad \Delta T_s = 0, \qquad \Delta T_R = 0, \qquad \Delta O_2 = 0$$

With reference to Fig. 14, where $\mathbf{PG} = \mathbf{A}$, the condition $Y_j = 0$, with $j \neq 1$, $Y_1 \neq 0$, implies $A_{j1} = 0$, *i.e.*

$$G_{h1} = \frac{\bar{P}_{1h} A_{11}}{|\mathbf{P}|} \tag{5}$$

where $\bar{P}_{1h}$ is the cofactor of P_{1h} and $|\mathbf{P}|$ is the determinant of $\mathbf{P}$ (if square).

A design based on eqn. (5) to start with would require knowledge of the plant transfer matrix $\mathbf{P}$. No need would arise, in principle, for other controllers acting on controlled variables other than P_e. However:

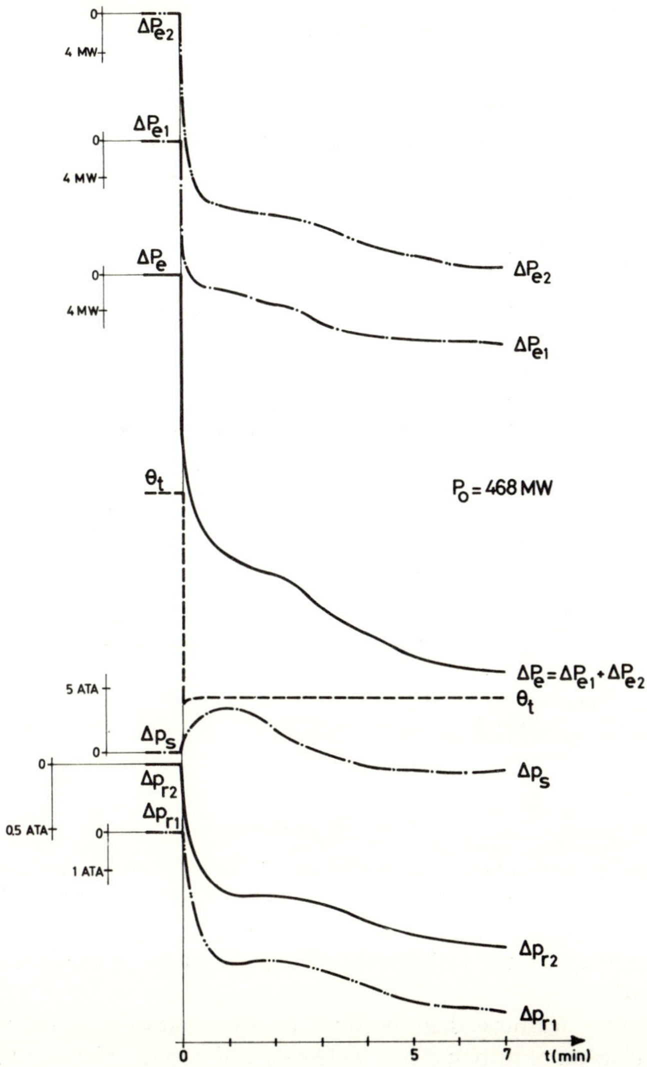

(b)

Fig. 8. Response to a power demand step. Field test on La Spezia 630 MW once-through boiler–turbine generator; speed governor on load-limiter: (a) positive step; (b) negative step. 1st shaft: supercritical stage + 1/2 LP turbine, electrical power P_{e1}. 2nd shaft: HP + 1/2 LP turbine, electrical power P_{e2}. (p_{r1}, p_{r2} first and second reheater outlet pressures.)

(1) disturbances as in §2.3 call for closed loops on p_s, T_s etc.;

(2) realizability difficulties are likely to exist, due to the form of some P_{ih}; and

(3) plant transfer function parameters are known with wide confidence limits and are strongly dependent upon the operating point.

Feedback loops on p_s, T_s etc. are thus necessary to take care of model inaccuracies and variations. Such variations may be very large with varying load, *e.g.* gains and

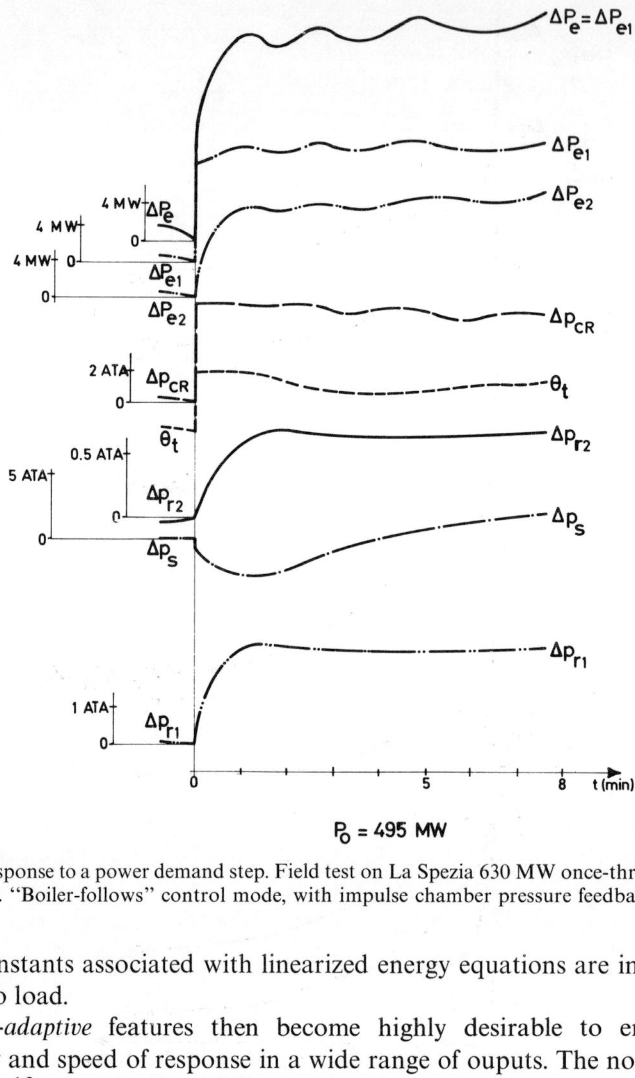

Fig. 9. Response to a power demand step. Field test on La Spezia 630 MW once-through boiler–turbine generator. "Boiler-follows" control mode, with impulse chamber pressure feedback. P_{e0} = 495 MW.

time constants associated with linearized energy equations are inversely proportional to load.

Load-adaptive features then become highly desirable to ensure optimum stability and speed of response in a wide range of ouputs. The non-linear control structure[13], Fig. 13, with inherent adaptivity, may show advantages here. In this respect, *digital computer control*, being typically suited to easy implementation of load adaptivity, has a point of preference. *Feedforward*, with adequate non-linear static "characterization" and dynamic compensation of manipulated-to-controlled variable transfer functions, might in principle serve exactly the same purpose in the same way as the above non-interacting structure. However, the practical merits of feedforward are considerably reduced by inaccuracy in static and dynamic characterizations and for the reasons mentioned earlier; hence, feedback loops are needed[14–16].

An *alternative design*, avoiding the pitfalls of non-realizability, is based upon the choice of a Kalman optimal state-feedback control matrix to minimize a quadratic

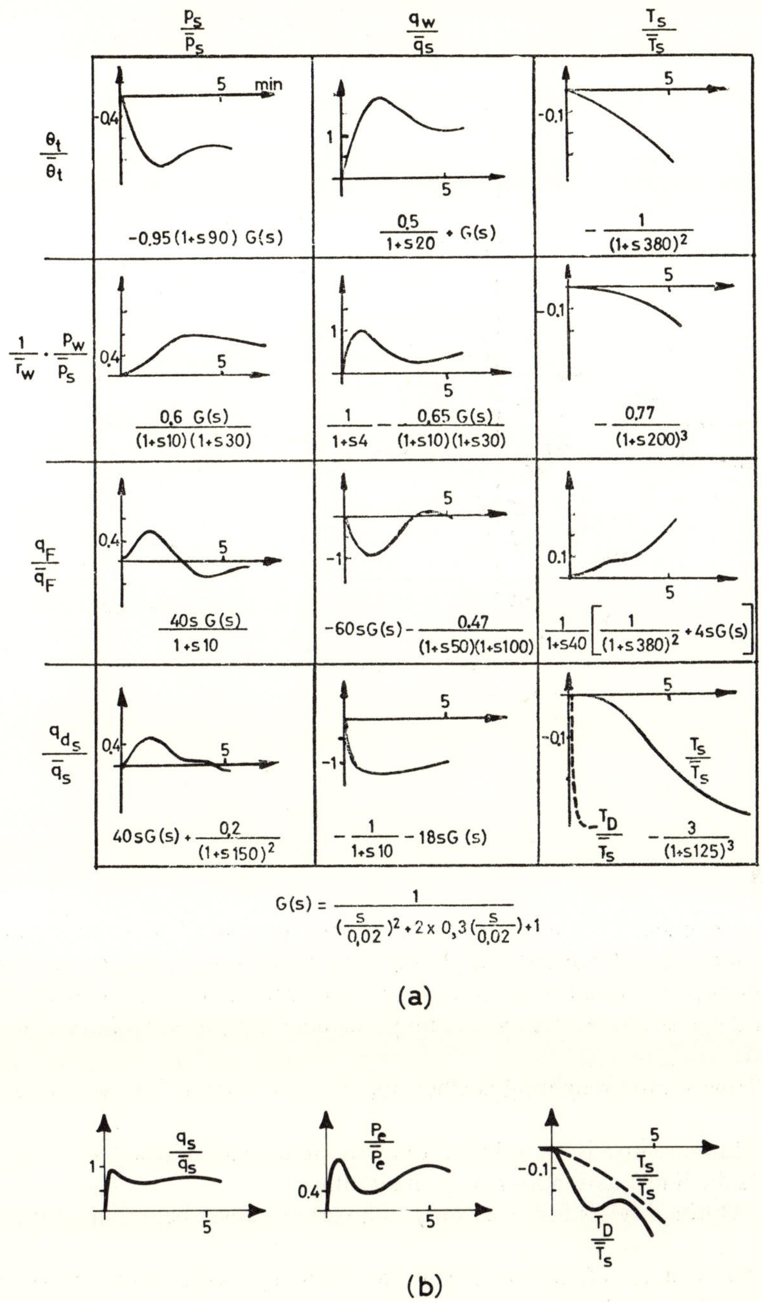

Fig. 10. (a) Once-through boiler, 250 MW, 1/2 load, transfer functions and unit step responses.
(b) Responses to a step in the turbine valve position.
(T_D : superheated steam temperature after desuperheat.)

References pp. 251–253

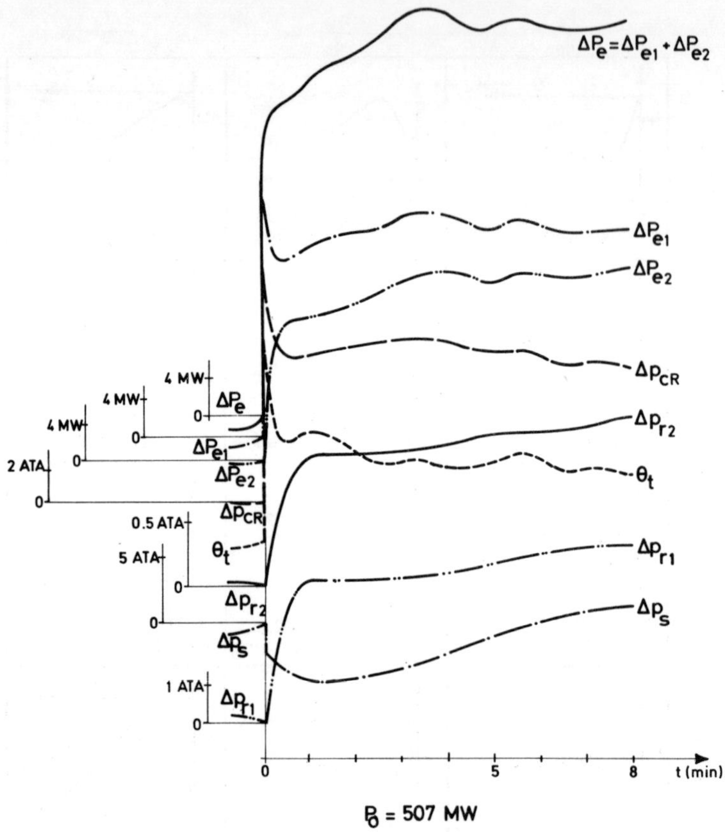

(a)

performance index[17,18]. Cases are reported in the literature[18] of optimal synthesis for the control of a 31st order drum-boiler model, with not too great deterioration of closed-loop performance when a sixth order accessible state reduced model replaced the 31st one. However, a number of practical problems have still to be answered:

(a) How should weighting coefficients on variable deviations be chosen (see §3)?

(b) How sensitive is the solution to plant model inaccuracies?

(c) Is the linear assumption a strong limitation?

(d) "Optimality" is achieved only in a certain sense; better structures may exist.

(e) Many states of the more elaborate (but necessarily only approximate) model are not accessible.

In any case, the optimal proportional state-feedback controller offers a good reference for the evaluation of alternative control structures by the same performance criterion.

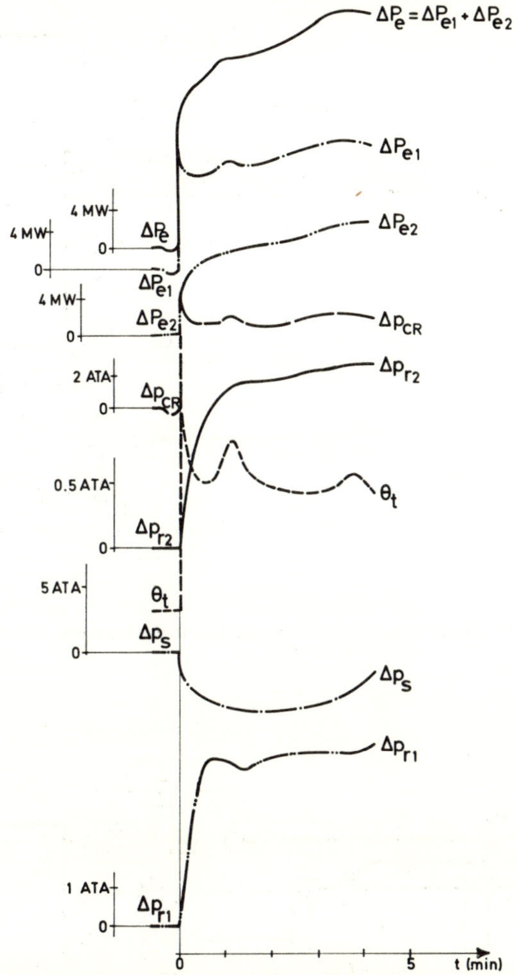

(b)

Fig. 11. Response to a power demand step. Field test on La Spezia 630 MW once-through boiler–turbine generator.

(a) Boiler–turbine coordinated control mode, with impulse chamber pressure feedback.

(b) Coordinated control, without impulse chamber pressure feedback.

A special mention should be made of temperature controls, not only because they pose the most difficult problems but because any improvement in them leading to reduced maximum transient deviations of T_s and T_R would allow increasing reference values without increasing material stresses; consequently, a better unit efficiency in steady state conditions might be achieved. More elaborate control structures, based upon accurate plant models, the use of suitable feedbacks from temperatures along boiler tubes and resort to adequate *dead-beat polynomial corrector digital control* algorithms[19] may therefore find here their economic justification. All designs should then be checked against the disturbances quoted

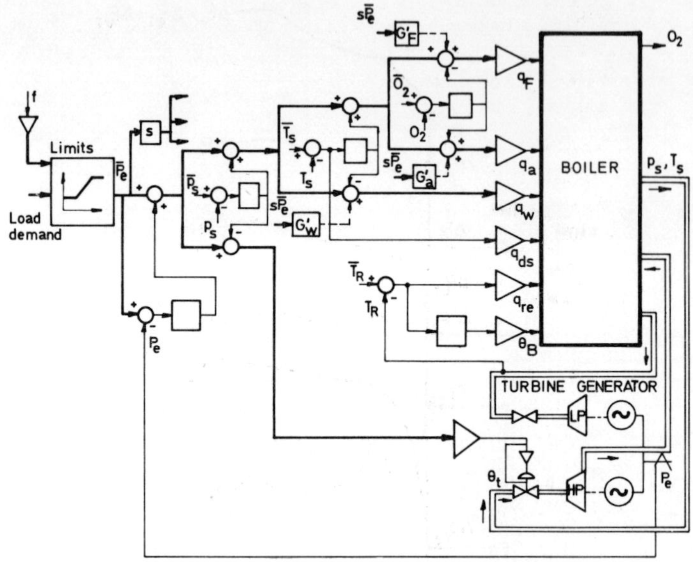

Fig. 12. Coordinated boiler–turbine control block diagram, 1st type.

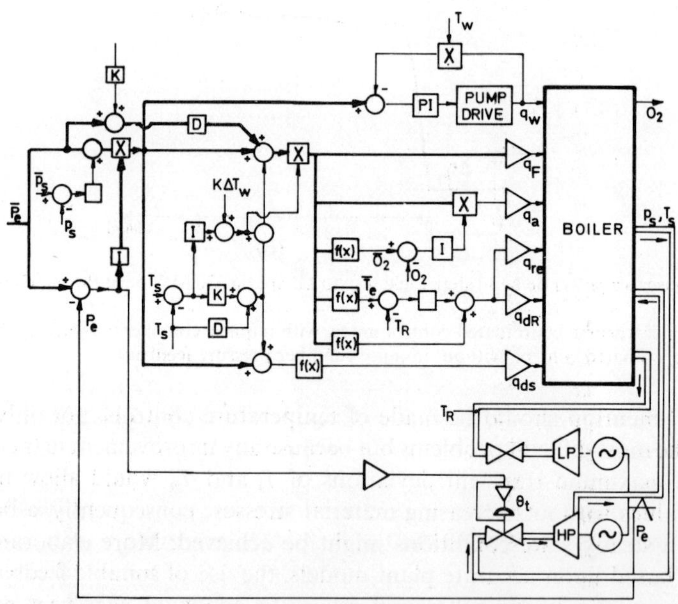

Fig. 13. Coordinated boiler–turbine control block diagram, 2nd type.

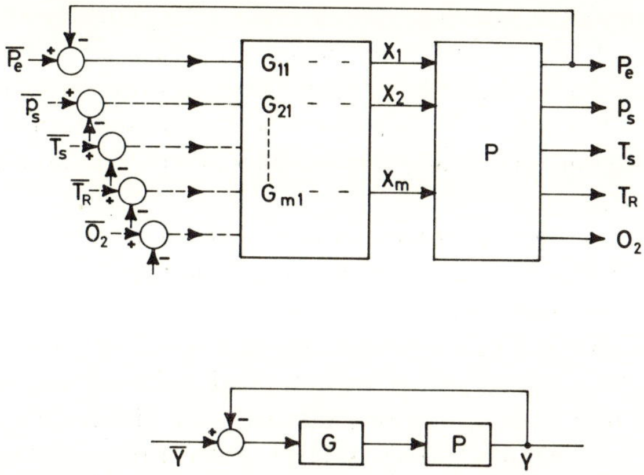

Fig. 14. Non-interacting boiler–turbine generator control.

in §2.6, and in general against *large perturbations.* "Thermal stresses" pose a major constraint upon maximum loading rates, magnitudes of abrupt load changes, start-up time and procedures, the time the unit can be kept on house-load after full rejection, etc.

Since the danger of permanent damage is associated with the integral deformation caused by repeated stresses, the constraint on each individual operation is more or less tight, depending upon the frequency of occurrence. For example, unit start-ups and loadings that should be repeated every day, or week, would certainly be subject to more severe constraints than load rejections hopefully occurring rather seldom. This explains why considerable effort has been spent in developing the application of *digital computers* or wired logic equipment to *automatic start-up*[20, 21]; the goal of easily repeatable optimally chosen procedures appeared to be worthwhile since the expected thermal stresses in all sections of the plant, including turbine, valves and superheater tubes, are definitely lower than in manual start-up. For hot restart of once-through supercritical units, another relevant argument in favour of using automatic computer control lies in the greater achievable speed in effecting all the necessary checks and control operations, with a higher probability of completing them in less than the nearly 10 min critical time for reaching maximum turbine blade transient stresses, beyond which a different and much slower start-up procedure must be followed. The advantage for the power system to have the unit back on service in a much shorter time is evident and important. While several types of recent big units accept lower repeatable loading rates than older units, although always admitting more than 1.5 %/min, most specifications today call for higher rates up to 5 %/min.

In *runbacks*, say from 100 % to 50 % load, the main limitations for drum boiler units do not come from thermal stresses but from drum level surges, a good performance of level controller being then essential. For once-through types, dangers may derive from imperfect calibration of water to fuel flow ratio, which may cause large temperature deviations after a few minutes; here again feedwater and fuel firing controls have a major role.

Full load rejection to house-load may cause heavy turbine stresses especially in supercritical units, due to the abrupt steam temperature decrease in the impulse chamber associated with the large pressure drop. In order to reduce them to safe values, the use of *interceptor valve* control has been proposed, to achieve a temporary increase in reheated steam pressure and consequently reduce the amount of impulse chamber pressure drop.

Another benefit of interceptor valve control is the achievement of a considerable reduction in resynchronization time after a turbine trip through smoother speed control and lower settling times.

Several other aspects of present experience and trends in power station controls might be worth at least a very brief mention, with reference to over-all system requirements and unit protection, for example: how drum critical low pressure limits[7] *versus* stresses in boiler tubes pose fast p_s control loop response requirements; how the power-frequency control participation band and average operating points are chosen with reference to material stresses, valve points, burner lighting sequence, mill, fan, pump margins of operating ranges; the trends to electrohydraulic governors to reduce dead-bands and avoid speed changer drive backlash; the wide use of *digital computers* for accurate and timely *supervision* in view of monitoring component deterioration; and on-line data processing for analysis of normal operation and faults, to the advantage of unit reliability and of improvements in design and operating procedures.

4.3 Nuclear plants

Because of their very high efficiency nuclear units are usually kept at base load, close to their maximum output, with no participation in system secondary control and often none also in primary frequency control.

However, their flexibility is in principle approximately the same as that of conventional steam units: full load rejection to house-load is admitted; the range of automatic control operation is from 15–20 % to 100 % load for PWR and AGR types; 10 % load steps and normal loading rates of 3 %/min for PWR, 5 %/min for BWR and more for AGR are tolerated.

Control structures[22–24] are, of course, different, since fuel firing is replaced by heat transfer control from reactor to coolant by gas or fluid flow rate modulation and by displacement of control rod positions in the reactor core. However, response speeds of the main output variables, including P_e and p_s, to θ_t are not very different from the ones obtainable for oil-fired boilers.

The experience so far gained on secondary control of nuclear units confirms that reactor behaviour would not pose any physical limitation to rather wide and fast output power variations. However, some evidence of deterioration in zircalloy fuel elements has lead to doubts whether they could stand repeated thermal stresses due to large and rapid rod displacements for reactivity control, while other more resistant material, such as stainless steel, has distinct disadvantages in neutron absorption. As a consequence, apart from efficiency considerations, reservations still exist about employing present nuclear plants in secondary control.

Space stability of neutron flux and temperature distribution in the core[24,25] with associated modulation of rod sector positions, together with unit protection and safety, are so wide and important subjects in themselves that even a superficial discussion of them would far exceed the limits of this paper. It may perhaps be enough here to say that existing protections and controls have ensured a nuclear plant availability which is now already quite comparable with that of conventional steam units. On-line digital computers have been systematically employed and are generally accepted for short-term plant monitoring and neutron flux distribution analysis, with clear merits in security assessment and in speeding up plant commissioning and the initial gradual increase in power output.

4.4 Hydroelectric plants

The different types of hydroelectric turbines

high head, large reservoir, impulse turbines
medium head, small pondage, Francis turbines
low head, run-of-the-river Kaplan and propeller units

for generating or pumping stations, or for pumping-and-generating plants, should be dealt with separately in view of their different features and responses. The discussion here is limited to a few general comments.

Firstly, "flexibility" of hydro units appears to be generally greater with respect to steam and nuclear units, although high-head turbines have loading rate limitations, usually through gate servomotor speed limitations, due to water-hammer in penstock; the normally acceptable 2–3 %/sec load pick-up rates are still relatively high. Unloading can be much faster, at the expense of some wasted water; start-up time is much shorter (except when the comparison is made with jet or gas turbines), with definite advantages for use as stand-by cold reserve. The participation band of reservoir units to system secondary control can be much wider and greater flexibility is added by the non-existence of short-term cost constraints.

Lower-head plants are currently employed for centralized power-frequency control if their water supply includes a reservoir, though small; otherwise, as in the case of run-of-the-river plants, since best water usage calls for using them at their maximum possible output, there is no participation in system secondary control except under very short-time emergency "boosting", as triggered by under-frequency-and-deceleration relays.

Secondly, hydro unit responses to dispatching commands or frequency deviations are not so fast, especially for high-head turbines, since frequency control bandwidth is limited by penstock resonant frequency. The effects of such limitations are especially evident at no-load before synchronization and in isolated network operation, when the controller transfer function, the power output-to-gate position static characteristic, the turbine generator inertia and the small perturbation hydraulic power-to-gate position transfer function[26]

$$\frac{\Delta P_h/\bar{P}_h}{\Delta\theta_t/\bar{\theta}_t} = G_w \doteqdot \frac{1 - 2\rho\tanh(s\tau/2)}{1 + \rho\tanh(s\tau/2)} \tag{6}$$

call for bandwidths ω_0 no greater than, say, 1/5 to 1/3 of penstock first resonant frequency $\omega_1 = \pi/\tau$ ($\tau/2$ is the sound wave transport lag from penstock outlet to surge tank; $\tau\rho = L\,\bar{V}/g\bar{H} = T_w$ is the water inertia constant, related to water speed V, penstock length L and head H). However, the above limitations exist also in normal operation, as all governor transfer functions add to determine the power system transfer function G_N (see Fig. 2).

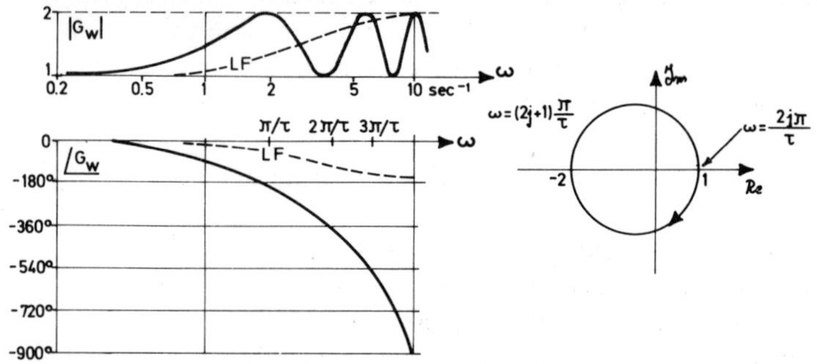

Fig. 15. Power output-to-gate position frequency response of hydraulic supply system (penstock).

Notice that eqn. (6) yields a phase lag of 180° at $\omega = \omega_1$ (see Fig. 15). The commonly reported medium-frequency approximation

$$G_w \doteq \frac{1 - sT_w}{1 + sT_w/2} \tag{7}$$

does not account for it; it holds only for much lower frequencies. As a proof of this statement, see Fig. 16 where the response of one of the Galleto hydroelectric units to a dispatching signal step is shown. The initial P_e *versus* θ_t behaviour is better described by a transport lag than by the sign inversion indicated by eqn. (7).

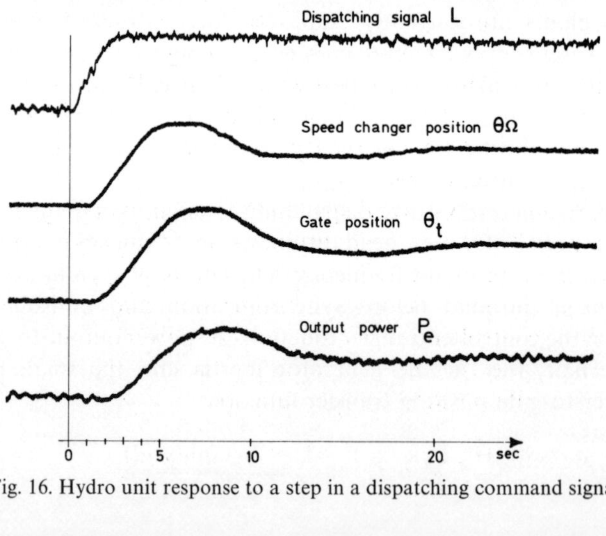

Fig. 16. Hydro unit response to a step in a dispatching command signal.

Notice that the above limitations in the bandwidth of the frequency control loop do not appear to be easily overcome by use of water pressure feedback, though minor improvements in bandwidth should be expected from it.

The output power-to-frequency transfer function of hydro units can be approximately expressed by

$$\frac{\Delta P_{rh}/\bar{P}}{\Delta f/\bar{f}} = G_{rh} \doteqdot \frac{1}{\sigma} \frac{1 + sT_a}{1 + sT_1} \frac{G_w}{1 + sT_2} \tag{8}$$

where $1/T_2$ stands for the wide bandwidth of the transient θ_t feedback loop in transient droop governors, or for a high frequency filter in accelerometric governors. If T_2 is neglected, at sufficiently low frequencies where $G_w \doteqdot 1$, expression (8) is similar to eqn. (1) (see also Fig. 2) and even the practical values of T_a ($\sim$1–6 sec, typically 2) and T_1 (4–25 sec, typically 8) are close to the corresponding values for the steam unit $\alpha\tau_R$ and τ_R.

Hydro unit controllers with suitable additional acceleration feedback signals have been occasionally employed[27] (Grand Coulee station) to introduce stabilizing effects for low frequency oscillations due to long tie lines, thus modifying the power system transfer function G_N (see Fig. 2).

It should be added that eqn. (8) does not describe the very low frequency behaviour of G_{rh} accurately. The water mass oscillations between surge tank and reservoir (periods are generally of a few minutes, with scarce damping at low load) call for a corrective term causing long tails in the power response and "noise" perturbations on system frequency. Furthermore, they set limitations on demands for fast large output power changes and especially on the timing of gate opening since the last closing[28].

Pumping units, which are generally not employed as continuously controllable loads, have however a relevant role in system-frequency control since they are the first loads to be shed in emergencies.

Pumping-and-generating units, although system operating economy would call them to follow a well-defined daily schedule, are often used for peaking and for secondary control.

In emergency, however, *e.g.* during the night, when they should be in the pumping mode, shift of each unit to generation may require minutes, which is too slow. Hence, in a station, it often happens that at night, when the need for avoiding waste of reservoir water calls for keeping only a few high-head units in operation, with less spinning reserve and big steam units on line and consequently greater danger, some units are kept in generating mode, no-load, as stand-bys for possible emergencies, even if this is against economy. The requirement of secure and fast control of start-up and changes of operating mode has led to the use of wired logic equipment or digital computers for automatic sequence control in pumped storage generating plants.

References pp. 251–253

5. VOLTAGE AND REACTIVE POWER CONTROL

5.1

As is well known, installed hierarchical systems for automatic voltage and reactive power control are as yet very few, almost all networks having only local controls in generating stations and in substations.

Recent trends[29–31] indicate considerable interest in the development of a three-level structure with a "secondary" control for set-point coordination of a.c. generators (or also other reactive power control means such as capacitors, shunt reactors, synchronous condensers, tap-changing transformers) within an "area" and a superimposed "tertiary" centralized control of areas for optimization. Whether manual or automatic, off-line or on-line, computer-aided or empirical, optimization should aim at ensuring acceptable quality of service in the face of highly varying demands of active and reactive powers during the day, with highest possible economy and security.

As a criterion for optimality, minimization of transmission losses (*i.e.* operating costs) subject to the constraint of keeping voltage variations at all significant busses within acceptable limits has been proposed. Alternatively, a reactive generation allocation to ensure that all generators are operated with the same "margin" with respect to the individual stability limits offers the advantage of better security, against a very slight increase in operation costs. In general, tertiary and secondary control demands may be expected to have some consequences on the type of information to be exchanged with generator regulators (for instance, position of the operating point with respect to stability limits) and also to affect the steady state performance requirements of generator voltage control. No matter what optimizing criterion is used at the "upper" level, or what more empirical provisions are decided by the dispatcher, their implementation at the power station or "area" level involves only a slow adjustment of set-points, with no appreciable demands on voltage regulator dynamic performance (*e.g.* the frequency bandwidth of the voltage control loop).

The situation may be somewhat different—but never to a large extent—with "secondary" reactive power control, since dynamic coordination of generator regulators, in order to avoid unnecessary exchange of power among generators with associated line losses and overloads, may suggest stricter specification of their desired responses. However, studies on secondary control are still underway and it is still too early to anticipate their conclusions.

Actually, network real requirements concerning dynamic response of voltage controls have not been well assessed quantitatively as yet. While it is apparent that correction of fast voltage fluctuations due to rapidly varying loads or voltage dips from other causes calls for fast response of excitation control (and also for enough ceiling in the field voltage)[32], further investigations are needed to arrive at a quantitative assessment of the desirable speed of response, and due account should be given to the danger that too large a bandwidth for the generator voltage control loop may imply a negative damping of the electromechanical oscillations as a consequence of the too high gain at their frequency.

This could be conflicting with the goal of fully exploiting alternators as sources of reactive power, especially for the cases of reactive power absorption when excitation control (and also governor action) is used to widen stability limits in the P–Q plane, as is described briefly below.

A reliable and rationally chosen limitation, in connection with thermal stresses and stability criteria, of the generator operating region in the P–Q plane is also a premise for a better automatic or manual coordination of different reactive power sources, with possibly lower "droops" in voltage regulation steady state characteristics and consequently better performance in terms of network voltage deviations.

6. STABILITY

6.1

The requirements posed by stability and by the damping of electromechanical oscillations concern both excitation controls and governors. Let us first consider *steady state stability*. It is known[33,34] that modern high-speed thyristor exciters with purely proportional transfer functions may lead to a narrower stability region in the P–Q plane, due to the higher gain of the voltage control loop at the electromechanical oscillation frequency and the already mentioned consequence of the lower or negative damping of such oscillations.

If some slowing down in the voltage response is acceptable for small variations (or for the "tail" of larger transients), the use of an adequate lead–lag network can reduce the voltage loop gain in a suitable frequency range and then restore or even broaden (see Fig. 17(a)) the stability region. Furthermore, it is also known[35–41,45] that the introduction of additional feedbacks from generator speed, frequency or output power may widen this region (see Figs. 17 and 18) and improve damping with less deterioration in voltage control dynamic performance.

Figures 19 and 20 are examples of experimental evidence of the improvement in damping obtainable by employing different additional feedbacks. In addition to speed and electric power (or "torque") supplementary signals (gains G_ω, G_c), a feedback from field current (gain G_f) is also considered, which may improve the damping without reducing the voltage control loop bandwidth.

If the schematization alternator–series reactance–infinite bus is acceptable, the speed signal and (if the governor action is neglected) the power signal are proportional respectively to the first and second derivative of angle δ; in the implementation of additional feedbacks the signal of the first derivative of angle may be obtained, with some practical advantages, by integration of the electric power deviation[42].

The damping action of additional feedbacks can be easily visualized (with some simplifications) in terms of elementary control theory, with reference to the open-loop transfer function of the "torque loop" (Fig. 21(b)), as derived from the linearized system block diagram shown in Fig. 21(a)[39].

	G_ω	G_c
1	-	-
2	$0.01\,\omega_s$	-
3	-	0.1

4 { Without add. feedbacks; With "lead lag" network in voltage control

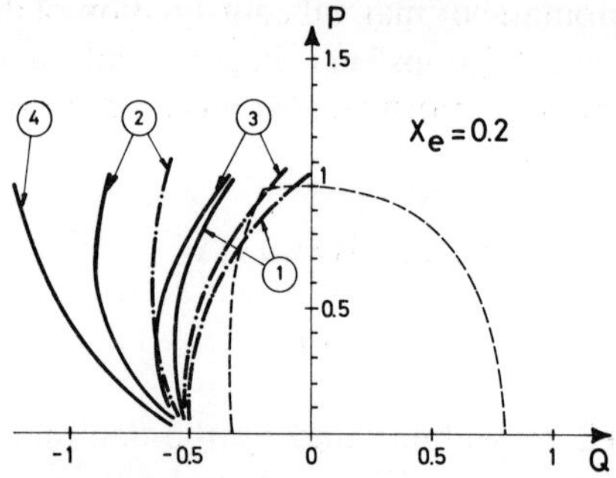

—·—·— Without dampers and saturations
——— With dampers and saturations
- - - - - Thermal limits

(a)

	G_ω	G_c
1	-	-
2	$0.01\,\omega_s$	-
3	-	0.1

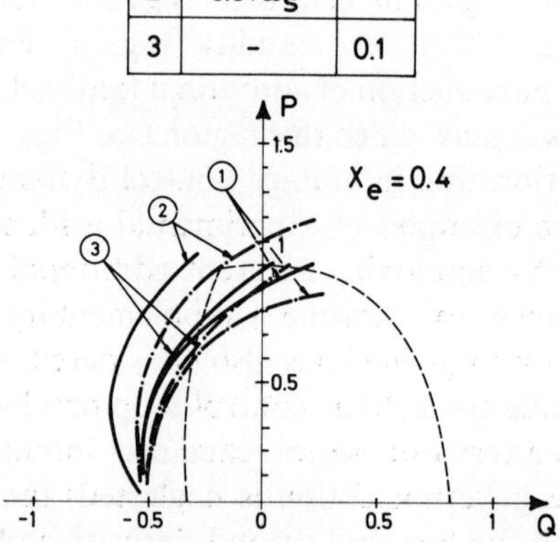

—·—·— Without dampers and saturations
——— With dampers and saturations
- - - - - Thermal limits

(b)

Fig. 17. Turbine generator connected to infinite bus through a series reactance x_e. Steady state stability limits with different additional feedbacks (governor action neglected). G_ω : gain of speed feedback (p.u.); ω_s : synchronous speed (rad/sec); G_c : gain of electric power feedback (p.u.). Alternator parameters: $x_d = x_q = 1.94$; $x'_d = 0.246$; $x''_d = x''_q = 0.15$; $T'_d = 1$ sec; $T''_d = T''_q = 0.027$ sec; $T_m = 8$ sec.

	G_ω	G_c
1	-	-
2	$0.01\,\omega_s$	-
3	-	0.1

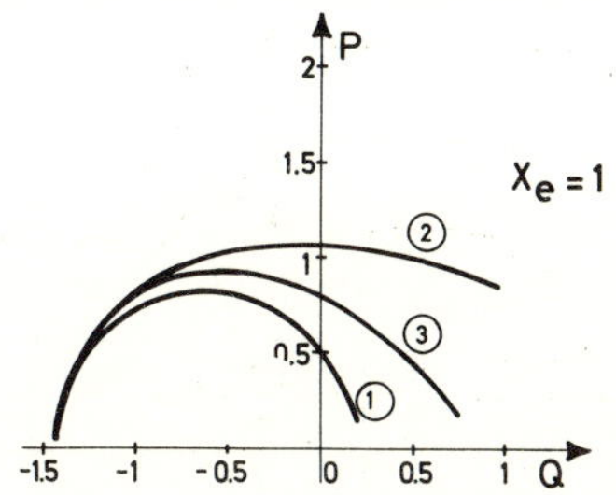

Fig. 18. Salient pole generator connected to infinite bus through a series reactance x_e. Steady state stability limits with different additional feedbacks (dampers, saturations, governor action neglected). G_ω: speed feedback gain (p.u.); ω_s: synchronous speed (rad/sec); G_c: electric power feedback gain (p.u.). Alternator parameters: $x_d = 1.27$; $x_q = 0.7$; $x'_d = 0.24$; $T'_{do} = 7.2$ sec; $T_m = 4.75$ sec.

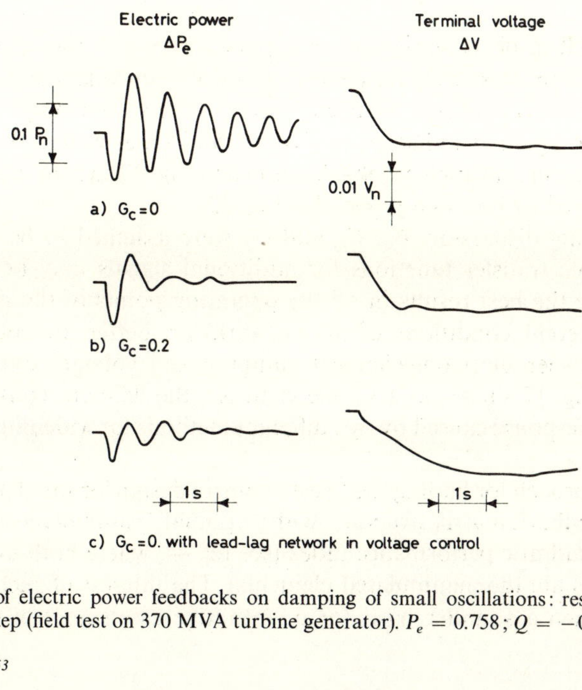

Fig. 19. Effect of electric power feedbacks on damping of small oscillations: response to a voltage reference step (field test on 370 MVA turbine generator). $P_e = 0.758$; $Q = -0.081$; $x_e \doteqdot 0.2$.

References pp. 251–253

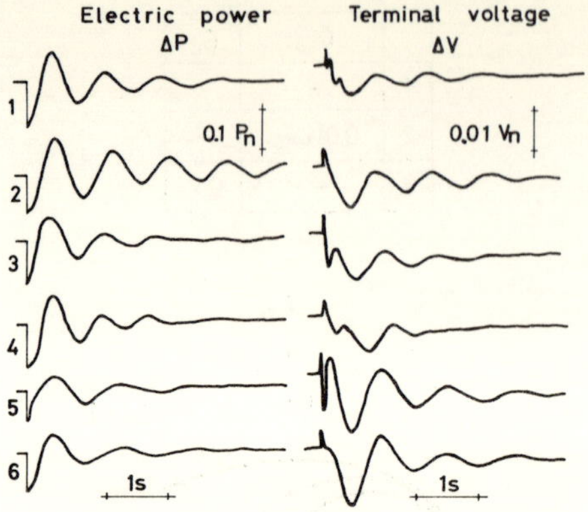

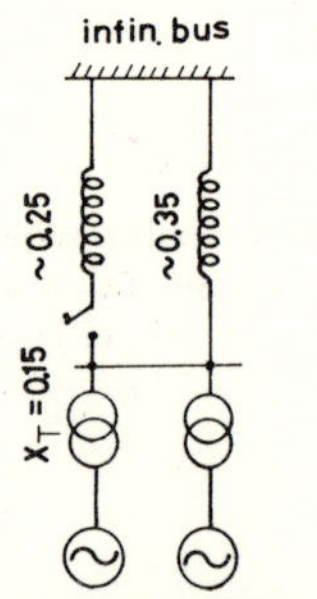

	μG_f	G_ω	G_c
1	7.5	—	—
2	2	—	—
3	7.5	0.01 ω_s	—
4	2	0.02 ω_s	—
5	7.5	—	0.2
6	2	—	0.2

Fig. 20. Effect of different additional feedbacks on damping of oscillations: switching of a line (field test on turbine generator). G_f : field current feedback gain (p.u.); ω_s : synchronous speed; G_ω : speed feedback gain (p.u.); G_c : electric power feedback gain (p.u.). $P_e = 0.75$; $Q \doteq 0$; $\mu \doteq 50$.

If account is taken of the variations of k_i parameters with the operating point, and the torque loop cross-over frequency *versus* the equivalent time constant $\tilde{T}$, the simplified analysis confirms the effectiveness of the second angle derivative in the underexcited region as indicated by Figs. 17 and 19, as well as the interaction between field-current and angle feedbacks shown in Fig. 20 and the worse damping that may be caused by too high μ (see also Fig. 22).

In the preceding discussion, G_c, G_ω and G_f were assumed to be constant; in general[44] suitable transfer functions for additional signals may be searched in order to achieve the best results in all the operating points of the machine (and possibly for different conditions of the network) or, better, to obtain the best compromise between electromechanical damping and voltage control dynamic performance. (Fig. 17 shows, as a simple example, the different consequences on voltage control response caused by two different methods for widening the stability region.)

A possible approach for finding the "best" control design for small perturbations uses the state-feedback matrix structure, with "optimal" parameters determined by minimizing a quadratic performance index (see ref. 44, where both excitation and admission valves are the manipulated elements). The interest of such studies will be enhanced when a sounder basis for quantitative assessment of the dynamic

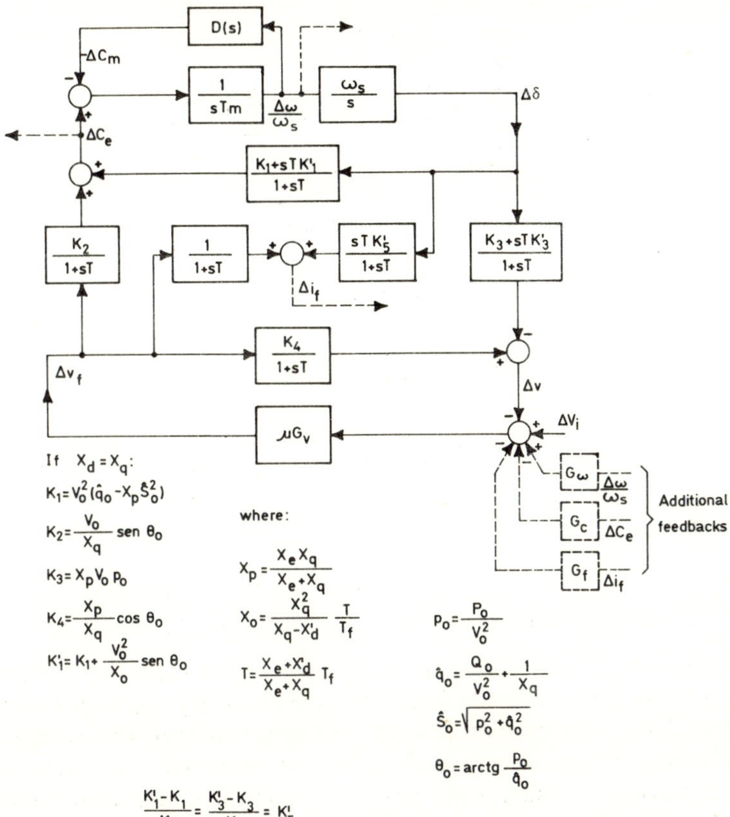

(a)

Fig. 21(a). Block diagram of a generator connected to infinite bus through a series reactance x_e (dampers and saturations neglected).

requirements of voltage control, and consequently less arbitrary penalties in the performance index, are available.

As another method, perhaps more suitable to the more important case of on-line optimal control of large perturbations than to operating-point dependent control design for small perturbations, the bang-bang technique has also been proposed or investigated[36,45–47]. However, not all the conceptual and practical difficulties in applying such a technique to a multi-machine system, where the reference angle cannot be uniquely identified, have been overcome as yet.

The introduction of angle derivative signals in excitation control makes terminal voltage, and therefore reactive power, sensitive to network frequency variation (and also to possible mechanical power variations). These generally undesirable effects can be reduced by a suitable low frequency cut-off in the transfer functions G_ω and G_c. In any case there appears to be a frequency range (which can also be determined on the basis of frequency variation spectra (see Fig. 1)) where these effects set an upper limit on the gain of G_ω and G_c. When a stronger damping action

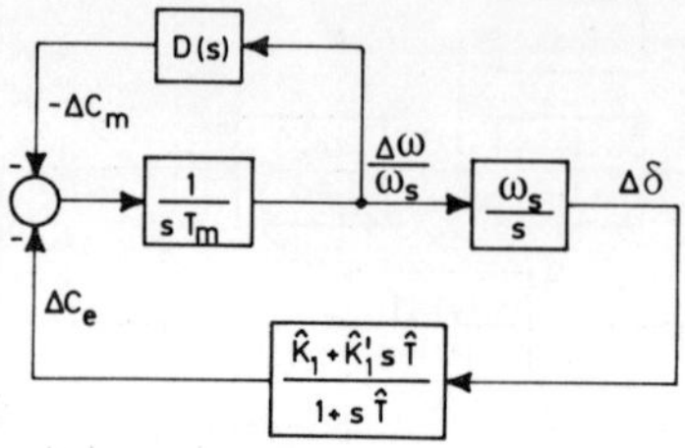

$$\hat{K}_1 = \frac{K_1(1+\tilde{\mu}K_4)+\tilde{\mu}K_2K_3}{1+\tilde{\mu}(K_4+K_2G_c)}$$

$$\hat{K}_1' = K_1' + \frac{\tilde{\mu}K_2G_\omega}{\omega_s\tilde{T}}$$

$$\hat{T} = \frac{\tilde{T}}{1+\tilde{\mu}(K_4+K_2G_c)}$$

$$\tilde{T} = \frac{T}{1+\mu G_f} \qquad \tilde{\mu} = \frac{\mu}{1+\mu G_f}$$

(b)

Fig. 21(b). Block diagram equivalent to Fig. 21(a) (G_c, G_ω, G_f constant; $G_v = 1$).

than is allowed by such limitation in the gains of the additional feedbacks is required, a more elaborate control may become necessary, *e.g.* with resort to parameter switching or to some other kind of adaptive control. This is especially true with reference to large disturbances, as will be described in the next paragraph.

In the previous considerations governor action has been disregarded ($D = 0$ in Fig. 21). However, its contribution to positive or negative damping may sometimes be relevant. If the gain of $D(j\omega)$ at the frequency ω_r of electromechanical oscillations is sufficiently high, the speed governor may improve, or deteriorate, damping depending upon the phase of $D(j\omega_r)$ (see Fig. 21(b)). Improving damping may call for a large frequency bandwidth of the speed loop, such as might be obtained from modern steam turbine governors; a check should be made whether so large a frequency bandwidth is acceptable in normal service—switching of parameters may be desirable instead.

Notice that the variation of ω_r which may be caused by the introduction of additional feedbacks (see Fig. 22) points to a possible justification for a coordinated design of voltage and speed controls.

6.2

The stability requirements for *large perturbations* are mainly related to the behaviour of the generating units in the case of transmission line faults. They include:

(1) for "first swing" stability the reduction, if possible, of the acceleration of units closest to the point of fault (caused by the reduction in the electric power output due to the fault) and

(2) a good damping of the subsequent oscillations.

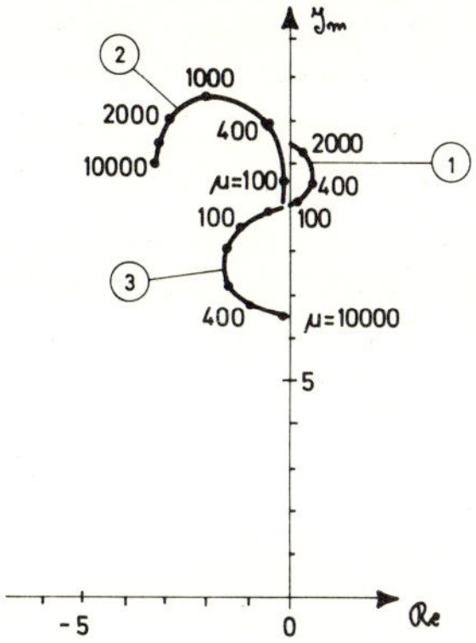

	G_ω	G_c	G_f
1	-	-	-
2	$0.01\,\omega_s$	-	-
3	-	0.1	-

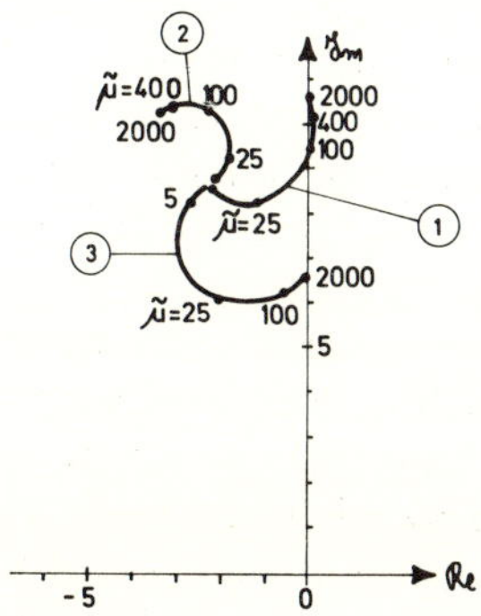

	G_ω	G_c	μG_f
1	-	-	7
2	$0.01\,\omega_s$	-	7
3	-	0.1	7

$X_e = 0.2$

$$\tilde{\mu} = \frac{\mu}{1 + \mu G_f}$$

Fig. 22. Turbine generator connected to infinite bus. Root loci (complex poles) with different additional feedbacks ($D = 0$; dampers and saturations neglected; alternator parameters as in Fig. 17; $P_e/V^2 = 0.9$; $Q/V^2 = -0.215$).

The first requirement has strongly influenced the design specifications for the excitation system (ceiling field voltage and, especially, highest field voltage gradient) in an attempt to increase the electrical braking torque.

Recently, steam turbine "fast valving" (especially with action on interceptor valves) has been given considerable attention[48–50] with a view to reducing the accelerating torque during the fault. Many problems arise with this solution, such as turbine and reheater stresses, valve wear-out, valve re-opening times, time lags in the valve opening-to-mechanical power transfer function and discrimination among different faults, especially in the case where many machines are involved. The choice of the most convenient specifications for excitation and admission valve control from the standpoint of first swing stability derives from a compromise between performance and cost.

A quantitative expression for this compromise is particularly difficult here, if fault probability and power system security are taken into account. Indeed, the pulling out-of-step of a unit for very unlikely contingencies can be accepted, provided resynchronization is fast and network security is not appreciably affected.

References pp. 251–253

As the first swing stability is affected by the generator operating point, these considerations may also affect the allowable operating region of generators, and therefore the total availability of reactive power (see the British experiments on two-field generators and the associated control problems[51,52,46]). The demand for rapid resynchronization, besides influencing the specifications concerning unit capability of fast re-excitation after a long fault[53] and operation on house-load, may affect governor requirements. A suitable modulation of excitation during asynchronous operation has also been studied[38].

6.3

Achieving good damping of oscillations after a fault is an obviously desirable goal that may influence the choice of the transfer functions on the additional feedbacks for excitation control (even for small variations, the best damping does not always coincide with the largest widening of the stability region) as well as governor transfer function. Damping may become a more stringent requirement when there are dangers for the absolute stability associated with possible out-of-steps occurring at subsequent swings or with negative damping of tie-line power oscillations among weakly connected areas[42,54,55].

As the analysis of these cases requires non-linear multi-machine models and, on the other hand, the damping can be carefully evaluated only if detailed machine models are employed, no general conclusion has so far been reached on the possibility of overcoming these problems by relatively simple means, such as the use of additional feedbacks with linear control in many generators. The elimination of scarcely or negatively damped tie-line oscillations may justify costly sophistications in control designs, at least for some "critical" generators. (The introduction of a two-field synchronous condenser in critical points of the network has also been suggested[56].)

In such cases, a centralized second-level control unit might be superimposed on the normal valve and excitation control loops, in order to adapt their parameters or to override them and assume the direct digital control of valves and thyristor firing devices on the occurrence of a specified event in the network (large perturbations) or of a received signal from an upper-level centralized control.

7. CONCLUSIONS

In conclusion, we realize that our review of the role of generating station controls in power system operation has been largely incomplete and unbalanced.

To digital computer control we have devoted only a few brief comments, and too little space has been left to the comparative analysis of alternative modern designs of coordinated boiler–turbine generator controls. We have emphasized the need for establishing performance criteria for generating plant controls in connection with system requirements, and have indicated the incompleteness of present specifications, but have not provided an answer.

However, we hope to have been able to show that the problems associated with rational design of station controls are interesting and challenging for designers and users and well deserve greater attention from automatic control engineers.

REFERENCES

1 IEEE Committee Report, Power plant response, *IEEE Trans.*, **PAS-86** (1967) 384–395.

2 U. Di Caprio, Identificazione sperimentale della densità spettrale di potenza delle variazioni aleatorie di carico agenti su reti elettriche, *LXXII Riunione Annuale AEI, Venezia, 1971,* memoria n. 6.26.

3 M. Cuénod and G. Quazza, Dynamic statistical analysis of electric power system controls, *IFAC Symp. on System Dynamic and Automatic Control in Basic Industries, Sydney, 1968.*

4 Ch. Concordia, S. Lalander and B. Favez, Rapport d'activité du Comité d'Etude 32, *CIGRE (Conférence Internationale des Grands Réseaux Electriques, 1966,* Rapport 334.

5 Ch. Concordia, Effect of prime-mover speed control characteristics on electric power systems performance, *IEEE Trans.*, **PAS-88** (1969) 752–756.

6 G. Quazza, Criteria for equitable participation of areas in tie-line power and frequency control of an interconnected power system, *Automazione e Strumentazione,* **14** (March 1966).

7 O. W. Durrant and R. P. Sigfried, Operation of drum and once-through boilers during electric power system emergencies, *ASME-IEEE Joint Power Generation Conf., Detroit, 1967,* paper 67-PWR-15.

7a H. Hale and J. H. Carson, Stored boiler energy gives quickly load response, *Electrical World,* **170** (Sept. 9, 1968).

8 H. T. Akers, J. D. Dickinson and J. W. Skooglund, Operation and protection of large steam turbine generators under abnormal conditions, *IEEE Trans.*, **PAS-87** (1968) 1180–1185.

9 R. H. Hillery and E. D. Holdup, Load rejection testing of large thermal electric generating units, *IEEE Trans.*, **PAS-87** (1968) 1440–1449.

10 E. Buzzi, Le facteur "temps" dans le démarrage et dans les variations de charge des turbo-alternateurs de grande puissance, *Journées Intern. D'Etude des Centrales Thermiques et Hydraulique modernes, Liège, 1972,* organized by AIM (Association des Ingénieurs de Montefiore), and *La Termotecnica,* **17** (1963) 206.

11 Ch. Concordia, F. P. deMello, L. K. Kirchmayer and R. P. Schulz, Effect of prime mover response and governing characteristics on system dynamic performance, *American Power Conf., Chicago, 1966.*

12 F. Läubli and F. Fenton, The flexibility of the supercritical boiler as a partner in power system design and operation: Pt. I—Theoretical relationships; Pt. II—Application and field test results, *IEEE Trans.*, **PAS-90** (1971) 1725–1733.

13 O. W. Durrant and J. K. Loeser, Boiler-turbine control system for application to universal pressure boilers, *IEEE Meeting, 1963,* report CP-63-1410.

14 D. J. Ahner, C. E. Dyer, F. P. deMello and V. C. Summer, Analysis and design of control for a once-thru boiler through digital simulation, *ISA Natl. Power Instrumentation Symp., Detroit, 1966.*

15 W. Friedenwald, Control and regulation in steam power plant, *Regelungstechnische Praxis,* **3** (1968) 98–104.

16 J. Debelle, Reglage automatique des générateurs de vapeur, paper presented at the *Seminar on "Projet Chance", Brussels, 1968,* organized by IBRA (Institut Belge de Régulation et d'Automatisme).

17 H. Nicholson, Dynamic optimisation of a boiler, *Proc. IEE,* **111** (1964) 1479–1499.

18 J. H. Anderson, H. W. Kwan and D. J. Leffen, Dynamic control of a 200 MW steam generator, *Proc. PICA (Power Industry Computer Applications) Conf., Boston, 1971,* organized by IEEE, paper XIV B.

19 H. Dang Van Mien, Etude des méthodes de régulation numérique, EDF (Electricité de France), *Bull. Direction Etudes et Rech.*, sér. A, Pt. I(1) (1968); Pt. II(2) (1968).

20 A. G. Montagna, Control system and automation of high capacity steam power units, *Intern. Seminar on Automatic Control in Production and Distribution of Electrical Power, Brussels, 1966,* organized by IBRA (Institut Belge de Régulation et d'Automatisme), Presses Academ. Européennes, Brussels; Dunod, Paris.

21 C. Agrò, Souplesse des centrales de grande puissance: démarrage, coupure de charge, variations

de charge, *Journées Intern. d'Etudes des Centrales Electriques Modernes, Liège, 1970*, organized by AIM (Association des Ingénieurs de Montefiore).

22 M. CALCAGNO, G. FOTI AND G. GARAGNANI, Evoluzione dei sistemi di supervisione, protezione e regolazione automatica nelle centrali nucleari di potenza, *LXX Riunione Annuale AEI (Associazione Elettrotecnica Italiana), Rimini, 1969*, memoria n. 4.1.04.

23 G. HASTINGS AND J. LOUIS, A control design development for HTGR power plant by digital computer simulation, *Proc. PICA (Power Industry Computer Applications) Conf., Boston, 1971*, organized by IEEE, paper XIV-A.

24 D. G. SCHULTZ, *Control of Nuclear Reactors*, McGraw-Hill, New York, 1961.

25 L. E. WEAVER, *Reactor Dynamics and Control*, Elsevier, Amsterdam, 1968.

26 G. QUAZZA AND F. SACCOMANNO, Considerazioni sul proporzionamento dei regolatori di frequenza per turbine idrauliche in relazione alle caratteristiche della rete, *LXII Riunione Annuale AEI (Associazione Elettrotecnica Italiana), Ancona, 1960*, memoria no. 160.

27 F. R. SCHLEIF, G. E. MARTIN AND R. R. AUGELL, Damping of system oscillations with a hydrogenerator unit, *IEEE Trans.*, **PAS-87** (1967) 438–442.

28 M. FANELLI AND F. SACCOMANNO, Performance of hydro-stations with dual-chamber surge tanks during cyclic operation, *Water Power*, **21** (1969) 421–427.

29 I. HANO, A. CHIBA, U. UMEZU AND ÖDE, Optimal control of voltage and reactive power in power systems, *Conférence Internationale des Grands Réseaux Electriques à Haute Tension (CIGRE), 1968*, Rapport 32-03, pp. 1–7.

30 K. EDWIN, Übergeordnete Spannungs-Blindleistungs-Regelung durch Prozessrechenanlagen, *Elektrotechnik und Maschinenbau*, (6) (1969) 229–233.

31 Z. SCZERBA, Réglage intégré de la puissance réactive et du niveau de tension dans une centrale électrique de grande puissance, *Conférence Internationale des Grands Réseaux Electriques à Haute Tension (CIGRE), 1970*, Rapport 32–09, pp. 1–9.

32 G. HOSEMANN, Die stromrichtererregte Synchronmaschine im Netzbetrieb, *Elektrizitätwirtschaft*, **5–8** (1963) 591–599.

33 F. REGGIANI AND F. SACCOMANNO, Considerations upon stability margins with special regard to the effect of generator, excitation system and voltage-regulation characteristics, *Conférence Internationale des Grands Réseaux Electriques à Haute Tension (CIGRE), 1966*, no. 330, pp. 1–30.

34 F. KATTELUS, Effect of excitation control on the steady-state stability of turbogenerators, *Sähkö-Electricity in Finland* (1966) no. 11.

35 V. A. VENIKOV, V. M. NOVITSCKII AND V. A. SHTROBEL, High-response regulation of excitation by the third and fourth derivatives of the absolute angle, *Elektrichestvo*, **3** (1964) 118–128.

36 P. DANDENO, A. KARAS, R. MCCLYMONT AND W. WATSON, Effect of high-speed rectifier excitation systems on generator stability limits, *IEEE Trans.*, **PAS-87** (1968) 190.

37 F. R. SCHLEIF, H. D. HUNKINS, G. E. MARTIN AND E. E. HATTAN, Excitation control to improve powerline stability, *IEEE Trans.*, **PAS-87** (1968) 1426–1434.

38 J. L. DINELEY, A. J. MORRIS AND C. PREECE, Optimized transient stability from excitation control of synchronous generators, *IEEE Trans.*, **PAS-87** (1968) 1696–1705.

39 E. FERRARI, R. FLORIS AND F. SACCOMANNO, Limiti di stabilità di turboalternatori con eccitazione statica e diverse strutture del controllo sull'eccitazione. Parte I: Analisi della stabilità per piccole variazioni, *LXX Riunione Annuale dell'Associazione Elettrotecnica Italiana (AEI), Rimini, 1969*, memoria n. 4.1.19.

40 R. T. BYERLY, F. W. KEAY AND J. W. SKOOGLUND, Damping of power oscillations in salient pole machine with static exciters, *IEEE Trans.*, Paper 69 TP 735-PWR, pp. 263–273.

41 D. OEDING AND P. NEMETZ, Stability and voltage regulation of large turboalternators in power systems; effect of machine data and excitation systems, *IEEE Trans.*, Paper 70 CP 199-PWR, pp. 1–10.

42 C. DUBÉ AND J. LEMAY, Improvement in system stability with effective static excitation control, *Power System Computation Conf. (PSCC), 1969*, Paper D. & C. 3, pp. 1–10, Appendix pp. 1–4 and 19 figs.

43 F. P. DEMELLO AND C. CONCORDIA, Concepts of synchronous machine stability as affected by excitation control, *IEEE Trans.*, **PAS-88** (1969) 316–329.

44 YAO-NAN-YU, KHIEN VONGSURIYA, L. N. WEDMAN, Application of an optimal control theory to a power system, *IEEE Trans.*, **PAS-89** (1970) 55–62.

45 H. GLAVITSCH, Möglichkeiten der Verbesserung der Stabilität und der Spannungsregelung von Synchronmaschinen mit Hilfe der Gleichrichtererregung, *Elektrotechnik und Maschinenbau* (20 April) (1966) 58–65.

46 S. Takada, E. Ohta and Y. Naganuma, Suppression of hunting in two-axis synchronous machines by control of field voltage, *Elec. Eng. in Japan,* **88** (1968) 28–37.

47 J. L. Dineley and A. J. Morris, Synchronous generator transient control: Part II—Theory and evaluation of new control techniques, *Proc. PICA (Power Industry Computer Applications) Conf., 1971,* pp. 191–200.

48 F. P. DeMello, D. N. Ewart, M. Temoshok and M. A. Eggenberger, Turbine energy controls aid in power system performance, *American Power Conference, Chicago, 1966,* pp. 438–445.

49 P. G. Brown, F. P. DeMello, E. H. Lenfest and R. J. Mills, Effects of excitation, turbine energy control, and transmission on transient stability, *IEEE Trans.,* **PAS-90** (1971) 1247–1252.

50 W. A. Morgan, B. Peck, D. R. Holland, F. A. Cullen and J. B. Ruzek, Modern stability aids for Calvert Clipps units, *IEEE Trans.,* **PAS-90** (1971) 1–10.

51 J. A. Soper and A. R. Fagg, Divided-winding-rotor synchronous generator—A comparison of simulated 30 MW conventional- and divided-winding rotor turbogenerators, *Proc. IEE,* **116** (1969) 113–126.

52 M. Rama Murthi, D. Williams and B. W. Hogg, Stability of synchronous machines with 2-axis excitation systems, *Proc. IEE,* **117** (1970) 1799–1808.

53 E. Andres and J. Lakota, Discussion of Special Report 11.00, *Conférence Internationale des Grands Réseaux Electriques à Haute Tension (CIGRE), 1968,* Group 11, Generators.

54 H. M. Ellis, J. E. Hardy, A. L. Blythe and J. W. Skooglund, Dynamic stability of the Peace river transmission system, *IEEE Trans.,* **PAS-85** (1966) 586–600.

55 O. W. Hanson, C. J. Goodwin and P. L. Dandeno, Influence of excitation and speed control parameters in stabilizing intersystem oscillations, *IEEE Trans.,* **PAS-87** (1968) 1306–1312.

56 P. C. Krause and J. N. Rowle, Synchronous machine damping by excitation control with direct and quadrature axis field windings, *IEEE Trans.,* **PAS-88** (1969) 1266–1274.

Discussion

J. P. Waha (*INTERCOM, S.A., Brussels, Belgium*)

Drum-type boilers are normally non-interactive. This is no longer true at low load and pressure conditions, like those of start-up. An example will clarify this. During start-up, there is interaction between the drum level and steam pressure, because enthalpy brought by feedwater in the drum is appreciable compared with the calories coming from the heat flux. This phenomenon disappears naturally at higher pressure. Could a coordinated boiler control, including multipliers, cope with such a situation?

G. Quazza

Although our direct experience with transient evaporation phenomena in start-up of drum boilers is relatively scarce, we agree that interaction effects may be more relevant during start-up than in normal operation. A coordinated and adaptive control system would then appear to be most suitable, provided of course that the mathematical model of the physical phenomena within the boiler is sufficiently well known. We can quote at least one case of once-through boiler start-up and low load operation when the control system structure and actions had to be substantially revised, by introducing non-linearities and changes of fluid paths through suitable valves, and altering controller transfer functions, in order to obtain stable and fast response. The new control design could be decided only after identifying the mathematical model of the thermo-hydraulic phenomena in the boiler, including fluid flow instabilities due to temperature effects on steam density. The choice of non-linear control actions was determined as a consequence of the non-linearity of the physical process to be controlled.

A. H. GLATTFELDER (*Eidg. Technische Hochschule, Zürich, Switzerland*)

You mentioned a successful application of Kalman state feedback to boiler power control. Could you give us some further information about it?

(1) How does this solution compare dynamically with the conventional feedforward–feedback approach in steady state operation?

(2) Have variations of transfer function coefficients been incorporated?

(3) How does this system behave on start-up or large, fast load variations compared with the above-mentioned, conventional approach, considering saturations?

A quantitative performance index approach being too complicated, control performance is specified by tolerances on the controlled variables. As controller settings are determined by one specified disturbance, the problem should be overdetermined by using more than one test disturbance.

Which disturbance is the most restrictive on the controller settings?

Would you suggest an adaptation of controller settings to disturbance?

G. QUAZZA

The applications of the Kalman optimal state feedback matrix to boiler–turbine control mentioned in our paper are described in refs. 17 and 18. Further work has been done on this subject elsewhere: see, for example, ref. 57 where a Luenberger state estimator has been used in connection with the synthesis of a once-through boiler optimal control system, and ref. 58 where a preliminary account is given of a study on a drum boiler optimal controller with a Kalman filter for state variable estimation.

The above papers supply only a partial answer to the questions posed by Dr. Glattfelder, and more generally to the ones included in our report. All of them refer to digital or analogue simulations, none having had as yet the support of an application to a real plant. References 17, 18 and 59 include performance comparisons with conventional controllers, showing some merits in the optimal state feedback approach—even when an accessible state reduced order model is employed[18]—but the conventional control structure used as a reference may not have been truly representative, at least because no feedforward was used, and the case of wide variations was not explored.

When Kalman optimal controllers, with due integral action, are actually implemented on a power generating plant, a more reliable assessment of their advantages will be possible. Digital or analogue simulation may, however, contribute to a reasonably complete comparison, provided the analysis is not limited, as in refs. 17, 18, 57 and 59, to small perturbations around a given operating point, but includes all significant working conditions, such as start-up, runback after loss of auxiliaries, and load rejection or when load changes are large. These studies, plus a sensitivity search to account for model parameter uncertainties and variations, are often done to check the validity of conventional control designs, and ought then *a fortiori* to be included in optimal control synthesis. To our knowledge, although the digital computers proposed for optimal control are ideally suited to parameter adaptation *versus* load-varying transfer function coefficients (see refs.

17, 19, 58 and 60) and also to boiler on-line identification, no analysis of desirable completeness is as yet available. Even the application to non-linear boiler models[61] should still be considered as somewhat academic. However, work under way in some research centres (see, for example, development of refs. 58 and 60) should contribute to filling this gap.

To the second set of Dr. Glattfelder's questions, concerning control system performance and specifications, we would first point out that giving tolerance ranges does not necessarily lead to overdetermining the problem of controller synthesis. Indeed, infinite solutions may still exist, some of them being, of course, more expensive than others, but all equally within tolerances. Although it may be true that in practice some conditions pose higher requirements than others, it should be recognized that, while faster but limited-amplitude load changes are more restrictive in specifications on maximum transient errors, the case of approximately constant load is more significant in limiting steady state errors and noise, and the fast wide load variations refer to stress-limited operation with the associated different control system structure or mode. Since tolerances on maximum errors in operating conditions 1, 2, 3 and 4 are different, specifications may then be consistent, none of them being in itself more restrictive, *i.e.* inclusive of any other.

As to controller setting adaptation to disturbance types, it can safely be stated that it is not only desirable, but already followed in existing designs. Indeed, in programmed loading, start-up and shut down, not only are controller settings adjusted as functions of load, but different control loops are switched on and off depending upon the operating point, or the rotor speed, or the time elapsed from the last shut down, etc. The same applies to runbacks due to loss of auxiliaries, or to load rejection and emergency trips.

Of course, controller parameter adaptation is possible and realizable only for given or easily identifiable types of disturbances, having known dynamic consequences on the plant.

57 R. BRESCHI, Applicazione delle tecniche di controllo ottimo a un modello di caldaia Benson, *Rendiconti 70 Riunione Annuale AEI, Rimini, 1969.*

58 M. BONGHI, Applicazione di un calcolatore al controllo digitale diretto di un gruppo termoelettrico, Presented at *Convegno ANIPLA "I calcolatori nei processi industriali", Bari, 1971.*

59 H. NICHOLSON, Dynamic optimization of a boiler–turboalternator model, *Proc. IEE,* **113** (1966) 385–399.

60 M. BONGHI AND G. A. MONTAGNA, Computer application to steam station control in Italy—Part II: Application to direct digital control, *UNIPEDE, Data Processing Conf., Lisbon, June 1971.*

61 H. NICHOLSON, Integrated control of a nonlinear boiler model, *Proc. IEE,* **114** (1967) 1569–1576.

K. N. STANTON (*Systems Control, Inc., Palo Alto, California, U.S.A.*)
What is ENEL experience in supplementary control signals?

E. FERRARI
Although relevant theoretical and experimental studies[39, 62] have been performed in this field, no angle derivative feedbacks have been employed in excitation control systems for Italian generators as yet. Their introduction in thyristor exciters for turbine generators has been planned for future installations. The above studies have shown that a field current negative feedback, as used by an Italian manufacturer, can improve the damping of electromechanical oscillations (see Figs. 20

and 22 of the report). In static exciters the electrical torque feedback has appeared to be useful for widening the stability limits in the underexcited region, with less deterioration of the voltage control loop (see Fig. 19). The speed feedback signal has also been studied; its damping effect may be improved if a field current feedback is added (see Figs. 20 and 22). Due to some difficulties in implementing the signal by a speed transducer, as evidenced by our direct experience, the integration of the electrical torque signal is probably a more convenient alternative than a direct measure of speed deviation. Studies on the most convenient feedback transfer functions are still under way.

Since with these feedbacks the voltage reference will be sensitive to random variations of frequency and mechanical power, the gain of the feedbacks will be limited (especially in the low frequency band) in order to avoid undesirable oscillations of reactive power.

Generally, the most suitable transfer function should be based on a compromise between voltage control performance and the damping of electromechanical oscillations. Switching of parameters has not been considered so far.

62 V. Arcidiacono, E. Ferrari and R. Floris, Limiti di stabilità di turboalternatori con eccitazione statica e diverse strutture del controllo sull'eccitazione—Parte II: Studi al calcolatore analogico e rilievi sperimentali, *Report 4.1.20/1969, Rendiconti della LXX Riunione Annuale AEI*, 1969.

J. L. Dineley (*University of Newcastle-upon-Tyne, England*)
Can you please comment upon the use of feedback signals derived from the time-dependent state of the generator and turbine, such as rate of change of rotor angle used in conjunction with the voltage regulator, or rate of change of rotor angular velocity used in conjunction with the governor, to improve the stabilizing behaviour of these controls.

Do you agree with my experience that, for example, rate of change of rotor angle feedback is found to increase the damping of rotor angle oscillations, whilst rate of change of rotor angular velocity reduces the amplitude of rotor angle swings?

E. Ferrari
Some information about our studies and experimental results on angle derivative feedbacks in excitation control has been given in the answer to Dr. Stanton's question. If the alternator–infinite bus model is assumed, and the governor action may be neglected at the frequency of electromechanical oscillations, the speed and torque signals represent, of course, the first and second derivatives of torque angle. Concerning the possible governor action for stability, we think that a large frequency bandwidth in the speed control loop of steam generators could give an appreciable improvement in the damping of electromechanical oscillations, and that a speed derivative signal could enhance this action, if the phase lag of the speed control transfer function (open loop) at oscillation frequency ω_r is sufficiently low, by reducing such phase lag. However, we have no experimental results in this field, except in one case when the band of the speed governor was larger than usual but not large enough to give a phase lag lower than 90° at ω_r, and therefore the stability conditions were negatively influenced by the speed governor.

Coming to the last question, the answer would depend on which transients (*e.g.* small variation or faults in the network) are referred to in the remark about amplitude of rotor angle swings. The first swing amplitude after a fault is hardly affected by additional signals. In the subsequent swings, both first and second derivative feedbacks may improve the oscillation damping, as in the case of small variations[62]. The discrimination between the amplitude reduction of (non-linear) swings and the oscillation damping was not evident in our results.

For small variation, the damping improvement given by the considered feedbacks can easily be seen in the case of turbine generators if it is noticed that, in the torque loop shown in Fig. 21(b), the ratio $\hat{K}'_1/\hat{K}_1$ of open-loop "pole" and "zero" increases if G_ω or G_c are increased. The effectiveness of G_ω or G_c may also be influenced by the value of the cross-over frequency of this loop, *i.e.* approximately the frequency of electromechanical oscillations, *versus* the value $\hat{T}$; thus, when comparing the experimental improvements obtained with these feedbacks by different investigators, the values of oscillation frequencies should be taken into account (see also the results of Concordia and De Mello[43] about the best phase of additional feedbacks *versus* the oscillation frequency). On the other hand, an additional feedback from the field current may improve the effectiveness of the first derivative signal, also through its effect on the $\hat{T}$ value.

The second derivative feedback may be especially effective in the underexcited operation. If $G_c = 0$, the value of $\hat{T}$ rises rapidly when the internal angle approaches 90° (see expression for K_4 in Fig. 21(a)); this rapid variation is avoided if $G_c > 0$, and thus the allowable underexcitation is increased.

All these discussions are to some extent qualitative, since here damper windings (which improve the damping) are neglected, but they have generally been confirmed by results of simulations and experiments. We have had no direct experience so far on the effect of these feedbacks in the case of poorly damped low frequency oscillations between weakly connected "areas".

Hierarchical System Theory and Electric Power Systems

FRED C. SCHWEPPE AND SANJOY K. MITTER

Electrical Engineering Department, Massachusetts Institute of Technology, Cambridge, Mass., U.S.A.

SUMMARY

Hierarchical, multilevel control theory is a rapidly expanding but still immature field. In some respects, the theory is actually "behind" practice as many "newly developed" concepts of hierarchical, multilevel theory are already being used in electric power systems. However, an understanding of hierarchical concepts, as they exist at present, can help to clarify some of the many possible trade-offs in the control and operation of electric power systems. Furthermore, the available theory can be directly applied to some existing problems and hopefully the class of applications will expand as the theory itself expands.

In this paper, we will discuss some of the basic aspects of hierarchical systems, namely

(1) types of decomposition,
(2) control structures,
(3) information structures, and
(4) data flow.

These basic ideas will be related to a limited class of bulk electric power problems associated with real-time control (computer and operator) for economic and reliable operation. No attempt will be made to survey all available theory and power applications, or to give detailed mathematical results and descriptions of specific applications.

1. INTRODUCTION

Most complex systems consist of many interacting subsystems. Due to economic, political and social constraints as well as consideration of complexity and reliability, some "decomposition" of the over-all system to achieve "decentralized" control is almost mandatory. Hierarchical system theory deals with problems associated with such a decomposition and the resulting need for coordination between the many individual controllers.

From a hierarchical system theorist's point of view, electric power systems are perfect examples of applications of the theory. However, the electric power system of today has evolved to its present stage on the basis of need and good engineering practice and was not really conscientiously designed using hierarchical theory. In fact, in some respects, the theory is still trying to catch up with practice. Nevertheless, an understanding of hierarchical theory can clarify many of the trade-offs that are possible in electric power system control and operation and can point out new applications of practical value. In this paper, we attempt to provide such understanding. In the interest of keeping the length of the paper within reason, we discuss only general aspects of hierarchical theory without including mathematical details. We should also emphasize that there is no widely accepted definition of what hierarchical system theory really is. Hierarchical theory is sometimes considered to include only decentralized optimization techniques for a deterministic situation but we are taking a broader point of view which also includes uncertainty and information concepts.

Because electric power systems have a multifaceted character, we are forced to confine our discussions to limited aspects of the over-all system. We have chosen to concentrate on electric bulk power system problems associated with the real-time control for economic and reliable operation. We will consider "system levels" ranging from individual power plants and substations up to an interconnected system. We will concentrate on problems associated with system response times varying from about one second to several hours.

Hierarchical system theory can in principle deal with both competitive and cooperative situations. Since we are concerned with interconnected power systems we have a cooperative situation in mind. In a cooperative situation there may be a single goal or many goals. For an interconnected power system we are really in a single goal situation, *viz.* to satisfy customer demands at lowest cost subject to the system being sufficiently reliable. However, since a certain amount of decentralization is mandatory for an interconnected power system, the goals of the individual subsystems may be somewhat conflicting. A certain amount of conflict may also be introduced by decomposition.

This paper is divided into seven sections. In §2 we discuss three different ways in which a system can be decomposed: by level, by time and by mode. In §3 we consider the concept of information structure. In §4 we discuss the "main" problem of this paper: coordination of "decentralized" controllers which control the individual subsystems of the decomposed system. In §5 the problems of data flow, routing and coding are considered. Section 6 is concerned with trade-offs in the design of hierarchical systems. Section 7 contains a brief summary.

2. DECOMPOSITION

Decomposition is a key concept in hierarchical theory and there are many types of decomposition depending on the system and problem of interest. Three types of decomposition are considered in this paper:

Level decomposition
Time decomposition
Mode decomposition

These appear to us to be the most important decompositions relative to electric power systems but other possibilities do exist, *e.g.* it is possible to decompose the control function into automatic and human control. We will discuss these three levels primarily by use of examples. Precise scientific definitions will not be attempted.

2.1 Level decomposition

Decomposition of power systems according to level is usually a geographically related or spatial concept. Level decomposition is decentralization as understood in classical economics. It is often motivated by a desire to retain local autonomy as much as possible. An example of a three-level decomposition is:

Level 1: Power plant–substation level
Level 2: Individual system level
Level 3: Interconnected system level

In this case, there may be, for example, one interconnected system level (Level 3) which covers a large geographical area (say 10^6 km^2). Within this single highest level there are many (say 10) individual power systems (Level 2) which cover smaller areas (say, on the average, 10^5 km^2). Within each individual power system there are very many (say 50) individual power plants and substations (Level 1), each of which covers a very small area (effectively a point). Obviously, still higher and lower levels could also be defined. Decomposition by level gives rise to the "pyramidal" structure illustrated in Fig. 1.

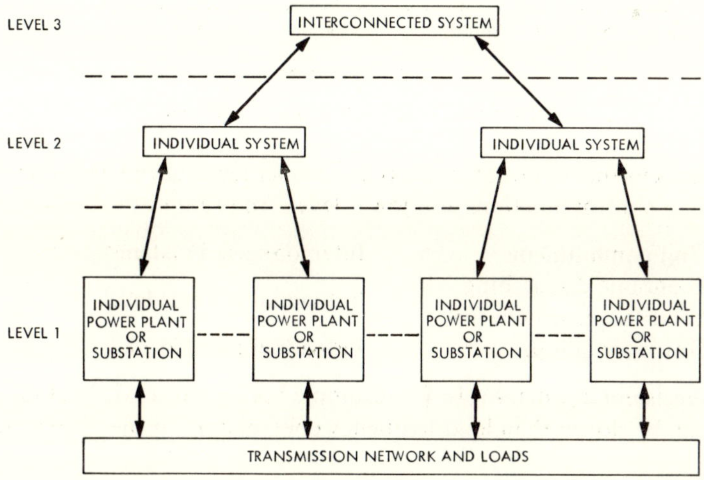

Fig. 1. Level decomposition (pyramidal structure).

Another example of level decomposition for power systems is in terms of the "bulk system" level (generators and EHV transmission networks, 230 kV and above) and the "distribution system" level (138 kV lines and below). However, we

References p. 276

are confining our discussion to bulk systems so this type of level decomposition is not discussed further.

2.2 *Time decomposition*

Decomposition by time arises naturally because of the extremely wide range of response times inherent in an electric power system. Time decomposition is almost always motivated by a desire to subdivide a difficult problem into smaller sub-problems.

An example of time decomposition is the control of individual power plants' real output as follows:

Control functions	*Time scale*
Unit commitment	hours
Economic dispatching	minutes
Load-frequency control	many seconds
Governor action	fewer seconds

There are still slower and faster functions than the above time decomposition, *e.g.* "maintenance scheduling" has a time scale of days while "relay action" is faster than governor action. Decomposition in time can be envisaged in terms of the sequential structure shown in Fig. 2.

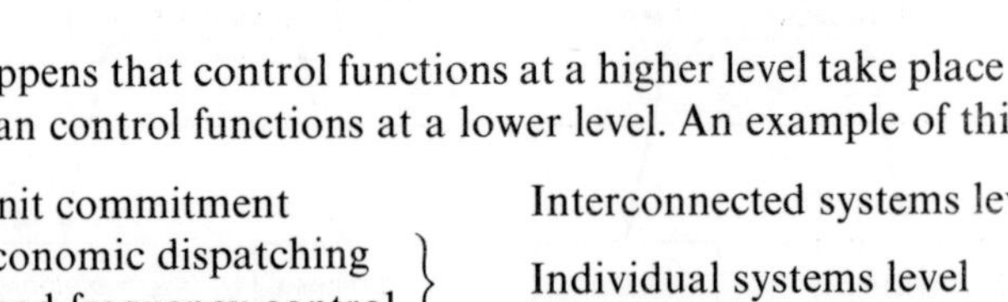

Fig. 2. Time decomposition (sequential structure).

It often happens that control functions at a higher level take place with a slower time scale than control functions at a lower level. An example of this is:

Unit commitment	Interconnected systems level
Economic dispatching Load-frequency control	Individual systems level
Governor action	Individual plant level

This, however, is not a general rule. For example boiler control done at the power plant level can be slower than load-frequency control done at the systems level.

2.3 *Mode decomposition*[1]

Analysis of system stability, load flow conditions, etc. yields certain "conditions of operation" (in the sense of continuous operation) for the power system. The need for decomposition by mode arises from a need to control the power system in all conditions of operation. Since the control strategies for different conditions of operation are likely to be qualitatively quite different, mode decomposition helps one to subdivide the over-all problem into subproblems dependent on the operating states. For power systems we consider the following mode decomposition:

(1) Normal mode
(2) Preventive mode
(3) Emergency mode
(4) Restorative mode

In the normal operating state the power system is being operated so that the demands of all customers are satisfied at standard frequency and voltage. Moreover the contingencies that are likely to occur in the immediate future are not likely to change the operating state to an emergency mode. An example of a control strategy for the normal mode is:

Keep frequency at approximately 60 Hz
Keep tie-line flow at about schedule
Meet power demand at minimum cost

If a contingency is likely to occur such that the system may not be able to return to the normal mode of operation, then the system goes to the preventive mode of operation. The assumption here is that by taking corrective action in anticipation the system can be made to remain in the normal operating mode. An example of a control strategy for the preventive mode is:

Keep frequency at approximately 60 Hz
Keep tie-line flows at about schedule
Meet power demand at minimum cost subject to constraint
on amount of spinning reserve

It is possible and often desirable to combine the normal and preventive modes into a single mode.

If a contingency has occurred such that the customer demand cannot be maintained at prescribed voltages and frequencies, then the system goes to the emergency mode of operation. Here economics is completely sacrificed. An example of an emergency mode control strategy is:

Try to keep frequency at approximately 60 Hz
Maximize the amount of demand being met

Once the system has gone into emergency mode (*i.e.* a contingency has occurred) then the system has to be brought back to the normal or preventive operating mode via the restorative operating mode. The mode decompositions and mode transitions are shown diagrammatically in Fig. 3.

Note that decomposition by mode does not have a pyramidal structure as in level decomposition nor a sequential structure as in time decomposition. In general, mode decomposition has no correlation with time and level decomposition.

References p. 276

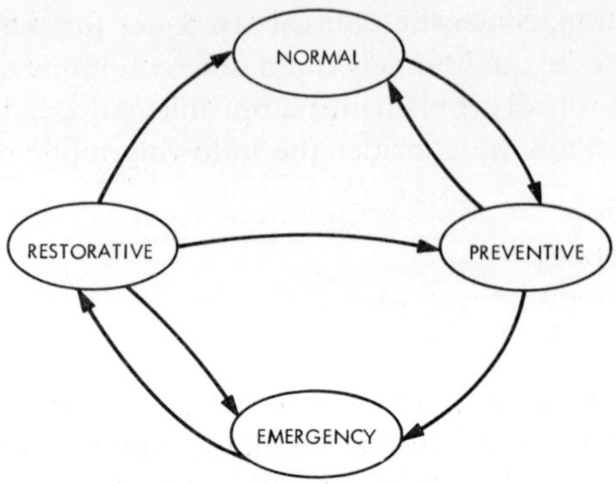

Fig. 3. Mode decomposition (interactive structure).

2.4 *Amount of decomposition*

The "amount" of mode decomposition is specified (by our definition). However, the amount of decomposition by level (*i.e.* number of levels and subsystems within each level) and time (*i.e.* number of time intervals) has not been specified. We will briefly discuss this subject because the choice of amount of level and time decomposition has a major effect on the over-all hierarchical system performance.

A fully integrated system is the limiting case where there is no level or time decomposition. For example, in a fully integrated system, instantaneous valve openings at the individual power plants would be determined at the interconnected system level by a computer program which explicitly considered unit commitment problems when computing turbine valve commands. Such a fully integrated system is not desirable (as it is not needed). Obviously, the other extreme of too many levels and time intervals is also undesirable.

There are many "natural" constraints on the choice of amount of decomposition. Level decomposition cannot ignore existing political, social and ownership conditions. The maximum amount of time decomposition is dictated by the natural response characteristics of the power system itself. Unfortunately we know no general rules for choosing from the various possibilities that lie within these constraints. All we can say is that the amount of time and level decomposition is one of the important trade-offs (see discussions in §6).

3. INFORMATION STRUCTURE

Consider a system which has been decomposed (by level, time or mode) into various subsystems. Each of these subsystems will have a certain "amount" of information of different "types" available to it for its decision-making process. The term "information structure of the hierarchical system" refers to both (1) the amount and type of information available to individual subsystems, and (2) the degree of coordination of information. In order to clarify these general ideas, we begin in §3.1 by considering "types of information". In §3.2 and §3.3 we then discuss the concepts of "amount of information" and "degree of information structure coordination". In §3.4 we point out the difference between usable information and inherently available information. Finally, in §3.5 we briefly discuss sources of information.

The term "information" is used here in a general sense with only limited correspondence to the formal mathematical definition used in communication–information theory.

3.1 Types of information

There are two basic types of information we will consider:

(1) Numerical values:
  system parameters
  state variables

(2) System structures.

After discussing these two types, we will consider the use of equivalents.

Numerical values of system parameters and state variables (and other variables) can be obtained either by direct observation or by estimation. Direct observation is essentially as its name implies—actual values that are the result of direct observation. For example, a watt meter provides a direct measurement of the real power flow on a transmission line. The "name plate" data for a generator provides a direct observation on inertia, etc. Estimation is a more general concept that involves the use of one or several direct observations to compute an estimate of the desired actual values. Such estimation can take many forms. For example, the real power flow on a given line can be estimated from direct measurements made by watt, var and volt meters located elsewhere on the transmission system. Future load demand can be estimated (predicted) by using past and present direct measurements. Response "time constants" of a boiler–turbine system can be estimated (identified) by using observations on how the system actually responds to inputs.

The second type of information is information on the structure of the power system. For example, the specification of a transmission network graph and the assumption of a "pi equivalent" for each transmission line constitutes a structural model (the values of the impedances are "actual values" and not part of the structural model). A special case of a structural model is structural information about the control strategies being used in different parts of the power system. Information on system structure is usually obtained by analysis (which can be viewed as a type of direct observation). However, it is also possible to obtain some information on system structure by estimation techniques using measured data.

It is important not to confuse the concept of the "information structure of the hierarchical system" with "information on the structure of the power system". One type of information in the "information structure" is "information on the power system structure".

The concept of an equivalent is very important when considering information on parameter-variable values and structures. Consider the level decomposition of

References p. 276

Fig. 1. It could be desired that each of the individual systems (Level 2) has equivalent models for the effect of the interconnected system as "seen by looking out through" the tie lines. Examples of equivalents are:

(1) network equivalents,

(2) spinning reserve equivalents (no dynamics),

(3) dynamic spinning reserve equivalents (effect on average system frequency behavior following major loss of generation) and

(4) transient stability equivalents.

3.2 *Amount of information: uncertainty*

As mentioned earlier, we are not using a formal mathematical definition of information so we have no quantitative measure of amount of information. However, in a general qualitative sense, the amount of information available is inversely related to the amount of uncertainty associated with the parameter-variable values and the system structure.

The amount of uncertainty can be measured by use of probabilistic–stochastic models or by use of unknown but bounded (set theoretical) models. Some possible approaches are briefly summarized in Fig. 4.

	Uncertainty in Parameter Values	Uncertainty in Structure
Probabilistic-Stochastic Model	Random vector with specific mean and covariance matrix	Hypothesize extra uncertain input modeled as stochastic process
Unknown but Bounded-Set Theoretic Model	Values lie within some set such as $2 \leq x \leq 4$	Hypothesize extra uncertain input modeled as unknown but bounded process; i.e. $2 \leq x(t) \leq 4$ all t

Fig. 4. Examples of modelling uncertainty.

In practice, models as in Fig. 4 for the uncertainty are usually not available. Often power system decisions treat (explicitly) values and structures as if they are exact and the uncertainty is ignored. This can be satisfactory if the actual uncertainty is small and its effect is implicitly considered when decisions are made. However, the uncertainty itself is always there and can have a major influence on the over-all performance. In the design of hierarchical systems, the amount of information available (*i.e.* the amount of uncertainty) is often a key and sometimes a limiting factor.

The use of the term "equivalent" in §3.1 is unfortunate as, except in rare cases, an equivalent is not truly equivalent. An equivalent such as one of the four examples of §3.1 usually introduces some uncertainty in addition to that in the complete representation itself. Sometimes this uncertainty can be completely neglected. However, in some cases the uncertainty can dominate even to the extent that the entire equivalent consists only of a model for the uncertainty. For example, one possible transient stability equivalent could consist only of false generators at each of the tie lines when the generator's real and reactive outputs are modelled by an uncertain process, either stochastic or unknown but bounded. Even when better (*i.e.* less uncertain) equivalents are available, the use of such uncertain equivalents can result in a major simplification in the computation and communication costs.

3.3 Degree of information structure coordination

Thus far we have discussed individual subsystem information structures, *i.e.* the type and amount of information available to the individual subsystems. The total amount of information in the whole system is the "sum" of the information available to all the subsystems. We now consider the "degree of information structure coordination" which determines how much of the total information is available to the individual subsystems.

Consider level decomposition as in Fig. 1. Each level and each subsystem at a given level has its own individual information structure. For example, if there is one subsystem at Level 3, 10 at Level 2 and 50 at Level 1 for each at Level 2, then there are $1 + 10 + (10)(50) = 511$ individual information structures. The information structure of the over-all system is said to be completely coordinated if all 511 individual information structures are the same, so that "everybody knows everything anybody else knows". The system information structure is completely uncoordinated if all 511 individual information structures are different. In general completely uncoordinated information structures cannot happen if the power system itself is interconnected. Even if there is no data flow between subsystems (see §5), the same power system is still "observed" and this provides some degree of information structure coordination.

The degree of information structure coordination must be viewed relative to the type of decision being made. For example, an almost completely coordinated information structure might exist for economic dispatch while the information structure might be completely uncoordinated relative to spinning reserve contingency evaluation.

The concept of equivalents is important when evaluating the degree of information structure coordination. For example, consider the problem of information (structure and values) on the network of the interconnected system. A completely coordinated information structure (relative to this problem) could exist if each individual power system (Level 2 of Fig. 1) had an equivalent impedance matrix looking out through its tie lines; complete detailed knowledge of the whole network is not required.

It is important not to confuse the concept of amount of information (*i.e.* degree of uncertainty) with the degree of information structure coordination. An increase in coordination will decrease the uncertainty but even a completely coordinated information structure could have a lot of uncertainty (*i.e.* a small amount of information).

3.4 Inherent versus *usable information*

There is an important difference between inherent information and the information that is actually usable.

References p. 276

Consider a transmission network with watt, var and volt meters on some but not all busses and lines. As a first example, consider a transmission line whose real power flow is not metered. The other metered values might be sufficient to allow the unmetered flow to be estimated so that information on the unmetered flow is inherently available. However, if the estimation is not done, this inherent information is not considered to be in a usable form. As a second example, assume the over-all network is an interconnected system containing many individual systems (decomposition by level). Assume each individual system has complete data on the structure and impedance parameter values of the whole interconnected network which could be used to compute an equivalent impedance network "looking out" from its tie-line interconnections. If some such computation is not made, the inherent information available in the network data is not considered to be in a usable form. As a third example, consider the same interconnected system but now assume data on the structure and impedance values of the whole network are not available to the individual systems. It is conceptually possible for one individual system to compute an estimate of an equivalent impedance network "looking out" by using just its own meter readings taken over a period of time when sufficient changes in tie-line flows have occurred. The inherent information available in these meter readings, of course, may not be used[3].

The earlier discussions on amount of information (uncertainty) and degree of information structure coordination have referred to the usable information rather than the information that is inherently available. Thus an information structure can be greatly changed by introducing extra computation facilities to convert inherent information into usable information.

3.5 Sources of information

Consider one subsystem of a system decomposed by level. Two sources of information for this subsystem are

(1) measurements made by the subsystem itself, and

(2) information externally transferred to it from other subsystems by a communication system.

These ideas are implicitly implied in the preceding discussions. We repeat them here to emphasize an important point. Even if there is no external information transfer by a communication system from other subsystems, the measurements made by the subsystem on itself provide some inherent information about the other subsystems because all the subsystems are interconnected. Of course this inherent information may not be converted into usable information but it is always there.

4. COORDINATION*

Thus far we have discussed decomposition and information structures. We now

* The concept of coordination has been considered by Mesarovic *et al.*[2] Their view is somewhat different in the sense that the role of information and information structure is not given the emphasis that we give it here.

discuss the main problem of interest: coordination (or coordination of control). Assume that for a given decomposition (by level, mode or time) the individual subsystems are controlled in a decentralized manner, so that each subsystem controller acts to some degree in an independent manner. The basic problem is to provide some degree of coordination between the decisions of these individual controllers.

4.1 Types of coordination

Two ways of improving the coordination between individual subsystem controllers are

(1) to increase the degree of information structure coordination, and

(2) to give one subsystem controller "superior" status so that it "coordinates" to some degree the other subsystem controllers.

As long as the individual subsystem controllers have a common goal, an increase in degree of information structure coordination will improve over-all performance. A prime example of a superior subsystem is in level decomposition where (Fig. 1) Level 3 (the interconnected system) assumes partial responsibility for coordinating the Level 2 (individual systems) controllers by furnishing them direct commands, goals or constraints. Examples of possible Level 3 coordination are when Level 3 tells one of the individual systems at Level 2 to

(1) increase base load generation by 500 MW (a direct command);

(2) increase base load generation by at least 400 MW but not more than 600 MW (a constraint);

(3) attempt to reach and maintain an increased base load generation of 500 MW (a goal).

Note that the concept of a superior controller furnishing goals or constraints to other subsystem controllers can be viewed as the transfer of another type of information that was not discussed in §3.1.

We now look at the coordination process in a little greater detail by considering Figs. 5–7. Let us consider level decomposition. Figure 5 represents the complete interconnected system. We now assume that the system has been decomposed by level and we consider two subsystems of the decomposed system and their control (Fig. 6). The final process of coordination is shown schematically in Fig. 7.

Figures 5–7 are for level decomposition but the same ideas hold for time decomposition.

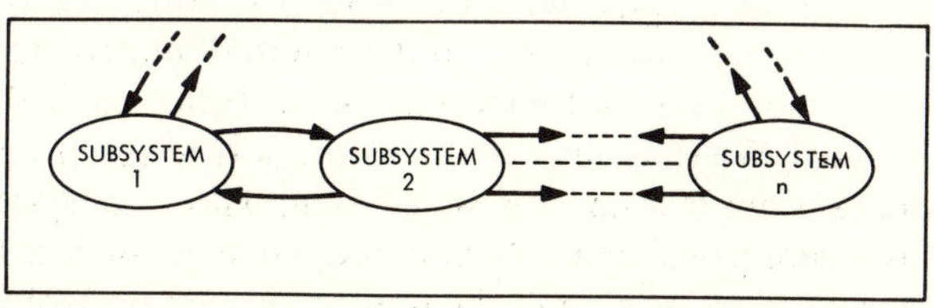

Fig. 5. Interconnected system.

References p. 276

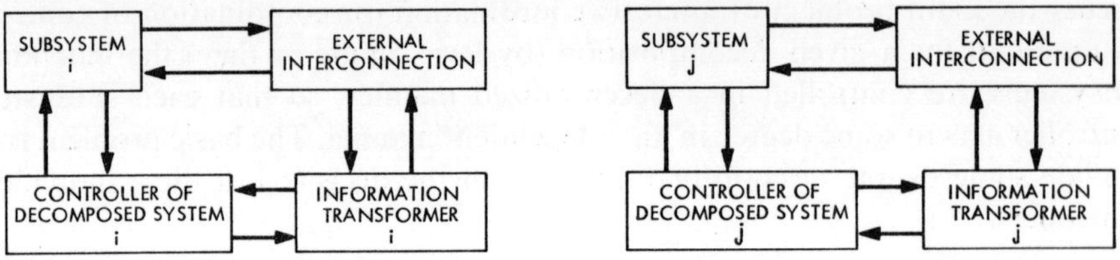

Fig. 6. Controllers of level decomposed subsystems.

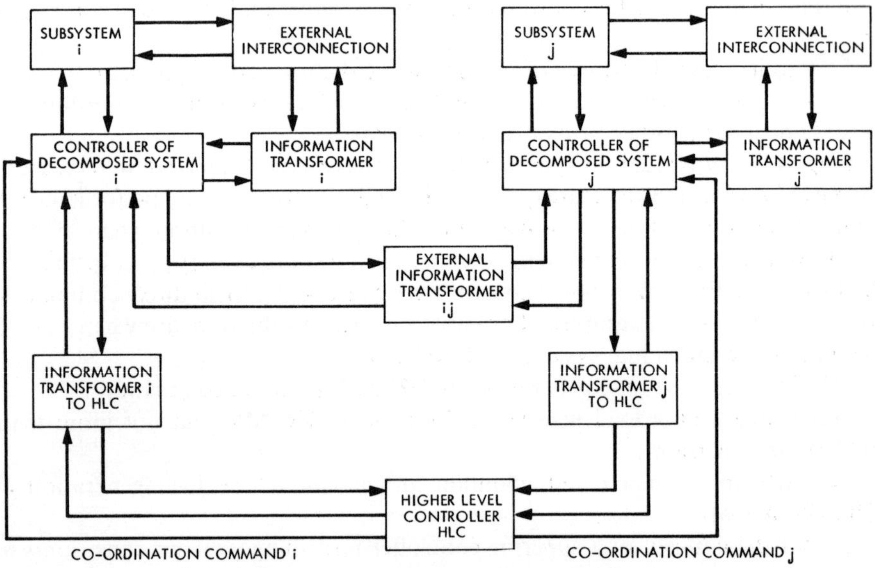

Fig. 7. Coordination.

4.2 Measures of coordination

Coordination can be measured on a scale of zero to full coordination. Consider the situation of Fig. 6. Assume that the information transfer ratio is zero (*i.e.* open circuit) and the external information transformer ratios are all zero in Fig. 7. This corresponds to *zero coordination.* The two subsystems are being controlled in a completely decentralized manner. Zero coordination is in general unattainable (unless the two levels physically separate) since by observing one subsystem we necessarily obtain some information on the external interconnection. Assume that the system has not been decomposed and is being controlled optimally relative to a single goal. We shall refer to this as the *fully integrated control system.* A *fully coordinated* system is a decomposed system whose subsystems are being controlled and the subsystem controllers coordinated in such a manner as to achieve a performance equal to that of the fully integrated control system.

From the point of view of reliability, complexity, etc., it is desirable to have a fully coordinated control system as opposed to a fully integrated control system (see also discussion in §6).

We should also point out the relationship between coordination as discussed here and information structure coordination. These are different concepts as

(1) a fully coordinated control system has a completely coordinated information structure, but

(2) a system with a completely coordinated information structure is not necessarily fully coordinated (if, for example, the individual subsystems do not have a common goal).

In the case of mode decomposition full coordination corresponds to a sequence of control actions whereby the system is brought back from the emergency mode to the normal mode in minimum time and zero coordination corresponds to the situation where the system operates in the various modes in an independent fashion and no coordinating action is taken.

4.3 An example: dynamic economic dispatching problem

To illustrate the concept of coordination we consider the dynamic economic dispatching problem for an interconnected system. The general problem may be posed as follows. Given the predicted uncertain dynamic load demands on two individual systems (of Fig. 1), schedule the dynamic generation patterns of each individual system and the power interchange program (including the pricing) such that the over-all demand is met at minimum cost (the cost function being an integral over one day) subject to the network and other dynamic constraints being satisfied. Thus the general economic dispatching problem is a dynamic optimization problem.

The steps in the solution of the problem by decomposition and coordination will now be indicated. We first apply *level decomposition* to decompose the over-all problem into two decentralized dynamic economic dispatching problems (one for each individual system) by fixing the interchange of power and the price mechanism.

Assuming the dynamic decentralized economic dispatching problems can be solved for a given level of interchange and given price mechanism, the *coordination problem* is to find the optimal price mechanism so that the system is guided towards the over-all optimum. The coordination in this case can be made either by information transfer between the two controllers or by means of a "superior" controller (at Level 3 of Fig. 1).

The dynamic decentralized economic dispatching problem is then further decomposed by applying time decomposition. The solution is divided into four parts:

Control function	*Time scale*
Unit commitment	hours
Economic dispatching problems	minutes
Load-frequency control	seconds
Governor action	fewer seconds

References p. 276

The basic coordination problem here is:

(1) to make sure that the unit commitment leads to feasible economic dispatching solutions;

(2) to update the unit commitment so that an improved solution to the dynamic dispatching problem can be obtained; and

(3) load-frequency control and governor action should attempt to maintain the optimal plant loading and also to maintain tie-line schedule and system frequency.

Finally, with regard to mode decomposition, in the emergency mode economics is sacrificed and one tries to obtain the maximum load solution subject to the system constraints being met. The coordination problem is to bring the system back to the normal mode using a dynamic restoration procedure in minimum time without violating constraints.

5. DATA FLOW

A decomposed system usually involves extensive data flow between the individual subsystems. This data flow can consist of direct commands, goals, constraints, direct measurements, estimated values, structural models, equivalents, etc. We will now briefly consider two aspects of this data flow question: (1) routing, and (2) coding.

To illustrate the routing problem, assume information is to be transferred between the individual systems of Level 2 of Fig. 1 so that they can make independent decisions. Two extreme methods of routing this information are:

(1) Pair-wise information exchange: the individual systems exchange information directly and independently with each other;

(2) Information exchange center: all information is channelled into and out of a central facility.

These two extremes are illustrated in Figs. 8(a) and (b). The information exchange center can perform a variety of tasks, such as:

(1) provide buffer storage as needed,

(2) assign message priorities,

(3) adjust data format to be compatible with individual systems' computers,

(4) maintain updated data files which can be accessed by individual systems, and

(5) convert the data in the data file into a form requested (needed) by an individual system (*e.g.* computed network equivalents).

Such tasks can be very important from a practical implementation point of view. However, by assumption in this discussion, the resulting information structure (uncertainty and degree of coordination) is the same for both Fig. 8(a) and Fig. 8(b) so data routing does not affect information structure.

In Fig. 8(b) the information exchange center can be viewed as a Level 3 (interconnected system) in Fig. 1. However, in this example it is being assumed that the individual systems are making independent decisions so, in Fig. 8(b), the infor-

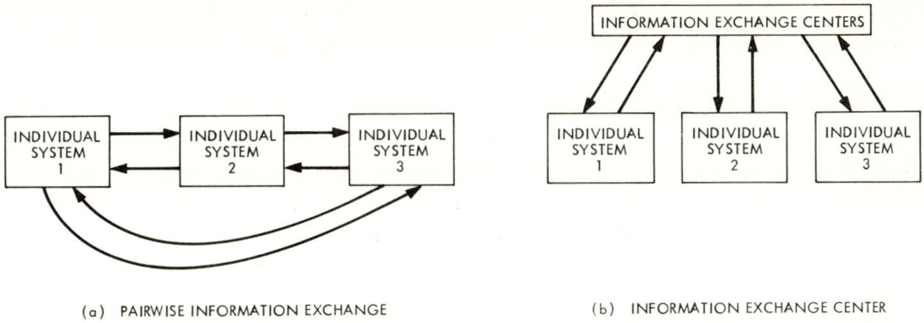

Fig. 8. Two methods of routing data flow between individual systems.

mation exchange center (at Level 3) is improving the degree of information structure coordination; it is not involved in decision making. From a conceptual point of view, it is important to emphasize the difference between higher level decision making and a higher level information exchange center which is simply one way of routing the data flow.

The preceding routing discussion points out the important point that a system may be decomposed one way relative to decision making and a different way relative to data flow.

The data flowing between, for example, the individual systems as in Fig. 8 can be "coded" in various ways. For example, instead of exchanging complete short-range load forecasts for every hour of a 24-hour day, the forecasted curve can be coded (parameterized) by a few parameters (such as Fourier coefficients). In this example, as in almost all cases, coding is simply a way of reducing the amount of data to be exchanged without appreciably affecting the information structure. In many cases there is no difference between the concept of equivalents and the concept of coding the information to simplify the data flow. In fact one of the main reasons for the discussion of the coding concept is to point out its close correspondence with the concept of equivalents.

The discussions on routing and coding of data flow have used level decomposition to illustrate the ideas. However, the ideas also apply to time and mode decomposition.

Since routing and coding problems do not affect the information structure or the actual decision-making process, data flow is not a "basic" aspect of hierarchical system theory. The main reason we have discussed it is because the choice of a data routing procedure and methods of coding can have a strong influence on the costs and reliability of the over-all system.

6. TRADE-OFFS IN DESIGN OF HIERARCHICAL SYSTEMS

Thus far we have discussed decomposition, coordination, information structure and data flow. The various possible combinations of these ideas lead to an almost infinite range of possible hierarchical systems. Since the "optimum" hierarchical

References p. 276

system is strongly dependent on the nature, needs and existing structure of the particular power system of interest, we concentrate only on the nature of the many trade-offs which must be considered.

The amount and type of level and time decomposition strongly influence the trade-off problem. However, we will assume these decompositions to be already fixed. For simplicity we will restrict discussion to level decomposition.

The discussion will consider only coordination and information structure trade-offs. Data flow problems are important in terms of cost and reliability but can, to some extent at least, be decoupled from coordination and information structure considerations.

One way to view the over-all trade-off problem is in terms of

(1) amount of coordination,

(2) amount of information in the information structure, and

(3) degree of information structure coordination.

All other things being equal, an increase in any of these three factors results in improved system performance. However, such increases cost money and the area which yields the most return per dollar spent depends heavily on the particular problem under consideration.

The reliability issue must also be considered. An increase in coordination or degree of information structure coordination is often obtained by some sort of centralization of facilities (such as a higher level controller or a regional information exchange center). Such centralization may increase performance but decrease reliability in the sense that the over-all system becomes vulnerable to the loss of the centralized facilities. An increase in coordination or degree of information structure coordination also usually implies an increase in complexity (computers talking to computers, etc.). A basic rule is that "complexity causes trouble" so as much simplicity as possible is desired.

A combination of cost and reliability arguments leads to the one general rule of hierarchical system design that we are willing to state.

> Minimize the amount of coordination and information structure coordination subject to the constraint that system performance is satisfactory.

Admittedly this rule is not quantitative. It is just a statement of a goal of good engineering design.

Another type of trade-off is between communication and computation costs. For example, it is possible to concentrate most of the computing capabilities at the highest level (the interconnected system). This requires extensive communication but prevents unnecessary duplication of computing facilities at the lower levels (say the individual systems). Another example is the difference between inherent available information and usable information as discussed in §3.4. The amount of information in the information structure can be increased by communication between levels or by increasing computation to convert more inherent information into usable information.

As an example of the complexities of these trade-offs, consider the problem of trying to use turbine valve openings and/or excitors to damp small amplitude, low frequency (2–10 second period) but long lasting system oscillations. For simplicity of discussion, we will naively assume there is only one individual system at Level 2 in Fig. 1 and concentrate on the interaction between the one Level 2 decision maker and the many individual power plants at Level 1. Three of many possible approaches to this problem are as follows.

(1) Central Direct Control: Level 2 receives all information available from Level 1. Level 2 sends direct commands to the individual turbine valves and excitors. Level 1 makes no decisions. This is a fully integrated control system with no decomposition.

(2) Central Supervisory Control: Level 2 receives some of the information available from Level 1. Level 2 identifies (estimates) the nature of the oscillation and then adjusts the Level 1 controller (governors, voltage regulators) by specifying the gains, time constants, etc. to be used. The Level 1 controller remains in direct control.

(3) Adaptive Local Control: Level 2 plays no role. There is no data flow between the individual Level 1 controllers. Each individual power plant at Level 1 adapts its own control laws (gains, time constants, etc.) depending on an identification of the nature of the oscillations made by direct measurements available at the power plant.

Central Direct Control is very expensive and complex to implement and is unreliable in that it is very vulnerable to the loss of the Level 2 controller. Note that, even when in operation, Central Direct Control is not guaranteed to work because the amount of information (structure and parameter-variable values) on what is actually causing the oscillations may be relatively small (large uncertainty). Adaptive Local Control involves no data flow between subsystems but still has some degree of coordination and information structure coordination because of the identification (estimation) done at the individual power plants. Adaptive Local Control is most reliable in that it involves no central facility. Both Adaptive Local Control and Central Supervisory Control involve adjusting the gain, time constants, etc. of the governors and voltage regulators; they simply use different information structures. The cost trade-offs between Central Supervisory and Local Adaptive Control are not obvious as Adaptive Local Control requires many individual small computer systems while Central Supervisory Control requires a communication system and one relatively large computer. The choice between these three approaches (and there are many other reasonable approaches) is not clear. However, using the general rule of minimizing coordination as much as possible, Adaptive Local Control is most desirable *if* it proves to be effective. If Adaptive Local Control cannot do the job, Central Supervisory Control is the next most desirable. Central Direct Control should be attempted only as a last resort. A final possibility, not yet mentioned, is to design the system (including voltage regulator and governor gains, time constants, etc.) so that oscillations are never important. This is the ultimate as far as minimizing control costs and maximizing reliability but the best way of doing this that we know is to increase the capability of the transmission system. This is very expensive.

References p. 276

7. SUMMARY AND CONCLUSIONS

In this paper we have discussed various concepts of hierarchical multilevel system theory and illustrated these concepts by considering the real-time control of bulk power systems for economic and reliable operation. We have shown how decomposition according to level, time and mode is a natural way to decompose the complex task of controlling an interconnected power system. We have then discussed the concepts of information and information structure. The concept of coordination was introduced and we discussed a scale of measurement for coordination. The relationship that necessarily exists between information structure and coordination (a concept different from that of information) was investigated. We then discussed data flow within the decomposed system. The final section was devoted to various trade-off considerations.

The goal of this paper has been to provide a general understanding of hierarchical system theory (as we view it) and its relationship to selected aspects of the real-time control of bulk electric power systems. The most important single point of the paper is that there are a huge number of possible trade-offs which must be considered before choosing a final design.

Hierarchical system theory is in its infancy. We hope that the concepts discussed in this paper and illustrated with reference to the real-time control of an interconnected power system will be useful in the future development of the theory.

REFERENCES

1 T. E. DyLiacco, The adaptive reliability control system, *IEEE Trans.*, **PAS-86** (1967).
2 M. Mesarovic, D. Macko and Y. Takahara, *Theory of Hierarchical Multilevel Systems*, Academic Press, New York, 1970.
3 T. G. DeVille, Identification for linear electrical power system models, *EPSEL Report No. 31*, MIT, Cambridge, Mass., May 1971.

Discussion

H. Glavitsch (*Brown, Boveri & Co. Ltd., Baden, Switzerland*)
I was a little bit surprised that the authors attached a measure of quality to the centralized computer as far as reliability goes. In the paper a comparison was made with the conclusion that the reliability is improved when the local adaptive system is used. If we look at this in the framework of hierarchical systems, neither one of the two extreme cases is really hierarchical. In one case we have direct digital control and in the other we have a completely decentralized control with a lot of observation capability. Would you please comment on this because it seems to me a contradiction.

F. C. Schweppe and S. K. Mitter
The term reliability was used in a general sense to emphasize the potential vulnerability of a highly centralized control system to the loss of the highest level controller, and to emphasize the authors' opinion that the best answer is the simplest

answer that works. The term was not meant to correspond directly to "power system reliability" as this expression is usually used, although there is obviously a relationship between the reliability of the hierarchical control structure and that of the power system itself.

In the example, three types of configurations were discussed:

(1) Central Direct Control,
(2) Central Supervisory Control, and
(3) Adaptive Local Control.

Central Direct Control is, indeed, direct rather than hierarchical control and was included only to reference the extreme case. However, the question of whether Adaptive Local Control is a hierarchical system is a matter of viewpoint. We consider it to be hierarchical because of the information coordination which results from processing observations from a common system. From our point of view, Adaptive Local Control is not completely decentralized, even though there is no external data flow between the controllers.

C. J. Eaglen (*Brown, Boveri & Co. Ltd., Baden, Switzerland*)

How do the ideas of multilevel control cater for batch and continuous systems?

Decomposition and allocation of control functions is to my mind necessary to realize real-time solution to an optimal control problem. How do the authors propose that the local optimization and decomposition can be structured to realize this?

What dissimilarity may exist between the processes at a level? If only a small dissimilarity may exist we may have too many levels.

F. C. Schweppe and S. K. Mitter

It is not possible to answer the questions raised by Dr. Eaglen adequately within the scope of this discussion. However, the following points can be made:

(i) There is no general theory of hierarchical systems, and we tried to stress this in our paper. Each case has to be looked at individually, although there are classes of problems in mathematical programming where some theory exists. Dr. Eaglen is trying to pose questions which are too general and, as such, ill-posed —at least in the present state of development of the theory.

(ii) The ideas of multilevel control are applicable to both batch (discrete-time) and continuous systems.

(iii) Dr. Eaglen is quite correct in stating that decomposition and allocation of control function is necessary to realize real-time solution to an optimal control problem. However, it should be pointed out that multilevel hierarchical system theory is not just a tool for solving optimal control problems. In fact, it is the authors' belief that far too much emphasis has been placed on optimality. In complex systems problems there usually does not exist a *single* optimality criterion.

The Role of the Human Operator in System Control

M. SYRBE

Institut für Informationsverarbeitung in Technik und Biologie, Karlsruhe-Waldstadt, Germany

AND

F. BACHER

Brown, Boveri & Co. Ltd., Department EDV, Baden, Switzerland

SUMMARY

In spite of the rapid development of research and technology the necessity for using human operators in system control has not been reduced. With the growth of power systems and the growing demands for the security of supply, increasing requirements are put on equipment and on staff also. A survey of this is given. An introduction to the methods of human engineering is presented showing the possibilities for solving the problems. Finally, conclusions are drawn concerning the design of the instrumentation of system control centres.

1. INTRODUCTION

Despite rapid progress in research and technology it is still necessary to use human operators in control systems. A typical example of increased efficiency in system performance can be seen when the successes of the manned Apollo spacecraft are compared with those of purely robot spacecraft. The success of the Apollo 15 mission was only possible by having a crew in command during difficult manoeuvres. The high adaptivity of human operators for cognitive and for manual processes makes it possible to cope with many unexpected states of operation. Therefore, the current trend of automating technical systems and of improving methods for information processing does not necessarily eliminate the need for human operators, but usually does modify their tasks.

In modern control centres, all the equipment for displaying the state of the system and selecting command inputs is concentrated together. Clear representation of all the data and clear organization of the command input is crucial for the proper functioning of these centres.

The planning of a control centre requires both experience and knowledge of currently available methods and equipment. In this paper, comments will be made on some points of the relations and developments mentioned above.

2. REVIEW

With the growth of power systems and the increasing demand for supply reliability, more stringent requirements are placed upon the equipment and the staff controlling these systems. This requires the installation of efficient information systems to give good support to the central control and to allow intervention into the system events. Considering that these power systems have a rate of growth of 7–15 % per year, the need for efficient running of the production and distribution can easily be recognized. The design of control centres has to include these conditions.

A consequence of this trend is that man is no longer able to register and utilize all the data with sufficient speed and reliability. This arises because of his limited ability to react and memorize (information storage capacity). Therefore, to relieve the operator, routine work has to be done with automatic control. Today this is possible with a particularly efficient tool, the digital computer. Through a hierarchical structure in data processing, it can be arranged that the operator receives only that relevant data which enables him to take action with sufficient speed and understanding. Only then will he be in a position to take decisions and to use his creative and adaptive abilities fully, including the ability to improvise. It may be mentioned in passing that these trends have social and labour policy implications.

The structure of electric utility systems divides them into national networks, regional networks with power plants and municipal networks with power plants. This requires a similar structuring to the system of control centres, as described in a previous paper[1]. Figure 1 shows this structure. The more difficult tasks of national interconnected networks, such as the supervision and control of generation and distribution of power in the area, are usually accomplished in two different centres: the coordination centre and the power pool centre. In the next lower level the tasks are divided into two categories: switching dispatch and generation dispatch. Frequently, these are also present in the third level of municipal networks.

Concern will not be given to the individual differences of the various centres, but emphasis will be placed upon the problems common to the design of advanced centres. It can be mentioned that electrical networks can be basically divided into the following states.

Normal state —the load is covered, adequate spinning reserve is available.
Alert state —the load is still covered, an adequate spinning reserve is not available.
Emergency state—the load cannot be covered any more.

During the normal state the operators will be mainly occupied with data registration, planning and similar jobs. However, their tasks will change abruptly when the network tends towards an emergency state. The operators have to initiate countermeasures quickly and, above all, correctly. These are derived from the available data. There are no common rules for any of these states that would allow complete automation of the necessary decisions.

In 1950 an operator in a typical power control centre with a controllable generator or network power of some 100 MW had to watch about 200 or 300 binary

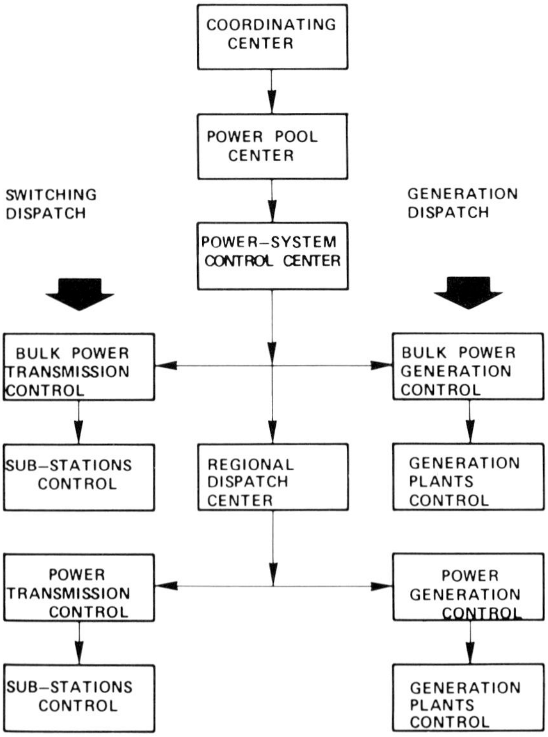

Fig. 1. Hierarchy of control centres.

messages (optical, acoustic), 50 analogue instruments and 10 plotters. Furthermore, he had to operate about 50 binary switches or press buttons and 10 analogue devices such as potentiometers etc. (Fig. 2).

By 1970 the controllable generator or network power had been increased to 1000 MW, and the requirements for ensuring operation of the power plant and supply to the customer had been increased, for the protection of life and material as well as for economic reasons. As a result, an operator today has to watch and to operate many more of these instruments (Fig. 3).

Since the number of operating errors and the reaction times have to be continually decreased as the systems become larger, methods have to be found which give the operator optimal support in carrying out his tasks, through the exploitation of the characteristics of process computers and of advanced display techniques. In other words, the man–machine interface has to be adapted to man.

3. MAN–MACHINE INTERFACE (INTERSECTION)

System components are separated from each other by intersections where the outputs and inputs of the components have to be made compatible. At the man–machine interface, man should influence adaptation by his

Fig. 2. Power control centre 1950.

Fig. 3. Power control centre 1970.

(1) body measurement,
(2) motor characteristics,
(3) mental–physiological characteristics and
(4) psychic and cognitive characteristics.

Man's body measurements are a very obvious characteristic and are therefore generally considered. Suitable standards have been drawn up, for example, in the GFR by the Fachnormenausschuss Ergonomie im DNA. Already, less attention is paid to motor characteristics such as muscle force, grasp velocity, grasp exactness, room for moving. For instance, many equally sized operating and display devices result in control panels which are too large. On the one hand, this puts too much demand on the human operator, and on the other, they do not employ him in an optimal manner (Fig. 2). In most cases, only the average sensitivity of the mental–physiological characteristics is considered. In this respect the eyes, ears and tactile organs are of particular importance. However, in many cases the time and spacial resolution limit of these organs, their frequency and distance dependence and their adaptability are crucial.

The psychic and cognitive characteristics have been least investigated. For instance, the registration accuracy and speed of instruments is decreased if task-alien, and therefore interfering, stimuli are presented simultaneously. This is illustrated in Fig. 4 by two typical displaying controllers for panel mounting. The left instrument has a lot of misleading information in the form of contrasting knobs, profiles, scales not necessary for regular operation and labelling which is too noticeable. The task-oriented information in the form of an indicator for the set-point and controlled variable—with the corresponding scale—does not

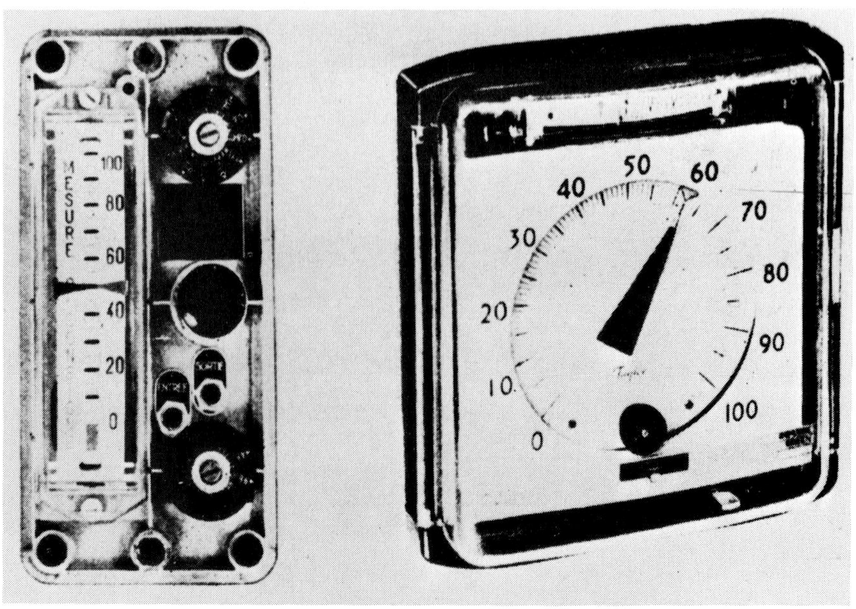

Fig. 4. Different lay-out of panel-mounted controllers.

References p. 291

dominate the over-all picture. The right instrument is ideal with regard to the shape of the indicator and scale. A good point in this design is the positioning of the, normally not important, controller output in a narrow horizontal instrument in the upper part. However, the noticeable button for set-point alterations and the equally noticeable name of the makers show poor design as both divert attention. Moreover, the button cannot be accessed directly and the glass handle is hardly visible.

Interfering information equal to signal noise can also be received by another organ. Thus, acoustic noise can hinder a registration, especially when it contains any sound indicating danger. In contrast, additional useful information received by a second organ can improve man's perceptual capacity. Examples of this are the coupling of warning lights and horn, or the improvement of manipulation elements in dark rooms by different handle shapes. Here, the standardization to a few control knob shapes hampers the differentiation for man.

In Fig. 5 the characteristics already mentioned of the man–machine interface and those of man are listed. The instrument field gives information to man's perceiving organs about the state of the machine. Some available designs are mentioned. Besides the common analogue instruments and signal lamps, digital display devices for accurate registration are now used. Furthermore, symbolic displays—known for binary signals as signs—are also applied to analogue variables. An example is a coloured bar for displaying levels of liquids.

Particularly important are the integrated displays which combine several quantities describing the state of the subsystems. Mosaic and illuminated panels are common. In these areas there are further possibilities, as will be shown in §5. The displayed information is also influenced by the arrangement of the instruments, by shape and colour, and by additional acoustic signals. The information processing by man is not only determined by the task, but also by his skill, motivation, mental load and environment. The result is an initiating action which is transferred to the control panel of the machine by muscular actions. Some remarks about design of control panels have already been made.

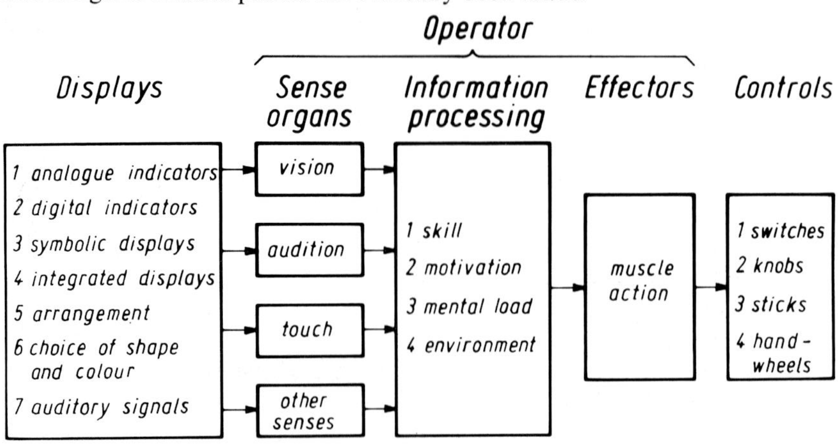

Fig. 5. Structure of human engineering problems in control centres.

4. HUMAN ENGINEERING FOR DEVELOPING METHODS

The above-mentioned intersection characteristics and the behaviour of human operators are studied in human engineering, which is still a young subarea of ergonomy. Existing knowledge has mainly resulted from problems of manufacturing rationalization, especially on assembly lines and also in aviation and space systems. For problems which occur frequently the optimal measures for seat, working places and control centres have been ascertained from those characteristics of the human body that are easily accessible to measurement[2]. The necessary power, shape and arrangement of control elements has also been studied. Good reviews of these results are published by McCormick[3], Neumann and Timpe[4] and, especially for network control centres, by Burgard *et al.*[5] The planning engineer will generally find sufficient information in these references. A less resourceful area is the definition and description of human ability (see ref. 4, Ch. 2.9), the quantitative consideration of its change with skills and motivation and the influence of working hours, noise, climate, illumination, room design and colours. In addition to the authors already mentioned, Murrell[2] gives an extensive review of the state of the art.

Surveillance and control tasks, which are the subjects of this paper, are mainly influenced by the complex psychic and cognitive characteristics of man. Without sufficient quantitative models, experimental studies will remain limited and unconnected. To date, quantitative studies only exist to deal with the design of display instruments using only one size of pointers or digits[2–4], and with the transfer characteristics of human operators working in closed-loop control when presented with fixed types of display and control devices[6,7]. In the course of planning, the engineer is first confronted with the problem of optimizing the multivariable display and multivariable control installations for the simultaneous solution of many operating tasks. At this point the engineer arrives at the limit of today's knowledge, allowing only the formulation of some qualitative fundamental rules.

(1) Consider the characteristics of the sensing organs (*e.g.* visual field, eye's angle resolution).

(2) The optimal shape of the instrument is task dependent (*e.g.* exact, tendency or limit-exceeding registration, importance of safety, speed or connection, or the frequency of use of the instrument).

(3) Choose solutions directly adequate to the task (*e.g.* accurate registration by digit and not by scales).

(4) Avoid information unnecessary to the problem (*e.g.* automatic change of display type (Fig. 1) for the decoupling of functions of complex instruments).

(5) Consider population stereotype expectations (*e.g.* turning potentiometers right results in higher values).

(6) Use corresponding designs for the display and corresponding control element (*e.g.* red pointer, red control knob).

(7) Instruments belonging to the same subsystem should have obviously common attributes (*e.g.* arranged closely, special colour).

References p. 291

Just how often these basic rules are violated was made clear by the comment to Fig. 4.

It may be pointed out that Burgard *et al.*[5] have divided these rules into about 100 individual rules for control centre design.

If the scientist has to deal with a problem which has only been solved qualitatively, he makes use of models derived from the problem's characteristics. These are based on functional and flow charts in which the elements are described by diagrams and tables, in the case of experiments, and by mathematical expressions, for the theoretical evaluation. The starting point of this procedure is shown in Fig. 6 for the example of human optical information perception. The display device of the machine sends pictorial information containing necessary and unnecessary interfering information for the operator's perception task. The pictorial information contains not only stochastic parts (*e.g.* pointer positions)—in contrast to the problems treated by C. E. Shannon—but also deterministic parts (*e.g.* scale). The latter are not negligible and they are observed at the beginning of the attentive fixation of the picture area. A recently published report[8] describes these facts mathematically.

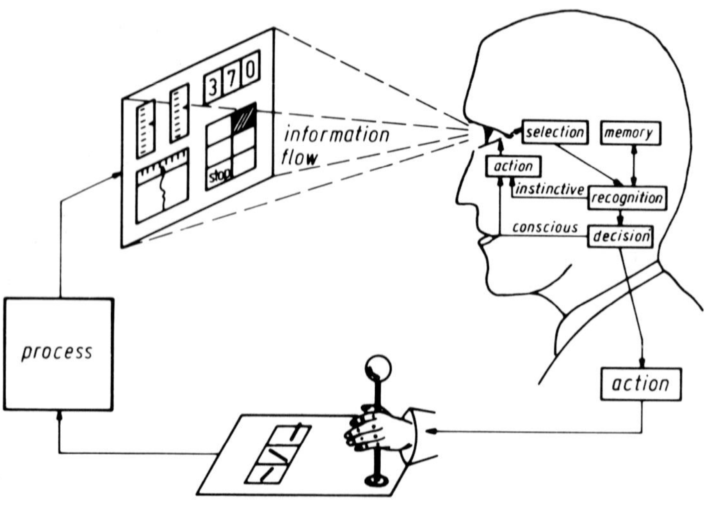

Fig. 6. Problem solving with human factors partial models.

An insight into the difficulties of the chained, task-oriented processing of the pictorial information is given in Fig. 7, published by Warr and Knapper[9].

The environment information entering through all sensory organs is selected relative to the accepted task (memorized information), *e.g.* stimulus information with input selector, see Fig. 7. When this information is completed with context information it is passed on for recognition and decision-considering ability, engagement, and charge of the percipient. The decisions are obtained by attaching

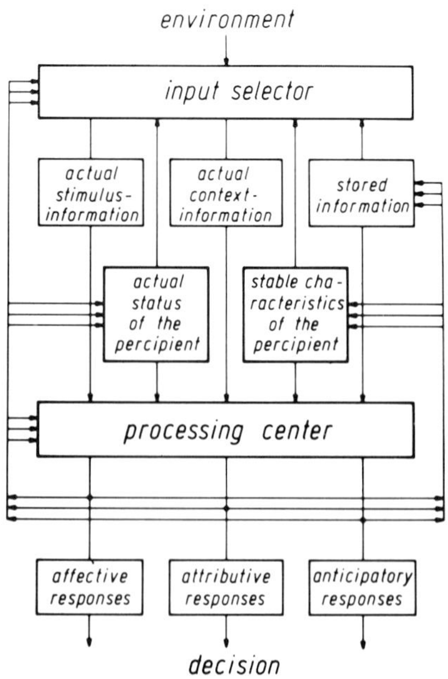

Fig. 7. Structure model of human perception.

them to a situation class (attributive response), but could follow anticipation at the beginning of information perception using past experiences (anticipatory response) or could be completed emotionally (affective response). Today, science is far from describing this very complex function scheme quantitatively.

5. MODERN CONTROL DESK DESIGNS

The foregoing sections indicate that incomplete knowledge demands a heuristic solution of the problem through a combination of experience, technical possibilities and plausible assumptions. In spite of this, it is possible to formulate some general rules for the processing of information in the control room as follows.

(1) All necessary past and present data are stored in the bulk memory of the process computer and are kept available either for the computation or for specific use by the operator.

(2) The parallel operation and parallel control possibilities have to be changed in a hierarchical ordered semiparallel or serial operation.

(3) The exclusive analogue display may be changed to a suitable choice of analogue, digital or symbolic display form.

(4) Use prediction techniques.

(5) Choose operation task-oriented display and control devices.

(6) Optimize the influence of the environment on the operator.

References p. 291

5.1 Display system

The particular significance of these rules is the fact that, through the organization of displayed information on the control board, an over-all picture can be obtained. The information can be divided into two categories:

Over-all view
Detail

For example, a mimic board gives the over-all view of network conditions and detailed information can be given by selection on the control desk (Fig. 8). This division is unnecessary in only a small system as the entire information can be displayed simultaneously.

In load dispatch systems the above-mentioned division of the displays is always necessary in practice. The operator's network interface equipment (the desk) temporarily displays that part of the network of interest to the operator. The display selected is for both received information and control outputs.

The selected part of the network display can be of different size, for example:

Part of the network
Group of stations
A station (Fig. 9)
Group of devices in a station
A circuit
An object or device

Fig. 8. Power control centre of the New York Power and Electric.

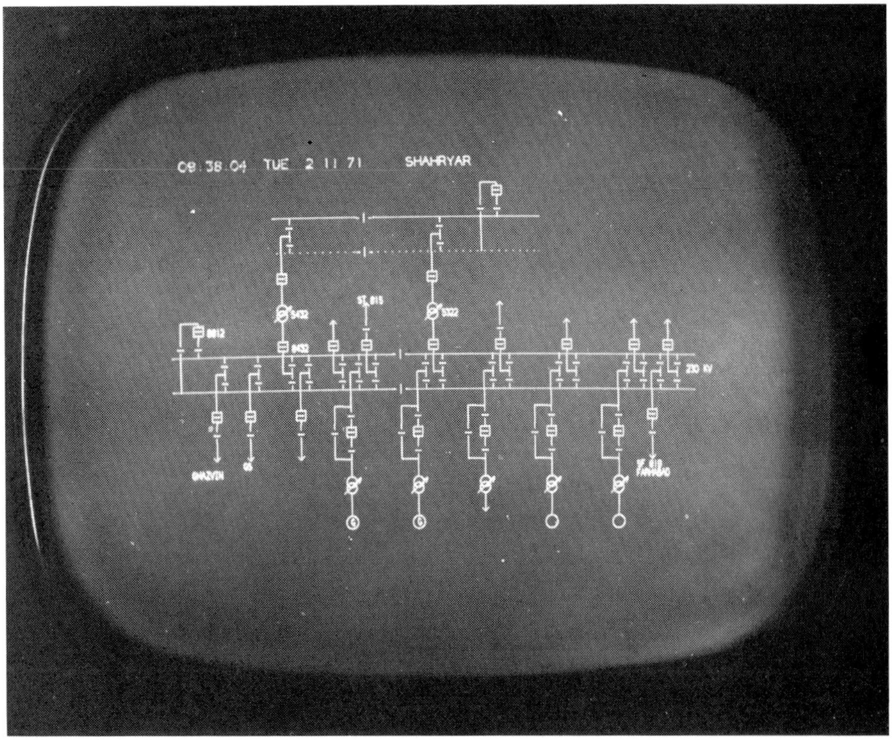

Fig. 9. Part of a network displayed on a CRT display.

The detail of the network part shown on the desk must not necessarily be of the same size as the range which is controllable by the control keys. In no case is it meaningful to make a display of smaller size than the area to be controlled. This would risk the operation of the plant.

The following examples can be mentioned.

(a) Display of transmitted information for a whole station on a mimic diagram, as displayed on a one-line diagram of a station. The command outputs are given through a device selector. The selector uses a unique code for each device on the display.

(b) Selectors with station and circuit keys effect the selection of the state of the devices in a circuit on the control board of the control desk. The control board contains the necessary control equipment for all controllable devices of the circuit. In this case, the selection of the whole of the transmitted information is given for the circuit.

5.2 *Display equipment (CRT display)*

Central control desks for controlling an electrical network are increasingly being provided with display apparatus. This allows different information to be generated by a process computer which can be output to the display device in any format.

References p. 291

At present, the most suitable display device available is the cathode ray tube.

In the control of a distribution network the following features can be used.

Network mimics divided into voltage level and area networks
Transformer stations and switching stations divided into size and type etc.
Geographical displays of supply networks
Plant outputs such as current, voltage, flow, load etc.
Output of measured variables in the form of lists
Output of stored curves and diagrams
Display of alarm messages, fault conditions, interconnection disturbances
Signalling of dangerous condition in clear text
Instructions and other communications between the operator and the data processing installation

Direct operational commands can be given from the display apparatus using either

a tracker ball,
a joystick or
a light pen

With the first two devices a tracker spot or a cursor mark can be moved on the screen of the cathode ray tube. The spot can be placed on or adjacent to a device to be controlled to identify the device. An operation key (*e.g.* push button) executes the required command.

The light pen contains a photocell in the point. By means of the photocell an interrupt is produced when the CRT spot coincides with the light pen position. Thus a controlled device can be identified.

Control commands can also be effected by the selection of points via an alphanumeric key board. The selection code can be taken from the display format. By suitable design, the code is quickly learned so that the displayed code is used as a verification.

Generally, two display units for a control position are included, whereby, for example, the station diagram is displayed on one unit and network conditions on the other. An additional unit is desirable as a standby in case of failure; the standby can be common for two or more control positions.

It is foreseen that the future control desks will be completely equipped with display units and the conventional display technique placed in the background.

5.3 Command execution

Two different methods of control can be chosen:

direct control
select control

A controlled device can be *directly* controlled from a control switch on the mimic diagram. The mimic diagram serves as the control board and the network display.

The control via *selection* is effected from a separate control desk and not from the mimic board. The control device is selected by the station circuit and device from the control desk. The control of the selected device is then executed by a common control switch on the control desk. The control selector is simply a coding of the command output. For practical purposes it is not important that the control operation takes place in steps.

With the selector it is necessary for the command output to the controlled device to be identified in order that the selection can be verified. A code can never have the same security as direct control. Thus, from the security aspect, direct control is preferable to select control. With the semiparallel display through a display screen system it is possible to have a combination of direct and select control systems.

6. CONCLUSION

In spite of rapid development of techniques further research is necessary. With the growth of energy supply networks and the increase in demand for security of supply, higher demands arise on equipment and operators. Here, a general review has been given. Through an introduction in methods of human engineering, a solution to the problem becomes possible and, moreover, suitable design criteria can be formulated. With these criteria, the design of the control and display systems can be made.

REFERENCES

1 F. C. Schweppe and S. K. Mitter, Hierarchical system theory and electric power systems, This Symposium, p. 259.
2 K. F. H. Murrell, *Ergonomics, Man in his Working Environment*, Chapman and Hall, London, 1965.
3 E. J. McCormick, *Human Factors Engineering*, McGraw-Hill, New York, 1964.
4 J. Neumann and K. Timpe, *Arbeitsgestaltung, Psychophysiologische Probleme bei Ueberwachungs- und Steuerungstätigkeiten*, VEB Deutscher Verlag der Wissenschaften, Berlin, 1970.
5 J. F. Burgard, R. J. Miller and H. E. Clauder, Human engineering consideration for power control centers, *Joint Meeting of the Communications & Power Systems Planning & Operating Section of the Canadian Electrical Association (CEA), Toronto, 1970.*
6 W. Oppelt, Der Mensch als Regler, *Regelungstechn.*, **13** (1965) 332–335.
7 W. Kreil and G. Schweizer, Der Mensch als Regler, ein Beitrag zum dynamischen Uebertragungsmodell, *Regelungstechn.*, **16** (1968) 49–56.
8 M. Syrbe, Anthropotechnik, eine Disziplin der Anlagenplanung, *Elektrotech. Z. ETZ-A*, **91** (1970) 692–697.
9 P. B. Warr and C. Knapper, *The Perception of People and Events*, Wiley, London, 1968.

Author Index

Page numbers in bold face type indicate full-length papers; all other page numbers refer to contributions to discussions.

Subject Index